Sensors for Health Monitoring

Advances in Ubiquitous Sensing Applications for Healthcare

Sensors for Health Monitoring

Volume Five

Series Editors

Nilanjan Dey

Amira S. Ashour

Simon James Fong

Volume Editors

Nilanjan Dey
Techno India College of Technology, Kolkata, India

Jyotismita Chaki
Vellore Institute of Technology, Vellore, India

Rajesh Kumar
Malaviya National Institute of Technology, Jaipur, India

Academic Press is an imprint of Elsevier
125 London Wall, London EC2Y 5AS, United Kingdom
525 B Street, Suite 1650, San Diego, CA 92101, United States
50 Hampshire Street, 5th Floor, Cambridge, MA 02139, United States
The Boulevard, Langford Lane, Kidlington, Oxford OX5 1GB, United Kingdom

Notices

Knowledge and best practice in this field are constantly changing. As new research and experience broaden our understanding, changes in research methods, professional practices, or medical treatment may become necessary.

Practitioners and researchers must always rely on their own experience and knowledge in evaluating and using any information, methods, compounds, or experiments described herein. In using such information or methods they should be mindful of their own safety and the safety of others, including parties for whom they have a professional responsibility.

To the fullest extent of the law, neither the Publisher nor the authors, contributors, or editors, assume any liability for any injury and/or damage to persons or property as a matter of products liability, negligence or otherwise, or from any use or operation of any methods, products, instructions, or ideas contained in the material herein.

Library of Congress Cataloging-in-Publication Data
A catalog record for this book is available from the Library of Congress

British Library Cataloguing-in-Publication Data
A catalogue record for this book is available from the British Library

ISBN 978-0-12-819361-7

For information on all Academic Press publications visit our website at https://www.elsevier.com/books-and-journals

Publisher: Mara Conner
Acquisition Editor: Fiona Geraghty
Developmental Editor: John Leonard
Production Project Manager: Anitha Sivaraj
Cover Designer: Greg Harris

Typeset by SPi Global, India

Contents

Contributors

Mohammad Affan
Department of Computer Science, Jamia Millia Islamia, New Delhi, India

Marcelo Antônio Alves Lima
Department of Electrical Engineering, Federal University of Juiz de Fora, Juiz de Fora, Brazil

Sanjiv Bansal
Department of Physical Medicine and Rehabilitation, Sawai Man Singh Hospital, Jaipur, India

P.K. Bhagat
Department of Computer Science and Engineering, NIT Manipur, Imphal, India

Thales Wulfert Cabral
Department of Electrical Engineering, Federal University of Juiz de Fora, Juiz de Fora, Brazil

Prakash Choudhary
Department of Computer Science and Engineering, NIT Manipur, Imphal, India

Vinay Chowdary
Dept. of Electrical & Electronics, UPES, Dehradun, India

Satya Ranjan Dash
School of Computer Applications, KIIT Deemed to be University, Bhubaneswar, India

Vindhya Devella
Dept. of Aerospace, UPES, Dehradun, India

Felipe Meneguitti Dias
Department of Electrical Engineering, Federal University of Juiz de Fora, Juiz de Fora, Brazil

Carlos Augusto Duque
Department of Electrical Engineering, Federal University of Juiz de Fora, Juiz de Fora, Brazil

Md. Rashid Farooqi
Department of Management Studies, Maulana Azad National Urdu University, Hyderabad, India

Akash Gupta
Dept. of Mechanical Engineering, UPES, Dehradun, India

Md. Shahriar Hassan
Department of Computer Science and Engineering, East West University, Dhaka, Bangladesh

Naiyar Iqbal
Department of Computer Science and Information Technology, Maulana Azad National Urdu University, Hyderabad, India

Vivek Kaundal
Dept. of Electrical & Electronics, UPES, Dehradun, India

F.S. Kazi
Department of Information Technology, Electronics and Electrical, VJTI, Mumbai, India

Mahdi Khosravy
Department of Electrical Engineering, Federal University of Juiz de Fora, Juiz de Fora, Brazil

Kriti
Thapar Institute of Engineering and Technology, Patiala, India

Vartika Kulshrestha
Department of Electronics, Banasthali Vidyapith, Vanasthali, India

Rajesh Kumar
Department of Electrical Engineering, Malviya National Institute of Technology, Jaipur, India

Debasish Kumar Mallick
School of Computer Applications, KIIT Deemed to be University, Bhubaneswar, India

Leandro Rodrigues Manso Silva
Department of Electrical Engineering, Federal University of Juiz de Fora, Juiz de Fora, Brazil

Amit Kumar Mondal
Dept. of Electrical & Electronics, UPES, Dehradun, India

Henrique Luis Moreira Monteiro
Department of Electrical Engineering, Federal University of Juiz de Fora, Juiz de Fora, Brazil

Prateeti Mukherjee
Department of Computer Science & Engineering, Institute of Engineering & Management, Kolkata, India

Amartya Mukherjee
Department of Computer Science & Engineering and BSH, Institute of Engineering & Management, Kolkata, India

Rayen Naji
Medical School, Federal University of Juiz de Fora, Juiz de Fora, Brazil

Ravi Kumar Patel
Research Centre, R&D Department, UPES, Dehradun, India

Sahar Qazi
Department of Computer Science, Jamia Millia Islamia, New Delhi, India

Atiqur Rahman
Department of Computer Science and Engineering, East West University, Dhaka, Bangladesh

M.G. Rathi
Department of Information Technology, Electronics, WCE, Sangli, India

Ratula Ray
School of Biotechnology, KIIT Deemed to be University, Bhubaneswar, India

Khalid Raza
Department of Computer Science, Jamia Millia Islamia, New Delhi, India

Ahmed Wasif Reza
Department of Computer Science and Engineering, East West University, Dhaka, Bangladesh

Abhishek Sharma
Research Scientist, Department of R&D, UPES, Dehradun, India

Nripendra Kumar Singh
Department of Computer Science, Jamia Millia Islamia, New Delhi, India

Gajendra Pratap Singh
Punjab Engineering College, Chandigarh, India

Yashmeet Singh
Punjab Engineering College, Chandigarh, India

Puru Lokendra Singh
Department of Electrical Engineering, Malviya National Institute of Technology, Jaipur, India

Kh. Manglem Singh
Department of Computer Science and Engineering, NIT Manipur, Imphal, India

S.P. Sonavane
Department of Information Technology, Electronics, WCE, Sangli, India

S.G. Tamhankar
Department of Information Technology, Electronics, WCE, Sangli, India

Seema Verma
Department of Electronics, Banasthali Vidyapith, Vanasthali, India

Samidha Mridul Verma
Department of Electrical Engineering, Malviya National Institute of Technology, Jaipur, India

Ankit Vijayvargiya
Department of Electrical Engineering, Malviya National Institute of Technology, Jaipur, India

Jitendra Virmani
CSIR-Central Scientific Instruments Organization (CSIR-CSIO), Chandigarh, India

About the Editors

Nilanjan Dey is an Assistant Professor in Department of Information Technology at Techno India College of Technology, Kolkata, India. He is a visiting fellow of the University of Reading, United Kingdom. He is a Visiting Professor at Wenzhou Medical University, China, and Duy Tan University, Vietnam. He was an honorary Visiting Scientist at Global Biomedical Technologies Inc., CA, United States (2012–15). He was awarded his PhD from Jadavpur University in 2015.

He has authored/edited more than 45 books with Elsevier, Wiley, CRC Press, and Springer, and published more than 300 papers. He is the Editor-in-Chief of *International Journal of Ambient Computing and Intelligence*, *IGI Global*, Associate Editor of *IEEE Access and International Journal of Information Technology*, Springer. He is the Series Co-Editor of *Springer Tracts in Nature-Inspired Computing*, Springer; Series Co-Editor of *Advances in Ubiquitous Sensing Applications for Healthcare*, Elsevier; Series Editor of *Computational Intelligence in Engineering Problem Solving and Intelligent Signal processing and data analysis*, CRC. His main research interests include Medical Imaging, Machine learning, Computer-Aided Diagnosis, and Data Mining. He is the Indian Ambassador of International Federation for Information Processing—Young ICT Group. Recently, he has been awarded as one among the top 10 most published academics in the field of Computer Science in India (2015–17).

Jyotismita Chaki is an Assistant Professor in Department of Information Technology and Engineering at Vellore institute of Technology, Vellore, India. She has done her PhD (Engg) from Jadavpur University, Kolkata, India. Her research interests include Computer Vision and Image Processing, Pattern Recognition, Medical Imaging, Soft computing, Data mining, Machine learning. She has published 22 international conferences and journal papers. She is the author of the book *A Beginner's Guide to Image Preprocessing Techniques (Intelligent Signal Processing and Data Analysis)* from CRC press. She has also served as a Program Committee member of 2nd

International Conference on Advanced Computing and Intelligent Engineering 2017 (ICACIE-2017) and 4th International Conference on Image Information Processing (ICIIP-2017).

Rajesh Kumar received a Bachelor of Technology Degree with Honors from National Institute of Technology, Kurukshetra, India, in 1994. He also earned a Masters of Engineering Degree with Honors from the Malaviya National Institute of Technology, Jaipur, India, in 1997; he earned a PhD Degree from the Malaviya National Institute of Technology, Jaipur, and University of Rajasthan, Jaipur, in 2005. He was awarded Post Doctorate Research Fellow in the Department of Electrical and Computer Engineering at the National University of Singapore (NUS), Singapore, from 2009 to 2011. He joined the Department of Electrical Engineering at the Malaviya National Institute of Technology, Jaipur, India, as a Lecturer in 1995. He is currently serving as Professor and Head. He is also adjunct faculty to Centre of Energy and Environment at Malaviya National Institute of Technology, Jaipur, India.

Dr. Kumar has carried out extensive research in various areas of theory and practice of intelligent systems, bio and nature-inspired algorithms, smart grid, power electronics, power management, applications of AI to image processing and robotics. He has published more than 450 papers in international referred journals and conferences. He has received and published 12 patents. He has supervised 15 PhD and 35 Master theses. He has delivered more than 100 expert talks in various conferences and workshops. Dr. Kumar has won the Career Award for Young Teachers, Government of India, in 2000. He received 6 best thesis awards, 5 academic awards, 12 best paper awards, 4 professional awards, and 30 student awards.

He is Vice Chairman, IEEE Rajasthan Sub Section, and Executive Member, IEEE PES-IAS Delhi Chapter and Computer Society of India, Rajasthan Section. He is Associate Editor of *IEEE ITeN* (*Industrial Electronics Technology News*); Associate Editor, *Swarm and Evolutionary Computation*; Associate Editor, *IET Renewable and Power Generation*; Associate Editor, *IET Power Electronics*; Deputy Editor-in-Chief, *CAAI Transactions on Intelligent Technology*; Associate Editor, *International Journal of Bio Inspired Computing*. He is an Editorial Member of more than 15 Journals. Dr. Kumar is also Senior Member IEEE (United States), Fellow IET (United Kingdom), Fellow IE (India), Fellow IETE, Life Member CSI, Senior Member IEANG, and Life Member ISTE.

Preface

With the growth of an aging population and chronic diseases, society has become more health conscious and patients have become health consumers observing for better health management. People's insight is shifting toward patient-centered, rather than the classical, hospital-centered, health services, which have been forcing the development of telemedicine research from the classic e-Health to m-Health and now to ubiquitous healthcare (u-Health). It is predictable that mobile and ubiquitous Telemedicine, integrated with Wireless Body Area Network, has a wide potential in developing the provision of next-generation uHealth. Regardless of the recent efforts and accomplishments, present u-Health wished-for solutions still suffer from shortcomings, hampering their adoption today. This book presents a comprehensive information of up-to-date requirements in hardware, communication, and calculation for next-generation uHealth systems. It compares new technological and technical trends and discusses how they address expected u-Health requirements. A detailed information on various worldwide recent system operations is presented. In particular, challenges in ubiquitous computing are highlighted. The purpose of this book is not only to help beginners with a holistic approach toward understanding u-Health systems but also present to researchers new technological trends and design challenges they have to cope with, while designing such systems. This book is concerned with supporting and enhancing the utilization of U-Healthcare in the several systems and real-world activities. It provides a well-standing forum to discuss the characteristics of the U-Healthcare systems in different domains. The book is proposed for professionals, scientists, and engineers who are involved in the new techniques of the U-Healthcare systems. It provides an outstanding foundation for undergraduate and postgraduate students as well. It has several features, including (i) presents an outstanding basis of the U-Health data analysis, and management tools in different applications with highlight on the sensor systems, (ii) highlights the Internet of Things-enabled U-Healthcare, and (iii) includes different applications and challenges with extensive studies for U-Healthcare system.

The book is organized as follows:

Chapter 1 gives the overview of advanced processing techniques and secure architecture for sensor networks in ubiquitous healthcare systems. This chapter discusses the applications and challenges in U-healthcare systems while examining secure architectures to address specific research problems. Advanced data processing methods are also studied to accurately label and predict possible medical problems based on past data collected from different resources, such as past medical records and incoming data via sensor networks.

Chapter 2 deals with wireless sensor networks toward convenient infrastructure in healthcare industry. This chapter seeks to shed light on WSN's applications in the healthcare industry. It is also shown that WSN offers wireless sensing technology and creates a sophisticated infrastructure for the use and affordability of multiple data collection.

Chapter 3 is devoted to a comprehensive dialogue for u-body sensor network. The chapter encourages novice and even expert users to use U-Health systems cases and proposes a model for choreographing the backend in the U-Body Sensor Network. This chapter examines currently existing U-Body Sensor Network technologies, communication protocols, and data capture methods. An experimental case study is broached with problems of network delay and latency that highlight the difficulties of ubiquitous computing, data handling, and data processing. The chapter focuses on the design perspective of the multichannel gateway consisting of streaming, surveying, and sampling data.

Chapter 4 discusses compressive sensing in medical signal processing and imaging systems. Compressive sensing in medical engineering, particularly in ECG, MRI, and CT, is briefly reviewed in this chapter. The advancement of electrocardiogram (ECG), magnetic resonance imaging (MRI) and computed tomography (CT) is an auxiliary tool for the diagnosis of different diseases. To evaluate the patient's condition, these tools use data processing. Such data must, however, follow a decent quality threshold in order not to undermine clinical judgment. Compressive Sensing (CS) therefore acts as a new approach that benefits the speed and quality of the diagnosis.

The focus in Chapter 5 is on nanopore sequencing technology and Internet of Living Things, which is the big hope for U-healthcare. The chapter focuses on the introduction of nanopore sequencing technology, its applications for U-healthcare, paradigms for the Internet of Living Things and their applications in U-healthcare, the convergence of concepts for U-healthcare between nanopore and the Internet of Living Things, the extent to which the promises, opportunities, and challenges are also discussed.

Chapter 6 provides an overview of Internet of Things-enabled virtual environment for U-health monitoring. This chapter is devoted to the introduction of platforms for Internet of Things, their technical insight, and their real-time implementation. In this chapter, Message Queuing Telemetry Transport platforms will be used as data networks and the reader will gain insight into the operation of Message Queuing Telemetry Transport Internet platforms, after which the reader can create a virtual Internet of Things communication platform him/herself. Raspberry Pi is a microcontroller with a built-in feature to connect to Wi-Fi, which is a prime requirement for virtual U-health monitoring for the Internet of things. This chapter offers all required insights to bridge the gap between sensor data and data networks using the Message Queuing Telemetry Transport.

Chapter 7 deals with behavioral relationship between Internet of Things and classification techniques in abnormal situation. The authors propose a method that includes Internet of Things in this chapter to predict better results from any diabetic dataset. Using wearable sensors, mobile phones, smart devices, or through hospital officials, this system sends the patient's diabetic information to the central database. If any abnormal or suspicious value is found after receiving the data from the user or operators, this proposed intelligent system will process these data using machine-learning techniques and properly predict the disease. In order to obtain a large number of data, the authors also propose a central database that all hospital and medical professionals, law enforcement agencies with different access control systems, can access.

Chapter 8 provides an overview of intelligent energy-efficient healthcare models integrated with Internet of Things and long-range network. Three case studies relevant to the context are introduced in this chapter. The first deals with a wearable asthma-monitoring system that can actually monitor the symptoms of an asthma attack by learning the patient's asthma pattern, which provides a basis for tracking events. The second study focuses primarily on improving the quality of life of patients with neurodegenerative diseases such as Parkinson's by developing a state-of-the-art wearable device that uses embedded sensors to detect fluctuations in neutral controlled movement and determines the dose of medicines. The third idea deals with the concept of an intelligent band-aid with different sensors that can ideally be positioned in the wounded area and the amount of drug to be provided can be monitored automatically by tracking the variances of the physiological parameters.

The focus in Chapter 9 is on wearable fitness band-based U-health monitoring. To analyze the effects of body metabolism, the system connects to the body and records one or more physiological parameters. The recorded data are then sent to the base station, which can be a PC, smartphone, or other embedded electronic device.

Chapter 10 deals with the role of trust in ubiquitous healthcare system. This chapter focuses on available low-power, secure technologies and challenges in all relevant aspects of smart healthcare embedded sensor systems with possible recommendations for future research.

Chapter 11 provides an overview of diagnosis of diabetic retinopathy and glaucoma using fundus images. To carry out a differential diagnosis between two of the most common eye diseases, diabetic retinopathy and glaucoma, a computer-aided classification system was developed in the chapter. These two diseases are the main reason for cases of visual loss in the world and a comparative method of differentiation between these two would therefore be of great assistance to ophthalmologists in proper diagnosis. The techniques involved in image preprocessing and image enhancement are used to build this model.

The focus in Chapter 12 is on different classifier for early detection of knee osteoarthritis. A comprehensive comparison was presented in this chapter for a knee osteoarthritis classification task using different Support Vector Machine kernels. It is observed that the Support Vector Machine Fine Gaussian Kernel with threefold cross validation gives the highest accuracy compared to other kernels.

Chapter 13 deals with a comparative study for brain tumor detection in MRI images using texture features. This chapter provides a detailed explanation of all methods of extraction of features based on texture and their importance in the detection of brain tumors. The chapter aims to explain how a particular method extracts the characteristics of the texture and builds a vector. This chapter has developed a common experimental protocol to determine and assess the efficiency of key feature extraction methods, as there is no thorough empirical comparison in literature. The proposed protocol contains a standard dataset and a range of classifications for evaluating the effectiveness of specific feature extraction methods over MRI images.

Nilanjan Dey, Editor
Techno India College of Technology, Kolkata, India

Jyotismita Chaki, Editor
Vellore Institute of Technology, Vellore, India

Rajesh Kumar, Editor
Malaviya National Institute of Technology, Jaipur, India

PART 1

U-Healthcare monitoring system using sensor networks

CHAPTER

1 Advanced processing techniques and secure architecture for sensor networks in ubiquitous healthcare systems

Prateeti Mukherjee[1], Amartya Mukherjee[2]

[1]Department of Computer Science & Engineering, Institute of Engineering & Management, Kolkata, India [2]Department of Computer Science & Engineering and BSH, Institute of Engineering & Management, Kolkata, India

1.1 Introduction

Ubiquitous Healthcare or U-Healthcare involves deployment of healthcare facilities on a wider scale than traditional healthcare systems. It seeks to build a communication platform through sensors, controllers, and servers between patients, physicians, and other healthcare workers. It is a pervasive healthcare structure that enables transportation of medical information accurately and securely, thereby supporting a wide range of applications such as patient monitoring, immediate response and emergency intervention, location-based delivery of medical services and equipment, telemedicine, etc.

The idea is said to have stemmed from the vision of Mark Weiser—the father of ubiquitous computing. According to M. Weiser, the most profound technologies are those that disappear as a fundamental consequence of not only technology but also of human psychology [1]. The concept of ubiquitous computing involves transparent computing hardware and software, enabling convenient interaction of the user with services. The definition of ubiquitous healthcare, in its simplest of form, involves two basic ideas—the domain of application of technologies enabling ubiquitous computing and the concept that integrates healthcare seamlessly into our daily lives. The essence of the vision lies in the creation of a computing environment where the users would become an integral aspect of computation, where technology would become pervasive and invisible, enabling the users to focus away from the technology toward higher goals. Realization of the concept requires multiple sensors, either mobile or embedded, in frequently used devices such as phones, wrist bands, cars, home appliances, clothing, and various consumer goods—all communicating through increasingly interconnected networks. In ubiquitous computing frameworks, the functionality is not based on each device separately but on the interaction of all of

Sensors for Health Monitoring. https://doi.org/10.1016/B978-0-12-819361-7.00001-4

them. The objective, according to M. Weiser, is that devices fit the human environment instead of demanding humans to enter theirs, and to surround a user with a large number of application-specific, network-connected appliances that unobtrusively provide them with information and services with respect to their environment and relevant to a context useful to the user.

A popular term that is frequently encountered while studying pervasive computing techniques is Internet of things (IoT), a concept that seeks to connect any device with the internet, thereby creating a network of interconnected devices. In Ref. [2], the term is aptly described as the technical equivalent of internet connectivity anytime and anywhere. Exploitation of IoT paradigms is expected to be a game changer in the healthcare industry [3, 4]. Reports [5, 6] suggest that in the period from 2017 until 2022, growth in IoT healthcare applications is poised to accelerate with various stakeholders stepping up their efforts. It is evident that IoT-based platforms formulate a key component in the digital transformation [7] of the healthcare industry. Interconnected devices provide more feasibility to patients, enabling continuous observation of heart rate, glucose levels, etc., while connecting them to specialists all over the world. There is an increasing consciousness and engagement of consumers with regards to their health, resulting in a boost in demand for remote, personalized healthcare services. This encourages various healthcare ecosystem players to come up with novel approaches and partnerships, thereby contributing to the development of more integrated and IoT-enabled eHealth approach.

The healthcare sector is a lucrative and appropriate domain for ubiquitous computing, since obtaining accurate information at the right time with negligible dependency on aspects such as location and time is of utmost importance in this industry. Ubiquitous healthcare frameworks, when appropriately deployed in practice, have the potential of saving large sums of money associated with care and delivery of medical services besides being instrumental in saving countless lives.

Healthcare is the largest service industry worldwide and is hence directly related to the global economy. However, even in the most developed nations, the exponential rise in healthcare costs is observed, possibly due to an aging population suffering from chronic diseases that require long-term treatments, continuous monitoring, and frequent hospitalization. Statistical data claim that the global population of adults over the age of 65 is likely to reach 761 million by the year 2025. With such huge numbers, providing quality healthcare through traditional methods of service results in increasing costs. At 1.02% of its gross domestic product (GDP), India's public health expenditure is among the lowest in the world. This figure has almost remained unchanged since 2009 and is lower than that of most low-income nations that spend 1.4% of their GDP on healthcare. Hence, the masses turn to the private healthcare sector, which is a much more expensive alternative to government services. Indians are the sixth biggest out-of-pocket (OOP) health spenders in the low-middle income group of 50 nations. According to studies [8–11], these numbers push approximately 32–39 million Indians below the poverty line every year.

Shifting from traditional healthcare methods to ubiquitous technologies is a possible solution to such terminal maladies faced by the healthcare industry.

U-Healthcare is not merely a technological innovation, it involves a paradigm shift in healthcare practices from managing illness to managing wellness while promoting evidence-based medicine and collaboration among physicians and their teams to provide best possible affordable healthcare to the patients. It heralds the advent of an age where healthcare will become increasingly digital and virtual and patients will play a more active role in managing their own healthcare through consumer-operated interoperable ubiquitous computing technologies. According to Ref. [12], ubiquitous healthcare would result in a switch from patient-centric systems to patient-centric operational models.

1.2 Challenges in ubiquitous healthcare systems

1.2.1 Research challenges

The fundamental concept of ubiquitous healthcare is characterized by a pervasive continuous collection of sensor data followed by real-time processing of monitored data to derive relevant physiological parameters, thereby enabling the system to draw effective, meaningful, context-aware, and patient-centric conclusions. The paradigm hence requires real-time processing of wirelessly collected vital signs and running complex models to analyze the processed information under the current circumstances, such as age, location, physical activity, and ambient conditions. Real-world biomedical sensor nodes are limited by their computational capabilities and are unsuitable for the execution of a series of complex tasks. Such problems are brought to light only when a theoretical research idea that existed solely on paper is simulated in a practical environment, having opened various avenues of research. The sequence of tasks is illustrated in Fig. 1.1.

The processing part involves tasks such as noise rejection, data disambiguation, and regular checks for data inconsistency since most inexpensive wearable sensing devices are prone to frequent failures. The sensors are present on multiple body sites on the patient, resulting in large chunks of raw data. Further, when deployed in a larger setting such as hospitals or retirement homes, several sensors are placed on multiple patients, increasing the dataset size even more. Hence, there is a need to prioritize the transmission of such huge amounts of preprocessed vital sign data in terms of their degree of importance for diagnosis to avoid network traffic congestion and to maximize reliability. Finally, the collected vital signs must be fed as input to an appropriate model to derive meaningful physiological parameters of interest.

The intelligence of a ubiquitous healthcare system and how well it can take into account the user's context and their current situation in its analysis is a direct measure for the success of the system in the real world. Knowledge of external and internal entities that cause changes in the user's state is necessary for ubiquitous healthcare technologies since it necessitates different interpretation of the data received at any given time. Such entities form a broader information set termed as contextual information and knowledge of such entities is termed context awareness. In ubiquitous

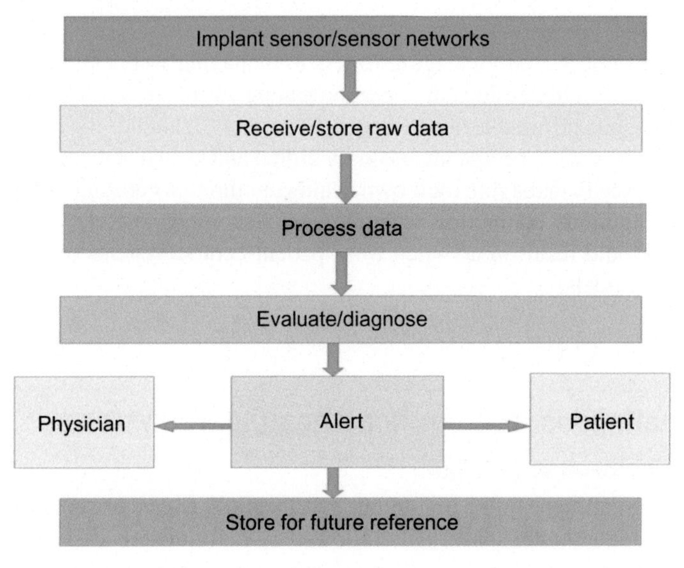

FIG. 1.1

A simple representation of tasks involved in ubiquitous healthcare systems.

healthcare, contextual information may contain parameters such as physical activity, location, as well as environmental conditions obtained by processing data from kinematic sensors, GPS, temperature/humidity sensors, etc.

The sequence of operations involved in ubiquitous healthcare architectures involves complex, computationally expensive tasks such as simultaneous execution of algorithms responsible for data processing and prioritizing tasks on the raw sensor data while running models that derive relevant physiological parameters for acquiring context awareness. Such processes are significantly limited by the computing capabilities of wearable sensor nodes. To overcome this limitation, researchers have suggested alternate electronic devices existing in the vicinity to act as elastic resource pools that process chunks of locally generated vital sign data in parallel [13]. The suggested architecture is shown in Fig. 1.2.

In this approach, the alternative devices form a wirelessly connected local mobile computing grid, and the collective computational capability is exploited to perform in-network tasks. In this framework, the sensor nodes only act as data providers, thereby offloading the task of executing compute-intensive algorithms and models to the computing grid. The vital sign data are transmitted via Bluetooth or ZigBee and a broker, existing either as a separate entity or as a logical role played by one of the service providers, is responsible for managing the elastic resource pool. The responsibilities of the broker include handling service requests from data providers and allocating tasks to the available service providers. The information thus obtained after processing in the local mobile computing grid can either be uploaded

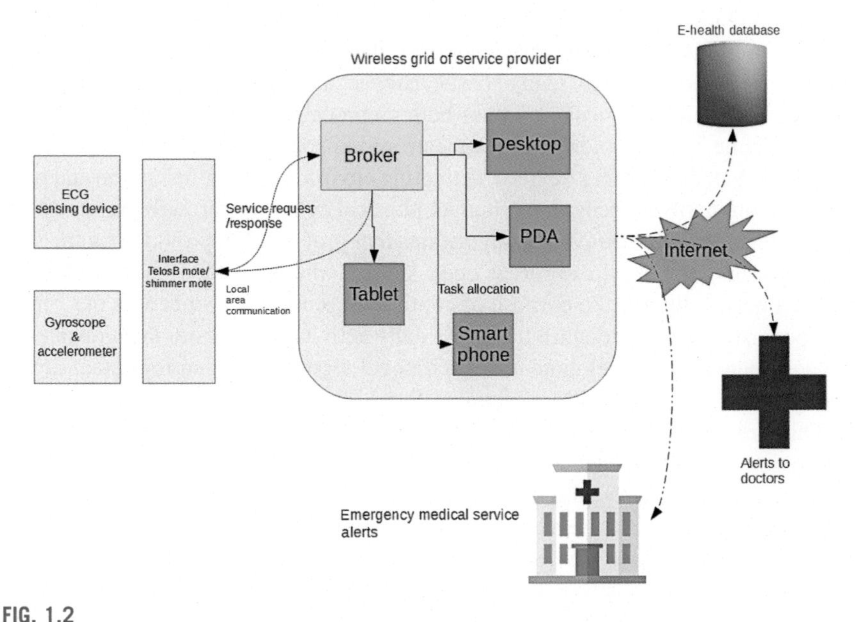

FIG. 1.2

Concept diagram illustrating the proposed solution.

to an electronic health record database or used to generate and send alerts to emergency medical service personnel over the Internet.

The advantages of the mentioned approach are as follows:

(1) The computational load on individual service providers is reduced since both data parallelism and task parallelism techniques are employed.

(2) Collective vital signs and processed contextual information can be stored in remote, secure electronic health record databases. Such historical data are advantageous for data and patient-centric decision making and form valuable datasets for research applications.

Numerous research challenges are associated with the realization of such seemingly robust frameworks. Discovery and provisioning of computing resources is an area of concern; however, the concept of the broker infused with a resource provisioning architecture for the mobile grids is a viable solution. Wireless sensor networks (WSN) also propose compatibility issues, with radio frequency interference produced by transmissions from multiple patients in homes and hospitals, along with the proliferation of numerous electronic devices using the same radio technologies.

Multiple body sensors also have biocompatibility issues that involve safe limits for various parameters. The radio frequency (RF) interference power is issued a safe limit of less than 1.6 W/kg, as issued by FCC Regulation, and is hence a limiting factor for the transmission of sensor data to the computing grids. The electronic devices implanted on a patient's body are bound to release heat when running for hours at a stretch. The heat dissipation of these sensors must be minimized as increased temperatures may lead to tissue damage and increase the risk of infections. The decision-making process involves both preprocessed data and contextual information. While in most ubiquitous healthcare systems, the responsibility of collecting data is delegated to sensing devices, extracting environmental information and physical location involves recognizing human physical activities through the design and development of novel distributed algorithms that process data provided by the kinematic sensors attached to different body sites. Computer-vision-based techniques may not be suitable for the purpose of continuous monitoring since it is not considered pervasive and may disturb the user's daily activities and result in an invasion of privacy. Hence, the development of such novel algorithms is an important area of research, discussed in detail in Section 1.5.

Other research challenges include the technical aspects of the devices involved. Integrating Ubiquitous Healthcare devices with expensive equipment renders them unpopular in the pharmaceutical and medical markets, thereby defeating the purpose of the architecture itself. Hence, reasonably priced sensors and boards must be used while considering reliability issues. The devices used and network grids employed must be robust and sensitive to erroneous data and respond appropriately. Further, the sensors must be versatile enough to operate in conditions not encountered by typical computing devices such as rain, sleet, hail, wild temperature variations, high humidity, saline, and other corrosive substances, and must have efficient heat dissipation techniques that allow prolonged use of these devices, without harming the patient. Hence, the product design and components used to play a crucial role in such healthcare architectures and are a prospective area of research for engineers.

1.2.2 Ethical challenges

Medical equipment supporting ubiquitous healthcare frameworks primarily comprise tiny sensors attached to the patient on multiple body sites that constantly monitor the patient's activities and the surrounding environmental conditions. This concept poses challenging ethical questions that range from individual trust issues to social and economic problems relating to security and financial sustainability of such architectures. In pervasive computing frameworks, the task of integrating computing infrastructure into everyday life while avoiding invasive behavior often brings about tradeoffs between flexibility and robustness, efficiency and effectiveness, as well as autonomy and reliability. Thus, security is of utmost importance in a pervasive environment where users share public resources to operate on sensitive private data. For this technology to be secure while involving limited human intervention requires advanced user authentication systems and transparent, reliable guidelines relating to resource management and privacy concerns. Controversies reduce the prospects of ubiquitous computing in healthcare and could greatly hinder the possibilities of ubiquitous healthcare for mass applications. Novel security architectures are discussed in Section 1.6 that strives to eliminate threats to the integrity of sensitive data in ubiquitous computing systems.

"The ethical challenges of ubiquitous healthcare" published in the *International Review of Information Ethics* [14] raises a relevant question in the context of pervasive computing methods in the medical industry. Medicine is essentially a controlled profession with government-appointed authorities monitoring and restricting the practices involved. When it comes to ubiquitous computing techniques, computer software and hardware development notoriously lack such regimes to check their performance and invasive tendencies, thereby giving rise to individual trust issues relating to such technology. The sensor motes provide continuous streams of either raw or processed data, which is eventually stored in public storage domains. The question of privacy is unavoidable in this framework. The question is, who owns the data? The EU Data Protection Directive [15] classifies medical information as sensitive data, which makes the concept of placing medical records onto single public systems that are vulnerable to unauthorized mass access questionable. When designing ubiquitous computing frameworks in the field of healthcare, close attention must be paid to the specifics of who controls the data, the extent of data gathered, and where the information is stored.

Societal questions relating to the ethics of ubiquitous computing often deal with forecasts of the future of humanity with the advent of such technology. In the works of Bentham and Foucault [16, 17], the question of whether technology holds the potential of gradually reshaping unhealthy behaviors is raised. Skeptics fear the advent of an era of robotic nurses in every hospital and care facility, carrying chunks of sensitive data, and prescribing textbook drugs and procedures, thereby making the process methodical and for many, unreliable. Development of frameworks that can mitigate such beliefs and win the trust of individuals and the society at large is necessary for ubiquitous healthcare to gain the approval of the masses.

The economy is also an undeniable factor in the healthcare sector, mainly when it comes to such advanced computing techniques with complex hardware and software platforms. The rift between the low-income groups and upper classes is forever widening in many developing and developed nations. Advanced healthcare requires monetary support, and the development of ubiquitous technologies might exacerbate the trend if executed poorly. The transfer of resources from traditional healthcare to ubiquitous devices could severely worsen the financial issues already troubling healthcare sectors worldwide.

1.3 Enabling technologies

The momentum for ubiquitous healthcare in the present era is created primarily by the abundance of technologies enabling ubiquitous computing and development of specific technologies related to modeling and measuring medical data. The technologies playing an enabling role in laying the foundation for ubiquitous healthcare include ubiquitous computing, ubiquitous communication, biomedical sensors, and monitoring devices. U-Healthcare does not involve off-the-shelf computing devices, rather novel holistic software and hardware technologies that seek to bridge the gap between the virtual and physical worlds through the incorporation of computing power and sensing into any electronic setup used on a day-to-day basis. Pervasive techniques in ubiquitous computing may also involve embedding microprocessors in everyday objects such as white goods, toys, plates, cups, glasses, houses, furniture, and even paint, as introduced by the concept of smart dust [18]. Such computing architectures form a critical evolutionary in a line of technological advancement that started in the 1970s. Fig. 1.3 illustrates the classification of technologies enabling ubiquitous healthcare.

The relatively recent advancement of grid computing in distributed computing paradigm has been discussed in Section 1.2, where file exchange goes hand in hand with resource access and transparency in hardware- and software-sharing methods. Grid technologies create dynamic and secure environments for computing and storing resources, which are very conducive for ubiquitous healthcare, enabling doctors from across the globe to access patient data and run complex simulations and virtual surgical procedures. The benefits of such shared network architectures translate to an improvement in the quality of the data, a decrease in development costs, and an efficient sharing of electronic patient-related information [19].

An approximate of 350 million mobile devices exists worldwide, with an expected increase in numbers beyond 1 billion in the following years. These statistics verify the importance of personal digital assistants (PDAs) and mobile devices in our daily lives. The cornucopia of such devices surrounding us offer a wireless mobile

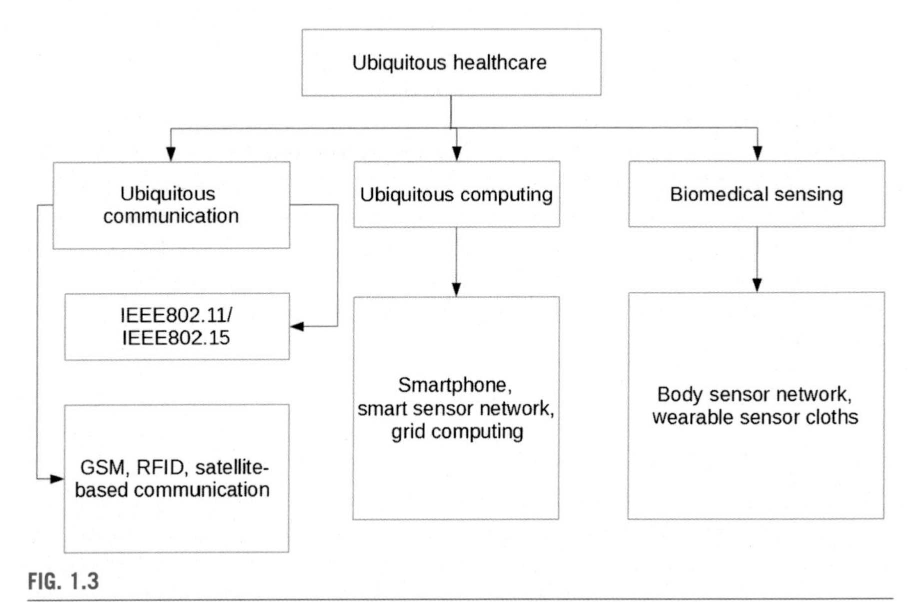

FIG. 1.3

Classification of enabling technologies for ubiquitous healthcare.

platform that may be used to run healthcare applications, user interfaces, and data-logging methods for monitoring systems, and provide a gateway to connect local devices collecting health-related information to global services such as hospital databases. radio frequency identification technology (RFID) allows simple, inexpensive wireless communication with nearby objects to obtain information such as product code, URL, or sensor data. Wearable biomedical sensors enable the collection and analysis of physiological patient data while the patient is mobile. Natural interactions between the user and computing devices are enabled by intelligent user interfaces (IUI) that is capable of interpreting speech and gesture while considering the context as well as user preferences. Such supporting technologies and their connections are discussed in detail as follows.

1.3.1 Wireless sensor networks

The sensors in ubiquitous healthcare frameworks predominantly act as data providers, sensing biological information from either outside or inside the human body. They communicate the acquired information to a control device worn on the body or placed in the vicinity. Then, the data assembled from the control devices are conveyed to remote destinations in a wireless body area network (WBAN) for the diagnostic and decision-making process to be executed. As mentioned in [14], the requirements of sensors used in wireless networks in healthcare systems are as follows:

1.3.1.1 Unobtrusiveness

The fundamental concept of pervasive computing lies in the unobtrusive nature of the technology involved. Therefore, the most essential requirement in the design of wireless medical sensors relates to their light weight and miniature size to allow continuous, noninvasive, and unobtrusive monitoring of the patient. It is experimentally known that the size and weight of any electronic sensor rely heavily on the size and weight of its batteries. However, simply decreasing battery size does not solve the issue, since a battery's capacity is directly related to its size. This is a relevant area of research in the development of microelectronics. Technological advances have recently led to the development of small-size, high-energy batteries [20] that seem promising for the purpose of wearable sensing systems.

1.3.1.2 Security

As discussed in Section 1.2, security and privacy concerns are a major challenge to this frontier of *E*-Healthcare. There are legal implications to data infringement activities, and the systems must fulfill privacy requirements provided by the law. Different security measures and complex architectures are discussed in Section 1.6 to resolve privacy issues. The ultimate goal is the development of frameworks with guaranteed security, where there is coordination between the system hardware and related security software components to establish reliable communication networks [21].

1.3.1.3 Interoperability

Data mobility and sharing of resources is a key element in ubiquitous healthcare systems. It seeks to reform the chaotic and often dysfunctional nature of information exchange among hospitals, physicians, and large care facilities. It is a testament of the extent to which various systems and devices can accumulate and interpret data, and record the information in a user-friendly, organized manner. Through this concept, data are made exceptionally mobile, since data exchange methods allow information to be shared across hospitals, pharmacies, labs, clinicians, and patients, regardless of the vendor involved. The concept minimizes cost through this resource-sharing mechanism, and the data entered once into the system becomes available to the patient for future use, whenever and wherever they should require it [22].

1.3.1.4 Reliability

This aspect of sensor networks has been briefly discussed as a challenge in Section 1.2. Although inexpensive sensing devices are a considerable threat to the reliability of the system, faults in the communication and transmission steps could lead to erroneous diagnosis and failure of the network. The communication constraint varies between nodes since the sampling rates required by each sensor are different. High demands on the communication channel, however, increase the requirements of the system and reduce battery life, thereby increasing the overall expenses. A calculated tradeoff between communication and computation is of paramount importance in the design of medical applications relying on WSNs. A traditional WSN structure is shown in Fig. 1.4.

WSN essentially comprise a network of typically small and portable wireless devices that are spread across an area, or in this case, in multiple body sites, to collect data in a cooperative fashion, and then forward the data to one or more collectors, commonly called sinks. A subset of such networks introduces another very important terminology in the field of ubiquitous healthcare—body area networks (BANs).

1.3.2 Body area networks

A body area network is simply a WSN where the wireless sensors are placed over, or inside the body of a patient or individual, to collect biomedical data. The biosensors generate data and transmit them to one or more sinks for storing or processing. A simple structure of such networks is illustrated in Fig. 1.5.

Designing a BAN involves classical WSN issues such as deciding the topology of the network and how the data are to be routed from the biosensors to the sinks, along with specific challenges relating to their presence on the human body. An issue briefly discussed in Section 1.3 deals with battery sources. The high-loss propagation behavior of wireless signals through and over the human body cannot be solved by increasing power emissions as in typical wireless networks. In BANs, power emissions must be contained to both avoids damages to human tissues due to overheating, and to preserve the charge of sensor batteries, whose substitution

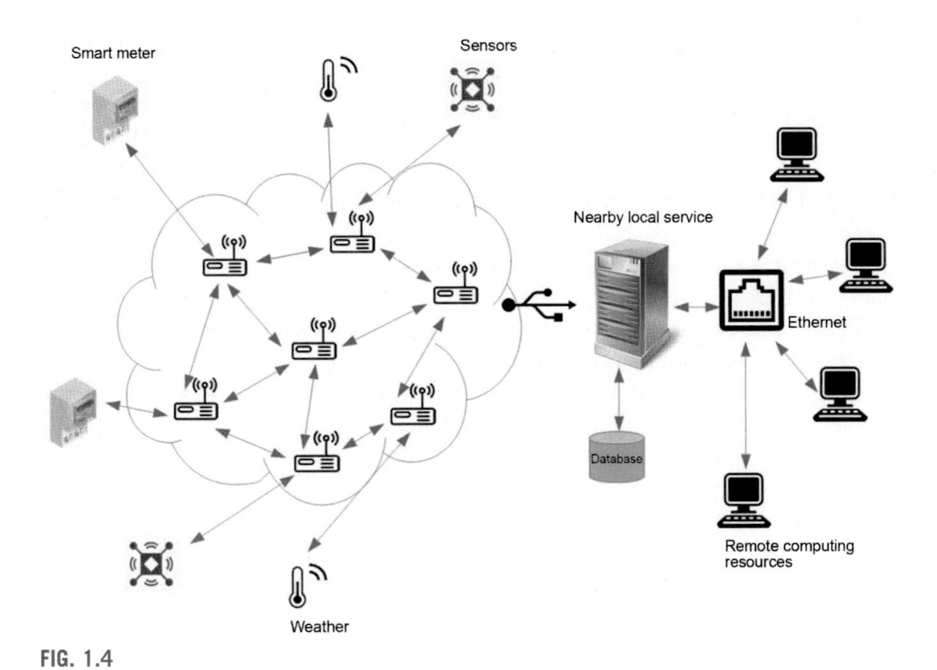

FIG. 1.4

Traditional WSN architecture.

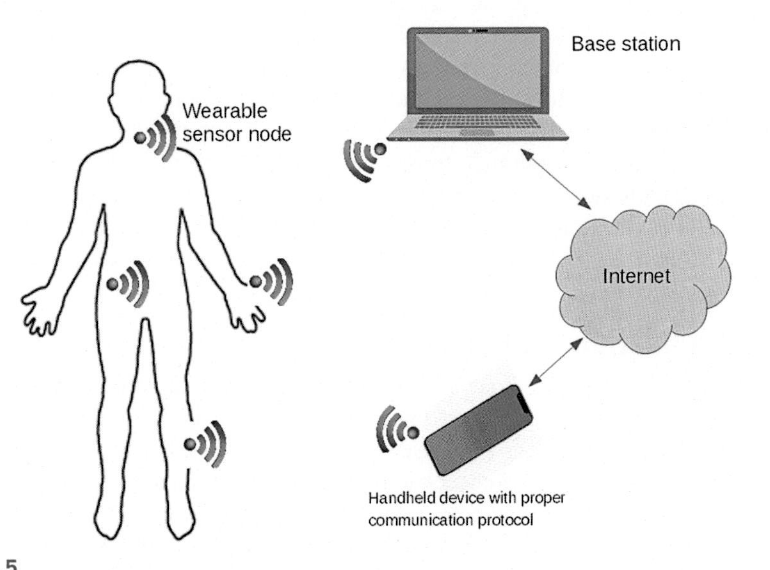

FIG. 1.5

Traditional WBAN architecture.

can be very uncomfortable and inconvenient for patients. Hence, controlling energy consumption is a major aim in BAN design, typically achieved through multihop routing [23], implemented through the addition of intermediate relay nodes between sinks and sensors. Several works exist in the field of energy-efficient routing protocols for BANs [23–27].

The design aspects of BANs could also be approached in terms of optimization models and algorithms. In Ref. [23], the concept of multipath routing and relay deployment is employed to minimize the total cost of BAN deployment by approaching the design as a mixed integer linear program. Mixed Integer Programming based concepts have been proposed [27] to solve complex optimization problems while tackling data generation uncertainty of BAN sensors. A recent research work [26] suggests a new integer linear programming (ILP) heuristic for solving the design problem, based on combining deterministic and probabilistic variable fixing strategies guided by peculiar linear relaxations of the robust optimization model.

Another energy-efficient network architecture involves multilayered topology based on the distance of the deployed sensors from the controller [28]. Each layer in the architecture is divided into multiple clusters, and cluster controllers from each layer are selected depending upon the residual energy of the sensors. Reclustering in the layers is avoided. The cluster collectors accumulate data from the cluster members and forward the aggregate data to the controller. The clusters in the individual layers are capable of communication with clusters of the immediate upper and lower layers. The problem of probable link failures in wireless network connections is addressed through cooperative communication among the controller units using amplify-and-forward and decode-and-forward techniques, and a fusion of all the information gathered from the different controllers is done at the fusion node. The architecture is illustrated in Fig. 1.6.

There is a plethora of research work dealing with the many problems prevalent in sensor network design for smart healthcare applications. Although the technology to support ubiquitous healthcare is still taking shape, the main challenge lies in interoperability among the multitude of devices and applications, and the short-range, low-power communication occurring between noncomputational devices (sensors) and parallel devices running healthcare applications.

1.4 Application scenarios

A multitude of wirelessly connected sensors enable spatiotemporal sampling of cognitive, behavioral as well as physiological processes in areas ranging from individual homes to large-scale settings such as care facilities, clinics, and hospitals. With the current advancements in sensing technologies, significant progress has been reported [29] in the manufacture of low-cost, high-quality sensors for use in both personal and public healthcare services. Further, constant developments in the field of machine-learning models [30] enable complex solutions to be inferred from massive bulks of sensory information. Pervasive communication techniques enable timely publication

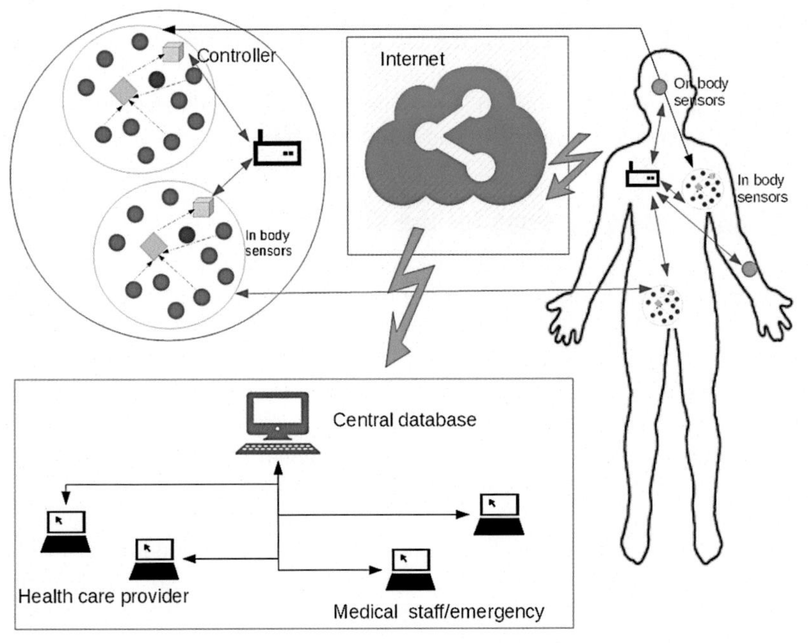

FIG. 1.6

The proposed communication architecture.

of the processed sensor information to caregivers, thereby opening a world of possibilities for ubiquitous computing in the healthcare sector.

1.4.1 Vital sign monitoring in hospitals and care facilities

In hospitals and clinics, a tangled mess of wires attached to a patient's body is a common scene. The restricted mobility is not only uncomfortable for the patient, but cases of increased anxiety levels are often encountered, and deliberate disconnections are fairly common. Failure to reattach sensors at appropriate body sites by the staff is another risk, since patients may be moved around in a hospital and carried through multiple units for different procedures. Ubiquitous healthcare networks help address such drawbacks associated with wired sensors, since their pervasive nature makes them less noticeable, and authorized access to backend medical record systems helps reduce the number of steps in a patient's diagnosis and treatment process.

In Ref. [31], the proposed system is capable of monitoring physiological parameters and making appropriate decisions based on data received from multiple patients. The coordinator node attached to the patient's body forwards the collected sensor data to a base station. The system detects abnormal conditions, issues an alarm to the patient, and sends text/e-mail updates to their physician.

Another work [32] considers the use of Zigbee technology to develop a home-based wireless ECG monitoring system for continuous real-time, noninvasive supervision of patients from the comfort of their own homes. Such smart systems take a portion of the load off the staff and physicians-in-charge, and enhance the quality of the patient's stay at care facilities and health centers.

1.4.2 Monitoring systems for the elderly

With age, several physical, cognitive, and social changes tend to challenge the health, independence, and quality of life of individuals. Their bodies become more susceptible to health issues like congestive heart failure, diabetes, arthritis, asthma, chronic bronchitis, memory loss, among others. The advent of any devastating episode can be quite erratic, and constant monitoring is crucial to provide emergency medical services to the patient.

Wirelessly networked sensors embedded in human's living spaces or carried on the person can collect precious information about personal, psychological, physical, and behavioral states and patterns in on the run real time. Such data, when correlated with contextual information such as social and environmental conditions, enable useful inferences to be drawn about health and status of the individual. Self-awareness in individuals is very useful as they can assist their medical team in early detection and intervention in health concerns. These procedures also provide effective and economic ways of monitoring and combating age-related illnesses.

A monitoring system code named UbiHeld (Ubiquitous Healthcare for elderly) [33] is an end-to-end healthcare monitoring device that combines pervasive computing platforms (primarily smartphones), employs skeleton-based activity detection and localization using depth sensors like Kinect, and analyzes the subject's social network feed to determine individual's mental condition for overall ubiquitous monitoring of the elderly. It makes use of IoT-based paradigms to support the back-end platform.

Similar systems designed to cater to the needs of the elderly have been developed exploiting embedded system technology and IoT frameworks [34–36].

1.4.3 Smart systems to assist the disabled

Wirelessly networked sensing devices are great tools when it comes to providing guidance and help to patients coping with declining motor and sensor capabilities. Intelligent assistive devices employing ubiquitous computing techniques coupled with IoT paradigms make use of the information about the patient's physiological state from sensors embedded in the device, worn or even implanted in the user or their surroundings, to perform appropriate meaningful actions.

The IlluminaStick is an original, unreleased prototype, designed to aid the visually impaired in their daily activities. The life of a visually impaired person is challenging, to say the least. Living in a world of almost complete darkness, safety of the individual is a pressing issue. Although through other sensory processes, a blind person may navigate through a known setting with ease, traveling places without assistance is never a viable option. Hence, this product aims to make the life of the disabled more independent, with amplified safety features to aid the concerned individual every step of the way. The design of the device is shown in Fig. 1.7.

Walking through the city is comparable to walking in a land mine when it comes to the visually impaired person. The smart blind-system grants complete freedom to such an individual. Users can easily navigate their way through busy, bustling streets without assistance. Although the device is robust and reliable, yet all control is not relinquished from the user and their loved ones. Since emergency situations and malfunctions cannot be ignored, the connectivity and tracking features ensure complete safety of the individual. Hence, the fear factor and subsequent dependence of the disabled is replaced by a smart, scalable, easy-to-use device that functions as the eyes of the user. This type of intelligent assistive devices can not only carry out their response to individual users and their current context, but also provide the user and their caregivers' important feedback for longer-term training. Embedded system platforms could be installed in traditional canes, crutches, walkers, and wheel chairs to provide personalized, context-specific feedback and guidance to the user at all times. Hence, such smart devices could go a long way in improving the life of the disabled or the recovering.

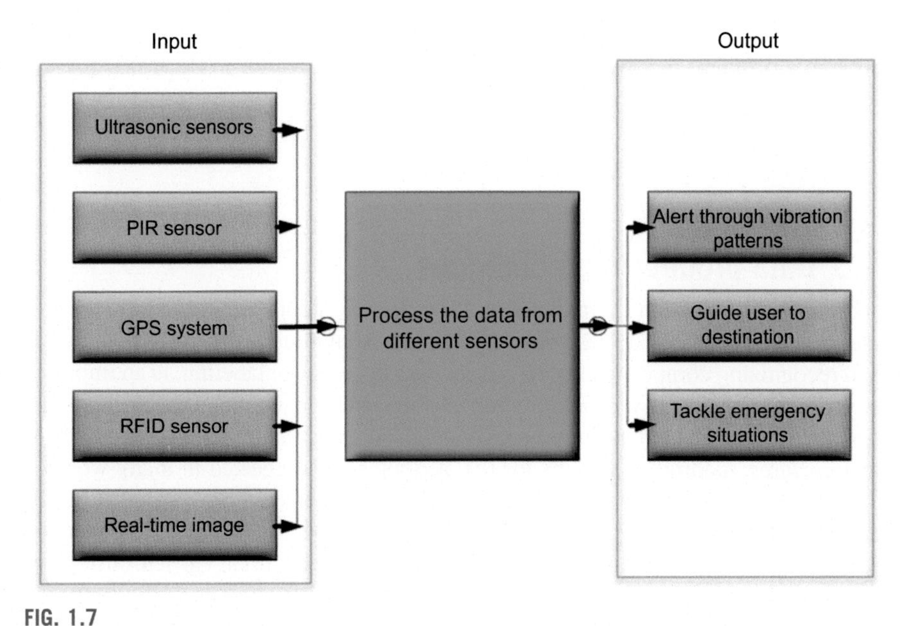

FIG. 1.7

Proposed architecture for IlluminaStick.

1.4.4 Expanding research prospective

Continuous collection of behavioral and physiological data from a large number of distributed subjects, going about their daily lives, through sensing motes,with their interconnected pervasive platforms, have begun to revolutionize medical and public health research studies across the globe. Controlled laboratory settings cannot replicate their ability to collect, process, and provide valuable insight to the subject's personal conditions, and cannot be measured from computer-assisted retrospective self-report methods. Hence, ubiquitous sensing systems are progressively becoming more and more important to medical, cognitive, and behavioral research.

The utility of distributed sensing technologies can be broadly classified into spatial and temporal scopes, relating to their data collection capabilities ranging from the scale of individuals to large populations. The spatial scope of distributed sensing can range from sensory observations of health status made when the subject is confined to a particular building, whether home, workplace, etc., or a larger, well-defined region, such as a particular disaster site for disaster management and rescue operations, to observations made when the individual is in motion, carrying out daily chores and activities. The temporal scope of sensing systems can range from specific monitoring for the duration of an illness, or continuous, long-term observations made for the purposes of early disease management, long-term health issues, or for public health purposes. Each spatial and temporal scope comes with its own set of challenges relating to network design, routing protocols, security measures, and different requirements on ergonomics.

There is a rich diversity of pervasive sensing technology with each component serving a different purpose and having different specifications. These sensing systems work harmoniously to enhance performance of a single platform tending to some specific application.

1.5 Data processing techniques

As suggested in Section 1.2, processing large chunks of data received from sensors at multiple body sites, and from multiple patients is a daunting task. Processing the data and recognizing composite data activities, with appropriate labeling, is an important area of research. For human activity recognition using sequential data from on-body sensors, human-labeled attributes are lacking. To solve this problem, deep learning architectures making predictions for the attributes are suggested in many works.

In Ref. [37], three attribute-based architectures are suggested, based on CNN, deepConvLSTM [38, 39], and CNN-IMU [40], followed by an empiric evaluation of random and learned attribute representations and networks, carried out on two datasets. These architectures have temporal-convolutional layers and fully connected layers in common. In the CNN-based architecture, the temporal local features are extracted by the convolutional layers, thereby creating an abstract representation of the input sequence. The fully connected units connect the local features, creating a global view of the input data. The activation function for the four temporal-convolution and two-fully connected layers is the ReLU function. The structure is illustrated as follows (Fig. 1.8).

The deepConvLSTM uses four temporal-convolution layers with ReLU activations functions, and two LSTM layers instead of fully connected ones. The global temporal dynamics of the inputs are captured by the LSTM layers. The network organization is depicted in Fig. 1.9.

In the CNN-IMU network, there are parallel temporal-convolution blocks, each with a fully connected layer. The parallels process and merge input sequences from the IMUs individually, creating an intermediate representation as per IMU. These intermediate representations are then combined by the network into a global representation, which is then forwarded to a softmax layer. Fig 1.10 illustrates the network design.

The work discusses the experimental results obtained by training and testing these architectures for attribute prediction and concludes that these networks present

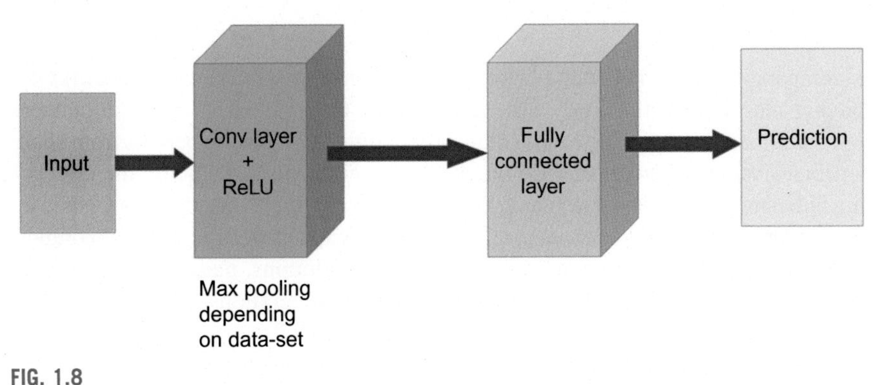

FIG. 1.8

CNN-based architecture.

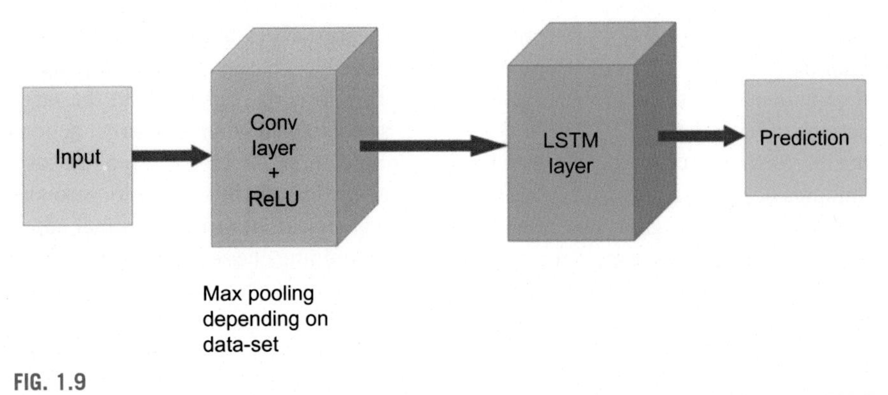

FIG. 1.9

Deepconvlstm-based architecture.

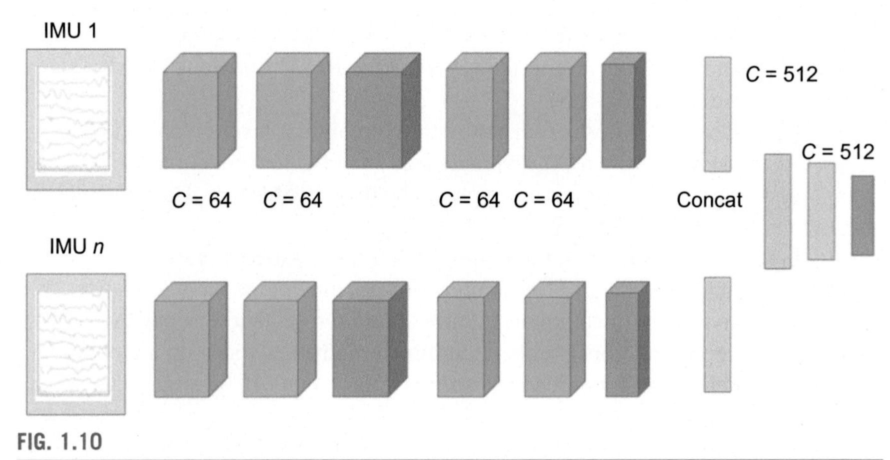

FIG. 1.10

CNN-IMU-based architecture.

a comparable or even better performance contrasting with similar networks that predict classes directly. They also suggest that the evaluation of these networks indicates that deep learning architectures for activity recognition benefit from shared attributes using information from the most frequent actions, reducing the effects of the unbalanced problem statement. Since the work suggests use of deep CNNs, it is important to note that such algorithms involve executing hundreds of compute-intensive functions such as multidimensional convolutions, matrix multiplications, re-LUactivations, etc., for hundreds and thousands of iterations. Hence, it is important to recognize techniques that optimize performance of such algorithms. Although user-controllable parameters are fairly easy to modify and are provided in the frameworks, they are not sufficient to ensure optimal performance [41] suggests the use of system-level optimizations in addition to framework-specific configuration parameters to achieve the best performance for CNN workloads on CPU platforms.

In another work [42], an entirely different approach is undertaken. The work identifies the drawbacks of deep neural network (DNN)-based methods in practical deployment, due to their inefficiency in adapting to certain changes in the target actions to be recognized. When a system user seeks to change or add target actions, the problem is nontrivial for DNN-based architectures, since they require large amounts of training data to learn about new target actions. In this scenario, zero-shot learning (ZSL) has great potential. ZSL methods are different from normal supervised learning frameworks in their dependence on training classes. In a regular supervised learning model, $y_{train} = y_{test,}$ where y_{train} is the set of class labels in the training dataset and y_{test} is the set of classes in the test dataset. In the case of ZSL models, this is not true, and the appropriate relation is given by $y_{train} \cap y_{test} = \varphi$ in some cases and $y_{train} \subset y_{test}$ in others. Hence, the test data include unseen classes, with no corresponding instances in the training set.

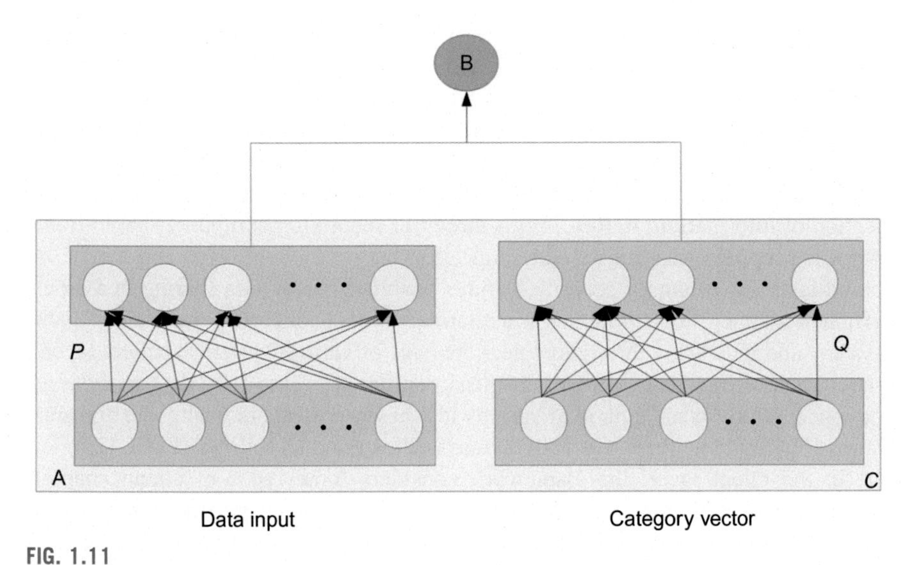

FIG. 1.11
Typical ZSL-based architecture.

Here, the unseen classes are classified using attributes together with a description of the class based on these attributes, which is usually given on the basis of external knowledge. The attribute represents a semantic property of the class. The presence of the attribute is judged by a classifier, learned from the training data. This idea has been demonstrated in related works such as [43–45]. A typical ZSL architecture is shown in Fig. 1.11.

Several other scholarly articles have been published in the field of data processing techniques for WSN in Ubiquitous Healthcare [46–48]. The works broadly fall into the category of advanced deep learning techniques with neural networks and the zero-shot approach.

1.6 Secure architectures

The pervasive healthcare paradigms employ wearable technology for Telemonitoring and diagnosis to provide inexpensive healthcare solutions. The sensing devices, either implanted in or attached to the body of the individual, are connected via the Internet, Bluetooth, or ZigBee, among other communication methods. Sharing, storing, and processing of public health data require secure infrastructure, with strict access rules. Privacy and security requirements for these applications are diverse as they are based on different usage scenarios ranging from prehospital, in-hospital, ambulatory, home monitoring, to database collection for long-term trend analysis. Further, resource constriction and constant topology evolving characteristics of

WSN aggravate the security challenges. This poses a major hurdle to smart healthcare frameworks and is a significant field of research.

Cloud computing frameworks have provided abundant storage and computational infrastructure for data analysis, facilitating the transition from desktop platform to virtual cloud storage. The open environment, storing heaps of sensitive health data, with shared assets, requires good security measures and guaranteed protection of information. In Ref. [49], a three-tier secure fog computing-based framework is proposed that allows communication between client layer, fog layers, and cloud layer for enhanced security features enabling health data sharing in a secure, efficient manner. Fog computing is a relatively recent concept, popular for reducing latency and increasing throughput near the edge of various systems at client layer. It requires less cloud storage and transmission power for long-term data analysis and has successfully been applied to various IoT-based projects in healthcare and smart cities [50–52]. The three-tier architecture is illustrated as follows (Fig 1.12):

In the client layer, the framework considers three types of clients, namely, mobile, thick, and thin. The clients are responsible for visualizing and analyzing the data, either operating on mobile devices, on standard web browsers, or on desktops. The cloud layer is responsible for overall storage and services executed on the servers. The fog layer works as a middle tier between the client and cloud tiers, typically characterized by low-power consumption, reduced storage requirements, and overlay analysis capabilities. Its primary responsibility is to assist and enhance the capabilities of the cloud. In doing so, the fog nodes process the data before transmission to the cloud layer, with added privacy benefits since the data are processed

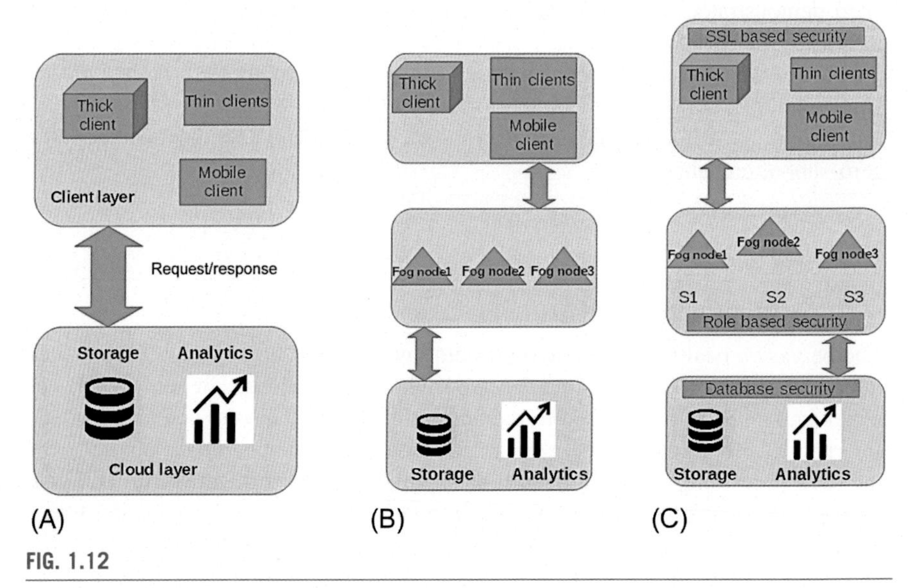

FIG. 1.12

SoA-Fog architecture.

locally at the fog devices and it is only the results of the analysis that is then sent to the open cloud framework. The proposed architecture is significant for the development of IoT-based frameworks, since fog computing is still in its infancy, and as a nontrivial extension of cloud, the security and privacy concerns prevalent in cloud computing techniques could impact fog computing as well.

In another work [53], a different, neural network-based architecture is suggested. Deep learning techniques play an important role in data processing and anomaly detection in pervasive body-sensing regimes but could suffer privacy issues considering users need to give out their data to the model, typically hosted on a server or cluster on the cloud, for training and prediction processes. The data could get leaked during the transmission process and could easily be misused when in the wrong hands. The proposed architecture comprises a deep network where the users don't reveal their original data to the model. Techniques such as feed-forward propagation and data encryption are combined into a single process and the first layer of the network is migrated to the users' personal, local devices. The activation functions are applied locally on said devices, and a dropping-activation-output method is employed to make the output noninvertible by randomly dropping some outputs from the activation function. Hence, the user's sensitive raw data are not exposed to the internet or other channels in this framework. The architecture is depicted in Figs. 1.13 and 1.14.

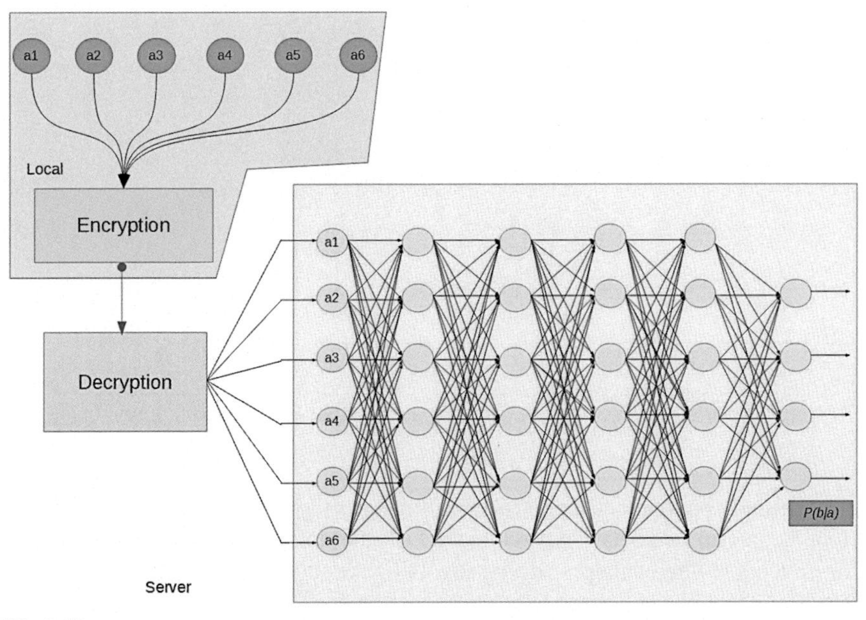

FIG. 1.13

Implementing feed-forward propagation in server.

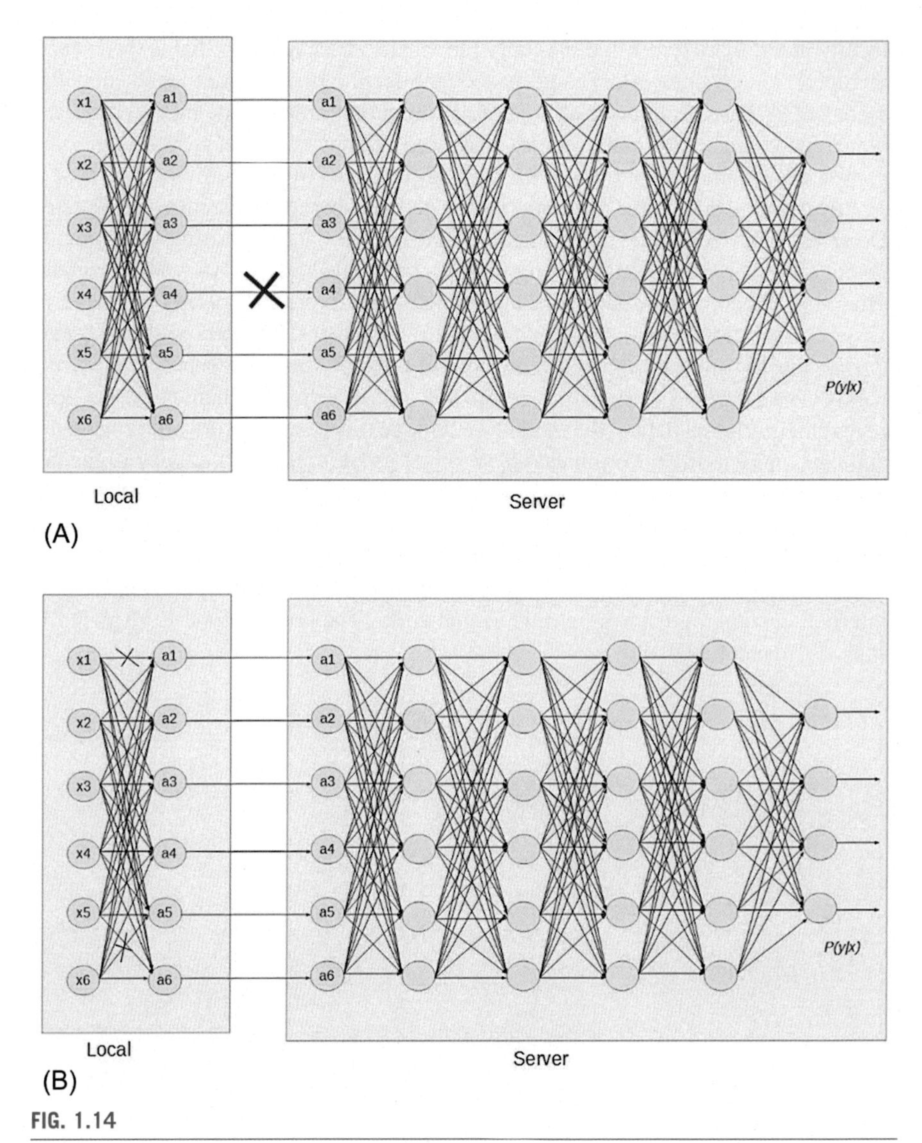

FIG. 1.14

The proposed architecture. (A) Dropping activation output and (B) splitting neuron network.

The study proves that their work ensures the original data cannot be recovered from the transmitted data. The architecture is beautifully constructed and is essential for protection when data processing involving sensitive information is executed in remote servers instead of the user's personal device.

Other methods include public key-based security architectures [54], and other neural network techniques [55–57].

1.7 Discussion

The present work identifies the hurdles to the realization of an efficient, low-cost, high-quality digitized healthcare system, and broadly classifies the same into two categories—research and ethical challenges. A thorough understanding of such challenges is crucial to the development of pervasive healthcare technologies and has spawned new areas of research.

The sequence of tasks involved in a ubiquitous healthcare framework is tricky and convoluting, since each seemingly simple process is faced with varied obstacles when operated in a practical setting. The task of data processing has long been a significant area of research, with contributions from multiple research works [58, 59]. The framework discussed in Section 1.2 is a notable improvement to the problem of limited resources and computational capabilities of portable sensor motes. With sensitive information traveling across nodes in a wireless network architecture, security is a major issue. For any technology to prosper in the real world, winning the trust of the masses is of utmost importance. With the notion of IoT gaining more devotion due to its wide range of applications, privacy-preserving data aggregation (PPDA) [60] has gained the attention of researchers in the past few years. The study discusses two very different approaches to the problem of data privacy and emphasizes on the realm of possibilities in this field. The theory behind information-sharing procedures in a pervasive computing platform is explained in Section 1.3, where the intricacies of wireless sensing networks as a primary enabling technology are discussed. The abundance of applications further affirms the enormous opportunities of ubiquitous sensing networks in the medical field.

The studies discussed in this chapter reveal crucial facts regarding pervasive computing techniques and their application in U-Healthcare. The financial conditions prevalent in the healthcare sector in India and all over the world make it absolutely necessary for the transfer of technology to pervasive platforms to be cost effective and affordable by the masses. At present, many systems described in the studies are in their prototype stages, with little reports regarding deployment issues. Hence, this chapter identifies the need of systematic evaluations of the effectiveness and efficiency of pervasive computing systems and records their performance in the real world not only in terms of technical efficiency, but in the context of user acceptance, societal acceptability as well as financial sustainability.

1.8 Conclusions

Ubiquitous healthcare is indeed a new frontier of smart, e-healthcare technology. We have already sailed away from traditional desktop-based computing techniques toward computing environments that progressively exist in our daily life activities. Pervasive technologies such as handheld mobile devices, wristbands, even shirts, and gloves have recently been woven into the realm of smart sensing networks and serve a variety of purposes. The possibilities for such smart healthcare methods range from simple home monitoring to high-end mobile and ubiquitous patient-monitoring systems using ad hoc wireless networks enabling profusion of services such as prevention, early detection, treatment and management of diseases, with a consequent reduction in a fatality.

The study discussed the concept of ubiquitous healthcare and the numerous challenges, both technical and ethical, prevalent in this emergent field. Further, enabling technologies and networks were explained in detail while reviewing newer topologies as proposed in multiple research works. The plethora of applications of U-Healthcare architectures and smart devices were studied, and one such design was presented to aid the visually impaired. Finally, the research aspects were reviewed to tackle the challenges in U-Healthcare systems. Advanced architectures were studied involving machine-learning methods and emergent cloud computing frameworks, to solve issues relating to the vast pool of raw data and privacy measures to secure said data. Research in these fields of sensitive health data processing and secure data transfer and storage is crucial to the success of this emergent field, and the authors encourage contributions from enthusiasts to help realize the ultimate goal of smart healthcare for all.

References

[1] M. Weiser, The Computer for the 21st Century, Scientific American, 1991.

[2] C. Bhatt, N. Dey, A.S. Ashour (Eds.), Internet of things and big data technologies for next generation healthcare, 2017 https://doi.org/10.1007/978-3-319-49736-5 ISSN: 9783319497365 (online).

[3] G. Elhayatmy, N. Dey, A.S. Ashour, Internet of things based wireless body area network in healthcare, in: N. Dey, A.E. Hassanien, C. Bhatt, A.S. Ashour, S.C. Satapathy (Eds.), Internet of Things and Big Data Analytics Toward Next-Generation Intelligence, in: J. Kacprzyk (Ed.), Studies in Big Data, Springer, Cham, 2018, pp. 3–20

[4] N. Dey, A.S. Ashour, C. Bhatt, Internet of things driven connected healthcare, in: C. Bhatt, N. Dey, A.S. Ashour (Eds.), Internet of Things and Big Data Technologies for Next Generation Healthcare, Springer, Cham, 2017, pp. 3–12.

[5] Internet of Things (IoT) in healthcare: benefits, use cases and evolutions; i-scoop

[6] The Internet of Things: today and tomorrow; Aruba

[7] Healthcare in digital transformation: digital and connected healthcare; i-scoop

[8] R. Shahrawat, K.D. Rao, Insured yet vulnerable: out-of-pocket payments and India's poor, Health Policy Plan. 27 (3) (2012) 213–221.

[9] E. van Doorslaer, O. O'Donnell, R.P. Rannan-Eliya, A. Somanathan, S.R. Adhikari, C.C. Garg, D. Harbianto, A.N. Herrin, M.N. Huq, S. Ibragimova, A. Karan, C. Ng, B.R. Pande, R. Racelis, S. Tao, K. Tin, K. Tisayaticom, L. Trisnantoro, C. Vasavid, Y. Zhao, Effect of payments for health care on poverty estimates in 11 countries in Asia: an analysis of household survey data, Lancet 368 (9544) (2006) 1357–1364.

[10] C.C. Garg, A.K. Karan, Reducing out-of-pocket expenditures to reduce poverty: a disaggregated analysis at rural-urban and state level in India, Health Policy Plan. 24 (2) (2009) 116–128.

[11] V. Roy, U. Gupta, A.K. Agarwal, Cost of medicines & their affordability in private pharmacies in Delhi (India), Indian J. Med. Res. 136 (5) (2012) 827–835.

[12] S. Sneha, U. Varshney, Ubiquitous healthcare: a new frontier in E-health, in: AMCIS 2006 Proceedings, 2006, p. 319.

[13] H. Viswanathan, B. Chen, D. Pompili, Research challenges in computation, communication, and context awareness for ubiquitous healthcare, IEEE Commun. Mag. 50 (5) (2012) 92–99.

[14] I. Brown, A.A. Adams, The ethical challenges of ubiquitous healthcare, Int. Rev. Inf. Ethics 8 (12) (2007).

[15] European Parliament and Council of the European Communities: Directive 95/46/EC of the European Parliament and of the Council of 24 October 1995 on the protection of individuals with regard to the processing of personal data and on the free movement of such data.

[16] Bozovic, M n.d.: The Panopticon Writings.

[17] Foucault, Mn.d.: Discipline and Punish.

[18] J. Preden, R. Pahtma, Smart Dust Motes in Ubiquitous Computing Scenarios, (2009).

[19] R. Draves, J. Padhye, B. Zill, Routing in multi-radio, multi-hop wireless mesh networks, in: Proceedings of the 10th Annual International Conference on Mobile Computing and Networking, ACM, 2004, pp. 114–128.

[20] P. Li, Y. Wen, P. Liu, X. Li, C. Jia, A magnetoelectric energyharvester and management circuit for wireless sensor network, Sens. Actuators, A 157 (1) (2010) 100–106.

[21] A. Sudarsono, M.U.H.A. Rasyid, H. Hermawan, An implementation of secure wireless sensor network for e-healthcare system, in: Computer, Control, Informatics and Its Applications (IC3INA), 2014 International Conference, Oct, 2014, pp. 69–74.

[22] H. Fotouhi, A. Causevic, M. Vahabi, M. Bj ̂rkman, Interoperability in heterogeneous low-power wireless networks for health monitoringsystems, in: IEEE International Conference on CommunicationsWorkshops (ICC), May, 2016, pp. 393–398.

[23] J. Elias, Optimal design of energy-efficient and cost-effective wireless body area networks, Ad Hoc Netw. 13 (2014) 560–574.

[24] G. Tsouri, A. Prieto, N. Argade, On increasing network lifetime in body area networks using global routing with energy consumption balancing, Sensors 12 (2012) 13088–13108.

[25] S. Yousaf, N. Javaid, Z. Khan, U. Qasim, M. Imran, M. Iftikhar, Incremental relay based cooperative communication in wireless body area networks, Procedia Comput. Sci. 52 (2015) 552–559.

[26] F. D'Andreagiovanni, A. Nardin, E. Natalizio, A fast ILP-based heuristic for the robust design of body wireless sensor network, in: G. Squillero, K. Sim (Eds.), EvoApplications 2017, Part I, LNCS 10199, 2017, pp. 1–17.

[27] F. D'Andreagiovanni, A. Nardin, Towards the fast and robust optimal design of wireless body area networks, Appl. Soft Comput. 37 (2015) 971–982.

[28] F. Afsana, M. Asif-Ur-Rahman, M.R. Ahmed, M. Mahmud, M.S. Kaiser, An energy conserving routing scheme for wireless body sensor nanonetwork communication, IEEE Access 6 (2018) 9186–9200.

[29] D.-S. Lee, Y.-H. Liu, C.-R. Lin, A wireless sensor enabled by wireless power, Sensors 12 (2012) 16116–16143.

[30] Y. Han, M. Han, S. Lee, A.M.J. Sarkar, Y.-K. Lee, A framework for supervising lifestyle diseases using long-term activity monitoring, Sensors 12 (2012) 5363–5379.

[31] M. Aminian, H.R. Naji, A hospital healthcare monitoring system using wireless sensor networks, J. Health Med. Inf. 4 (2013) 121.
[32] N. Dey, A.S. Ashour, F. Shi, S.J. Fong, R.S. Sherratt, Developing residential wireless sensor networks for ECG healthcare monitoring, IEEE Trans. Consum. Electron. 63 (4) (2017) 442–449.
[33] A. Ghose, P. Sinha, C. Bhaumik, A. Sinha, A. Agrawal, A. Dutta Choudhury, UbiHeld—ubiquitous healthcare monitoring system for elderly and chronic patients. in: UbiComp 2013 Adjunct—Adjunct Publication of the 2013 ACM Conference on Ubiquitous Computing, 2013, pp. 1255–1264, https://doi.org/10.1145/2494091.2497331.
[34] S. Monicka, C. Suganya, S. Nithya Bharathi, A.P. Sindhu, A ubiquitous based system for health care monitoring, Int. J. Sci. Res. Eng. Technol. 2278-08823 (4) (2014).
[35] Y. Hata, H. Yamaguchi, S. Kobashi, K. Taniguchi, H. Nakajima, A human health monitoring system of systems in bed, system of systems engineering, in: SoSE '08. IEEE International Conference on 2–4 June, 2008, pp. 1–6.
[36] M. Nambu, K. Nakajima, A. Kawarada, T. Tamura, The automatic health monitoring system for home health care, information technology applications in biomedicine, in: Proceedings. 2000 IEEE EMBS International Conference on 9–10 Nov, 2000, pp. 79–82.
[37] F.J. Ordóñez, D. Roggen, Deep convolutional and LSTM recurrent neural networks for multimodal wearable activity recognition, Sensors 16 (1) (2016) 115.
[38] J. Yang, M.N. Nguyen, P.P. San, X. Li, S. Krishnaswamy, Deep convolutional neural networks on multichannel time series for human activity recognition, IJCAI (2015) 3995–4001.
[39] R. Grzeszick, J.M. Lenk, F.M. Rueda, G.A. Fink, S. Feldhorst, M. ten Hompel, Deep neural network based human activity recognition for the order picking process, in: In Proc. of the 4th Int. Workshop on Sensor-Based Activity Recognition and Interaction, ACM, 2017.
[40] F.M. Rueda, G.A. Fink, Learning Attribute Representation for Human Activity Recognition, 2018.
[41] Vikram Saletore, Deepthi Karkada, Vamsi Sripathi, Kushal Datta, & Ananth Sankaranarayanan; n.d.Boosting Deep Learning Training & Inference Performance on Intel® Xeon® and Intel® Xeon Phi™ Processors
[42] H. Ohashi, M. Al-Naser, S. Ahmed, K. Nakamura, T.S.A. Dengel, Attributes' importance for zero-shot pose-classification based on wearable sensors, Sensors 18 (2018) 2485.
[43] H. Larochelle, D. Erhan, Y. Bengio, Zero-data learning of new tasks, in: Proceedings of the National Conference on Artificial Intelligence (AAAI), Chicago, IL, USA, 13–17 July, 2008.
[44] Y. Fu, T. Xiang, Y.G. Jiang, X. Xue, L. Sigal, S. Gong, Recent advances in zero-shot recognition: toward data-efficient understanding of visual content, IEEE Signal Process. Mag. 35 (2018) 112–125.
[45] C.H. Lampert, H. Nickisch, S. Harmeling, Learning to detect unseen object classes by between-class attribute transfer, in: Proceedings of the IEEE Conference on Computer Vision and Pattern Recognition (CVPR), Miami, FL, USA, 20–25 June, 2009.
[46] J. Wang, Y. Chen, S. Hao, X. Peng, L. Hu, Deep learning for sensor-based activity recognition: a survey, arXiv: 1707.03502, 2017.
[47] A. Frome, G.S. Corrado, J. Shlens, S. Bengio, J. Dean, M.A. Ranzato, T. Mikolov, DeViSE: a deep visual-semantic embedding model, in: Proceedings of the Advances

in Neural Information Processing Systems (NIPS), Lake Tahoe, NV, USA, 5–10 December, 2013.
[48] J. Liu, B. Kuipers, S. Savarese, Recognizing human actions by attributes, in: Proceedings of the IEEE Conference on Computer Vision and Pattern Recognition (CVPR), Colorado Springs, CO, USA, 20–25 June, 2011.
[49] R.K. Barik, H. Dubey, K. Mankodiya, SoA-Fog: secure service-oriented edge computing architecture for smart health big data analytics, in: 5th IEEE Global Conference on Signal and Information Processing GlobalSIP, 2017 (November 14–16, 2017, Montreal, Canada).
[50] N. Constant, et al., Fog-assisted wIoT: a smart fog gateway for end-to-end analytics in wearable Internet of Things, in: IEEE HPCA, 2017.
[51] R.K. Barik, H. Dubey, A.B. Samaddar, R.D. Gupta, P.K. Ray, Foggis: fog computing for geospatial big data analytics, arXiv preprint arXiv:1701.02601,(2016).
[52] H. Dubey, N. Constant, A. Monteiro, M. Abtahi, D. Borthakur, L. Mahler, Y. Sun, Q. Yang, K. Mankodiya, Fog computing in medical internet-ofthings: architecture, implementation, and applications, in: Handbook of Large-Scale Distributed Computing in Smart Healthcare, Springer International Publishing AG, 2017.
[53] H. Dong, C. Wu, Z. Wei, Y. Guo, Dropping Activation Outputs with Localized First-Layer Deep Network for Enhancing User Privacy and Data Security, 2017.
[54] M.M. Haque, A.-S.K. Pathan, C.S. Hong, Securing U-healthcare sensor networks using public key based scheme, in: 2008 ICACT, Feb. 17–20, 2008.
[55] L. Deng, D. Yu, Deep Learning: Methods and Applications, Now Publishers Inc., 2014.
[56] D.J. Bonde, S. Akib, et al., Review techniques of data privacy in cloud using back propagation neural network, Int. J. Emerg. Technol. Adv. Eng. 4 (2) (2014) 15.
[57] T. Graepel, et al., ML confidential: machine learning on encrypted data, in: ICISC, Springer, 2013, p. 121.
[58] H. Kim, Y. el Khamra, I. Rodero, S. Jha, M. Parashar, Autonomic management of application workflows on hybrid computing infrastructure, Telecommun. Syst. 19 (2–3) (2011) 75–89.
[59] E.K. Lee, H. Viswanathan, D. Pompili, SILENCE: distributed adaptive sampling for sensor-based autonomic systems, in: Proceedings of the International Conference on Autonomic Computing (ICAC), June, 2011.
[60] W. He, X. Liu, H. Nguyen, K. Nahrstedt, T. Abdelzaher, PDA: privacy-preserving data aggregation in wireless sensor networks, in: Proceedings of the INFOCOM, 2007, pp. 2045–2053.

Further reading

[61] N.R. Rishani, H. Elayan, R.M. Shubair, A. Kiourti, Wearable, Epidermaland Implantable Sensorsfor Medical Applications, 2018.

CHAPTER

2 Wireless sensor networks towards convenient infrastructure in the healthcare industry: A systematic study

Md. Rashid Farooqi*, Naiyar Iqbal[†], Nripendra Kumar Singh[‡], Mohammad Affan[‡], Khalid Raza[‡]

*Department of Management Studies, Maulana Azad National Urdu University, Hyderabad, India**
Department of Computer Science and Information Technology, Maulana Azad National Urdu University, Hyderabad, India[†] Department of Computer Science, Jamia Millia Islamia, New Delhi, India[‡]

2.1 Introduction

Over the last two decades, information technology has been broadly used in medical science and health care. The wireless sensor network (WSN) is one technological advancement that has made an impact on the healthcare industry. It allows users to access data and services electronically irrespective of their geographical locations. WSNs can be effectively applied in the medical industry to improve the health of patients as well as the healthcare services they receive. A WSN can be placed in any kind of environment and the information it collects can be stored, processed, and sent to any location as per requirements. It can be used to collect data that patients and healthcare providers can use to monitor health and make decisions regarding medical treatment. WSN technology is cost-effective and widely accessible [1].

Traditionally, people monitor their health by going to regular checkups with their healthcare providers. The advent of WSN technology, however, has brought about a drastic change in how health is monitored; now it can be done remotely. It enables large-scale studies of human chronic diseases and behavior. WSNs differ from traditional wireless networks such as mobile ad hoc networks (MANETs), wireless local area networks (WLANs), and cellular networks. In fact, in these networks mobility management and organization routing are used to enhance the quality of services. With the adaptation of a WSN a patient performing his routine activities at work or home can monitor his health more efficiently and in a cost-effective manner [2].

Sensors for Health Monitoring. https://doi.org/10.1016/B978-0-12-819361-7.00002-6

A WSN is a multihop network comprised of a variety of nodes. It comprises tiny wireless computers that process and communicate environmental stimuli, including vibrations, temperature, and light. The technological advancements in wireless communication and microelectromechanical systems (MEMS) allows for creating low-power, large-scale, low-cost, multifunctional networks that are more robust than traditional sensing methods. WSN technology opens new avenues in the field of health care and has several benefits over traditional sensor networks, such as easy deployment, low risk of infection, little discomfort, improved mobility, and better quality of care at lower costs. A WSN can replace thousands of wires connected to the machines in hospitals.

2.2 Background

In order to understand the evolution of the modern WSN, it is helpful to know its origin and evolution. Like much of today's advanced technology, WSN technology was born from military and industrial applications. The sound surveillance system invented by the United States military in 1950 was the first wireless network used for detecting and tracking Soviet submarines in the Pacific Ocean.

The distribution sensor network (DNS) program was started by the US Defense Advanced Research Projects Agency (DARPA) in 1980. The governmental and academic institution began using WSN for forest fire detection, health care, natural disaster prevention, and structural monitoring. In a very short span of time, WSN technology became widely recognized and found a place in scientific research and academics. In fact, the market value of WSN was very high and demand was continually increasing. Previously, these sensor networks placed a premium on functionality, while other factors such as high deployment cost, low networking standard, and more power consumption prevented their widespread application. In fact, there is a long history of applying sensors in medical and public health care. The emergence of information technology now makes it easy to use WSN to interconnect medical sensors with other devices, whereas in the past medical sensors were isolated.

Today's WSNs incur low deployment and maintenance costs, bringing new sensors of information, convenience, and control to our professional and day-to-day lives. These WSNs enhance productivity and reduce costs.

2.3 Wireless sensor network: Technical architecture

WSN technology is being enhanced to fix real-life problems. WSNs hold up enormous sensor nodes that act as overseers to the changes occurring in the environment and collect the information related to it. Compared to traditional nodes, WSN sensor nodes are remunerative in terms of disbursement efficiency, finite power, efficient computation, and action of resources. Sensor nodes are endowed with the competency to communicate with each other for the sending of gathered data to a sink node, which is a central region bound to other sensor nodes of the WSN. To make the interaction possible a network is forged using sensor nodes to bring computing and sensing actions. WSNs are popular for their output-giving potential, easy deployment, lack of cables, and facile troubleshooting. WSN innovations were used to build up military-based applications for combat area surveillance. Further innovations using WSNs to monitor natural phenomenon like humidity, ferocity, temperature, and pressure, as well as physical health have attained popularity [2, 3].

2.3.1 Categories of wireless sensor networks

Structured WSNs: In a structured WSN, there are confined sensor nodes that are dispersed in a preordained fashion. Therefore the basic cost for this type of WSN is comparatively low and its preservation is simpler.

Unstructured WSNs: An unstructured WSN contains a large number of sensor nodes that are randomly dispersed in the network. It possesses an ad hoc nature due to which its preservation is difficult. Furthermore it is susceptible to connection breakdown. The architecture of a typical WSN is shown in Fig. 2.1.

2.3.2 How the process flows

In a real environment, sensor nodes are dispersed to sense data and gather information by observing their surroundings. Sensed data is required to be circulated in succession through a gateway node to a host computer wirelessly. The collected data goes to the host computer for evaluation and conversion into useful information.

2.3.2.1 Battery

WSN gadgets/devices are subject to battery regulation. A node can't execute for more than a month in a complete active mode, as an alkaline battery contributes around 50 Wh of power. In a real-world scenario, it has to be assured that the gadget will work without replacement for one or two years. Efficient scheduling algorithms

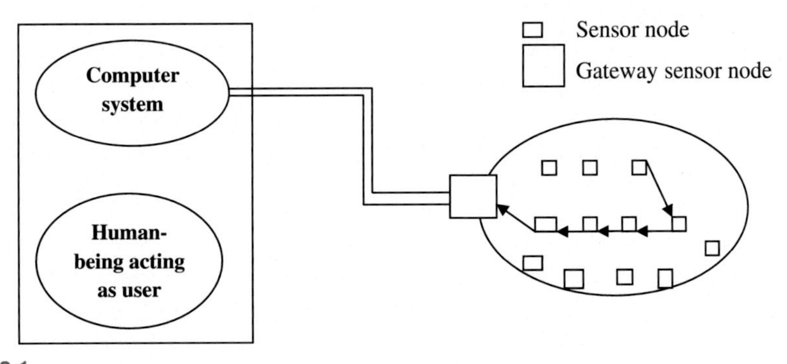

FIG. 2.1

Architecture of a typical wireless sensor network.

and energy management are necessary for dealing with energy issues [1]. There is a huge effort in modeling low-power sensors to manage these issues [2]. Wireless sensors worn on the body cannot be used for solar cells because solar cells cannot yield under direct sun up to 15 m. Therefore power techniques based on body heat [3, 4] and motion [5] must be reconfigured for use with the healthcare system.

2.3.2.2 Sensors

Static and dynamic peculiarities for choosing sensors must be contemplated. These peculiarities perform a vital role in the efficient performance of sensors. Veracity, period and assessment, fault bands, assessment resolution, lifeless band, reactivity, hysteresis, and uncurled and untwisted nature are the static peculiarities of the sensor. The dynamic peculiarity can be determined as potential to handle hasty variation or transformation in the input. When building a sensor module or its network, not all peculiarities are considered significant. Relying on the application and some environmental consideration, a selected subset of properties of sensors are acknowledged by designers.

The WSN is a sensing instrument which connects one or more transceiver nodes and base stations. It applies a broad scale of RF communication tools. The configuration of the sensor can be performed through software to determine and regulate disparate natural yardsticks like temperature, force, movement, and so on. Wireless sensors accompanied by RF communication combine to make a WSN. WSNs administer and gather data, determine relevant quantities, analyze and gauge information, establish purposeful user displays, enable decision making, and set and send alarms.

Wireless sensors have implications in various fields, including automation, manufacturing, military combat, aerospace, architecture, engineering, medical devices, research and development, and retail and wholesale services [6]. Embedded systems for computing are advancing, leading to the evolution of WSNs that include tiny, battery-operated "motes" equipped with confined computation and radio communication capacities. A sensor network allows the collection of data and information to be highly embedded in the natural environment.

2.3.2.3 Memory

Large-bit computation is not performed due to lack of sufficient memory that the biosensors possess. Therefore biosensors do not possess much computational power. We can conclude that when the sampling frequency rises, the surplus memory size rises and vice versa [7]. Also, it is inferred that sampling frequency is straightly equivalent to the memory usage.

2.3.2.4 Typical node

In a real environment, the distribution and placement of sensor nodes may take many forms. Nodes can be distributed randomly or placed deliberately at selected spots. The deployment of a node can be a one-time action where placement and use of the sensor-based network are disjointed tasks. Alternatively, the placement of the node can be a continual procedure where many nodes are placed at any time when the network is being used. The placement affects significant characteristics such as node locale, a regular arrangement in node locale, predicted the degree of node dynamics. Sensor nodes can change their locale after placement. Mobility can come from environmental impacts like wind and water; nodes can be adhered to or transported by mobile entities and nodes can acquire automotive capacities. Relying on the real needs of the applications for a single node, the size obtained can range from large to microscopically small. The price of a single device may fluctuate from a few dollars to hundreds of dollars. Sensor nodes are unbounded, self-controlled equipment; their power and equipment are confined by shape and price restrictions or limitations. Fluctuating price and size limitations result in corresponding fluctuating limits on power available. Therefore a sensor node may differ greatly from system to system. An amorphous computing project [8] considered that sensor nodes were identical.

2.3.2.5 Network

A sensor network incorporates a massive number of sensor nodes. Nodes are placed either inboard or very near to the phenomena they are designed to sense. A network designer must consider the following in order to construct and execute a sensor network:

- *Wi-Fi:* Radio technologies such as IEEE 802.11x are being used by Wi-Fi networks to choose and transfer wireless data [9]. Wi-Fi is advantageous when it comes to ad hoc wireless network implementation.
- *Bluetooth:* Bluetooth is acceptable for concise limits, low energy, and low-priced digital radio wireless communication [10]. It can be used for small-range application to be deployed in a WSN.
- *Zigbee:* Built on IEEE 802.15.4, Zigbee is a new wireless technology used to create personal area networks (PANs) with a transmittal range of 100-plus meters [11]. Communication machines established for Zigbee use very little power and therefore battery life of a thousand or more days is not uncommon. Compared to Bluetooth, Zigbee has expanded coverage area, lower power consumption, and secured networking.
- *Network topology:* Network topology is the arrangement of the elements of a communication network, including transmitted power, transference economic cost, packet loss, information delay, and so on. Some examples of topologies include Star, Bus, and Ring.
- *Communication protocols and routing:* Routing techniques and disparate communication protocols can be implemented for a WSN. The protocol being employed depends on the sensor to be used in the application. A handshaking protocol begins when one device sends a message to another device indicating that it wants to establish a communications channel. The two devices then send several messages back and forth that enable them to agree on a communications protocol. CodeBlue is the latest technique implemented to make use of acutely

confined resources efficiently in the wireless network so as to handle sensor nodes that have confined computation competencies.

- *Power management:* Time-division multiple access (TDMA) is helpful for power saving, as a node can easily have stopped for the stipulated time slots in between to them, alerting to pick and transfer messages in time. The majority of power is being used up in the procedure of radio-frequency (RF) communication. Therefore power management techniques related to software can highly reduce the power dissipated by RF sensor nodes.
- *Network coverage:* A sensor's coverage area is related to its range, which is determined when the sensor is attached to its sensor node. Broader coverage makes the system more powerful.
- *Data management:* A WSN can be seen as a distributed database that collects physical measurements, indexes them, and replies to all the queries raised by its users. Each sensor node is capable of generating a stream of data that is received from the sensing devices on the node. In general, each sensor consists of a small node with computing, sensing, and communication abilities. Sensor nodes are treated as small computers. WSNs are now enabling applications in health care that were not practical previously. Nowadays we can realize the wide scope and various implementations of distributed database management in wireless sensing devices [12].

This approach will bring better health care facilities to those patients who do not want to stay in the healthcare centers or hospitals; however, the efficiency of such a system depends upon the capabilities of the decision support system (DSS) integrated with it. The development of DSS architecture provides support to data collection and processing from multiple WSNs. The real-time data received from multiple sensors are analyzed for various diseases. Finally, the results are stored and sent to the concerned person.

2.3.3 How it works

A WSN is comprised of geographically dispersed, self-directed sensors used together to monitor variables or factors such as heat, sound, vibration, pressure, and so on. The latest progress related to MEMS technology has given birth to sensors. Sensory data is obtained from sensors placed in dispersed locations. For sensing and executing the data WSNs are liable, relying on the requisite of the network. For a sensor node, the frequency of its sensation may rely on an instance of an event or it can be cyclic, relying on the application related to the linked corresponding node being used. Fig. 2.2 is a diagram of a WSN. To execute the sensing operation there is a sensor unit that the WSN possesses, an inbuilt battery for power, an implanted processor for data execution, and a device with a confined range antenna (i.e., transceiver) responsible for transmittance and acceptance of data. For the sensor node, the memory brought is of confined size as the node is responsible for passing rather than storing the data that is sensed. The battery is equipped with commencing energy and has

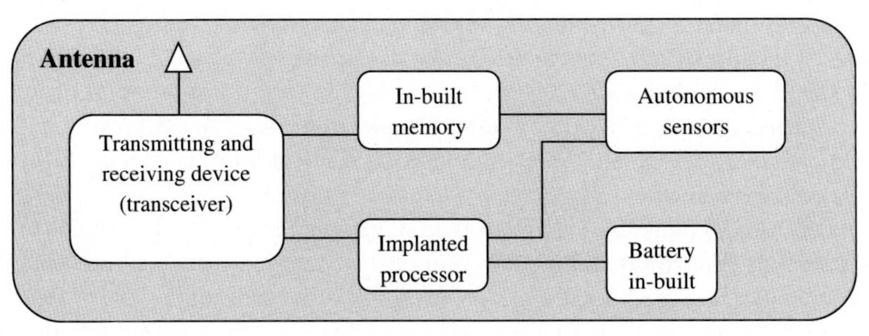

FIG. 2.2
Wireless sensor node structured diagram [13].

Table 2.1 Different types of sensors for various physical indexes [14].

Indexes	Sensor
Glucose testing	Blood feature may be analyzed. Requires an invasive test. Everyday diabetic surveillance population.
Heart/pulse rate (ECG)	An electrode that can be worn on the skin. It can detect electrocardiogram trace and pulse rate.
Motion perusal	Requires amalgamation of various sensors for EMG.
Muscle contraction/ tightening (EMG)	Electrode is same as that of an ECG electrode. It can give the situation related to muscular and correlated motor neuron.
Respiration and temperature	Plethysmograph and thermistor are used for determining breathing and body temperature, respectively.
Blood flow (SpO_2)	To determine pigmentation changes, photocell can be taken. It is done so as to send back oxygen in the blood stream.
Cancer	No significant applications of WSN have been seen for cancer diagnosis, management, and prediction.

confined range. As these five mandatory units, a sensor node may be rigged for additional units like a power generator, a locale searching unit, and so on. For circulating information from origin to destination distinct routing protocols are used, including opportunistic routing, which enhances reliability, and extended transference range. Different types of sensors used for various physical indexes are presented in Table 2.1.

2.4 Wireless sensor network applications in healthcare industries

2.4.1 Monitoring of patients and detection of diseases

Alzheimer's disease: Alzheimer's disease (AD) is a neurodegenerative disease without any cure. It is the main cause of dementia in which the patient exhibits several neurocognitive and motor problems, such as memory loss, delusion, disorientation, depression, hallucination, thinking and language problems, and so on. Overall a patient's quality of life deteriorates as the disease progresses. Olaniyi and Nadine [15] stated that individuals 60 years of age or older are more prone to developing AD. According to the World Health Organization (WHO), 35.5 million people worldwide were affected by dementia in 2010. This number is expected to double in the next 20 years.

AD is usually diagnosed at an advance stage when the individual begins to show symptoms of behavioral and memory deficits. Recent technological advancements have enabled clinicians to detect AD early. Several procedures and assessments can be used to determine possible or probable AD, including the following:

- questionnaires about a person's daily activity, clinical records, and behaviors
- MRI or CT scan
- memory, problem-solving, language processing, and psychometric tests

WSN technology can be used to detect early AD via various sensors designed to monitor patient movement and activity. Varatharajan et al. [16] proposed a foot movement monitoring system for early detection of AD by implementing sensors with multicast IOT devices that communicate with integrated machines with the help of a WSN. A dynamic time warping (DTW) algorithm is used to compare the many shapes of signals collected from individuals [17, 18]. DTW is responsible for similarity classification of multivariate time series data to classify walking patterns, speech recognition, and handwriting classifications of patients without AD and those with AD.

Asthma: Asthma affects almost 300 million individuals globally and is the cause of one of every 250 deaths [19]. Asthma is a chronic medical condition involving the respiratory tract. It is characterized by constriction of airways and bronchospasm due to inflammation and excessive mucus production. Major symptoms of asthma include coughing, tightness, chest pain, and shortness of breath. An acute asthma attack can be fatal. So far the disease has no cure. Treatment involves medications (corticosteroids, adrenergic agonists, and antiinflammatory drugs) and management of the condition.

As management of asthma is crucial for patient care, any method for live monitoring of a patient's physiological conditions and risk factors plays a very important role. Hosseini et al. [20] developed a device called mHealth (mobile health) to assess risk factors in pediatric asthma. The mHealth system is comprised of a wireless sensor for detection of physiological and environmental risk factors and an application program interface (API). Data obtained from wireless sensors are sent to the cloud through a connected smart watch that ultimately performs data analysis and sends back the results. Ho-Kyeong et al. [21] designed a WSN device called AsthmaGuide for elderly patients that is designed to collect various physiological data (such as lung sound, pulse, and inspired air volume) from sensors and instruments and environmental data (such as air quality index, pollen count, dust levels, etc.) from the Internet. Physiological and environmental data are integrated and analyzed over the cloud and the results sent to the user and the user's healthcare provider.

Preventive medicine: Innovation in mobile healthcare has made it easier to prevent certain diseases and conditions (e.g., asthma, anaphylaxis, type 2 diabetes, and seizures) and has helped reduce healthcare visits and allowed doctors to reach more patients. Appelboom et al. [22] found that smart wearable sensors and WSN are effective and reliable in preventing medical conditions. Fig. 2.3 represents patient monitoring using WSN technology.

Several researchers [14, 23, 24] have provided a solution for integrated health using IoT, big data, and the cloud, which interconnect all available medical resources and provide a comprehensive, effective, and reliable healthcare system.

Cancer: Cancer is one of the biggest burdens on global health. According to the GLOBOCAN 2018 database, 18.1 million new cases of cancer and 9.6 million deaths occurred from cancer worldwide [25]. Cancer is the state of unchecked division of any cell type to form a mass of cells that becomes malignant. There are a number of agents that can cause cancer, for example, chemicals, radiation, oncogenic viruses,

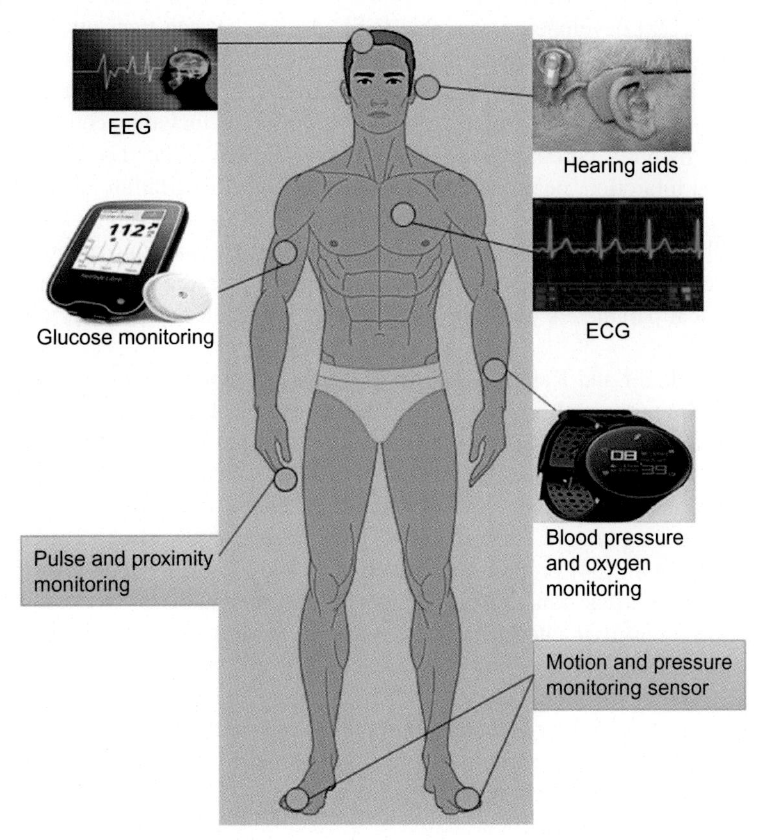

FIG. 2.3

Patient monitoring using various WSN-enabled devices.

and so on. Prevention and treatment entirely depend on the severity of cancer. Late-stage, metastatic cancers are difficult to treat and ultimately lead to death. In order to control tumor growth and alleviate a cancer patient's condition, combinatorial radiotherapy and immunotherapy (e.g., immune checkpoint inhibitors) have been proven to be the best treatment modality in recent years [26].

There are very few wearable WSN devices designed to predict, diagnose, and manage cancer. Researchers from Purdue University have developed a risk-on-a-chip device to assess risk factors of breast cancer. This dye-based, micro-fluidic device allows in vitro (3D cell culture) simulations to assess oxidative stress and tissue stiffness, both of which are major risk factors for developing breast cancer. Ionizing radiations are used in various industrial and clinical setups, and people operating in such systems are vulnerable to the deleterious effects of the radiation and highly prone to develop cancers. Yoon et al. [27] developed a wearable device that serves as an indicator of ionizing radiation and can alert personnel to the same. This device is based on measuring the metabolic activity of yeast to ferment glucose, which is hindered in the presence of ionizing radiation due to DNA breakdown and mitochondrial damage. Singh et al. [28] developed a wearable WSN device that can transmit information from a tumor during ultrasound hyperthermia treatment of the tumor. This WSN might serve as a way to control treatment conditions for patients in remote areas.

Heart diseases: There are many types of heart disease, including congenital heart disease, arrhythmia, coronary artery disease (heart failure), dilated cardiomyopathy, myocardial infarction, pulmonary stenosis, and so on. Heart failure or heart attack can occur due to high blood pressure, which is something that can be tracked very early [29]. There are many wearable wireless devices for real-time monitoring of a patient's cardiac health. These devices measure blood pressure and pO_2 and send the physiological parameters to the hospital servers for further analysis by physicians [30, 31].

Kappiarukudil and Ramesh [32] developed a wearable wireless sensor system (WWSS) for remote real-time monitoring and detection of heart failure. The WWSS continuously senses the ECG of the patient and transmits it to a mobile phone via Bluetooth. The mobile control system (MCS) receives the ECG signal and generates a warning signal to the patient. Kakria et al. [33] included more parameters like blood pressure and temperature for heart attack monitoring in an effort to reduce false alarms. Ahn et al. [34] proposed a new device called Titan for long-term wireless monitoring of cardiac function. It consists of an implantable pressure sensor and an extracorporeal user interface. The researchers performed a trial on 40 patients for approximately 560 days and showed accuracy and safety in standard medical treatment. Dey et al. [35] developed a WSN in-house ECG monitoring device relying on Zigbee protocol similar to Bluetooth for data transmission. This device is a sleek, low power consuming, and low-cost option that provides real-time measurement and transmission of ECG data to physicians.

Artificial retina: An artificial retina is an electronic device consisting of a miniature camera mounted in eyeglasses that captures images and sends information wirelessly to a microprocessor that converts visual information to an electronic signal and transmits it to microelectrodes (retinal stimulus) implanted in the retina.

These electrodes are stimulated to generate pulses to send information through the optic nerve to the brain.

The artificial retina is mainly for those blind people with retinal disease due to macular degeneration (AMD) and retinitis pigmentosa (RP), which causes damage to the light-sensitive cells (photoreceptors). In practice, the artificial retina differs from traditional WSNs [36, 37]. Schwiebert et al. [38] developed a smart sensor package to restore vision from the damaged retina. It consists of an epi-retinal implant an Artificial retina prosthesis chip wirelessly receive information from the sub-retinal part. Miura et al. [39] proposed a wireless power-driven artificial retina using a thin-film transistor (TFT). It works correctly even in with an unstable power source. Fig. 2.4 represents the artificial retina's processing system.

Code blue: Code blue is an emergency medical situation that refers to cardiopulmonary arrest. Most hospitals have a team on hand to deal with code blue crises. Use of a wearable WSN can be of great help to the code blue team, allowing them to easily identify, locate, and get vital physiological data of the patient. To address code blue situations, Malan et al. [40] prepared a wireless sensor called CodeBlue that is capable of monitoring vital physiological information, including heart rate and blood oxygen saturation (SpO_2). These vital signs are sent wirelessly from CodeBlue to the base station [40].

Diabetes: Diabetes is a common disease that affects many people. Complications from diabetes include stroke, high blood pressure, blindness, and kidney failure. The standard treatment for diabetes includes blood glucose monitoring, insulin injections, exercise, and diet. Wireless biomedical sensors present an effective and efficient way for managing diabetes. For example, a biosensor can monitor a patient's glucose level and send the data to the patient's smart watch [41].

Preventing medical accidents: Presently one of the most important issues in the medical field is preventing accidents caused by human error. Approximately one hundred thousand people die in hospitals every year due to human error. The E-Nightingale project is a nursing risk management system that incorporates wearable environmental sensors in a sensor network to monitor and predict medical accidents. The aim is to reduce frequency of accidents and thus save patient lives [41].

Home monitoring: Using WSNs at home confers multiple benefits on patients, including convenience, privacy, and dignity [41].

Stroke and post-stroke: Strokes are very common worldwide and recovery from them is a time-consuming process that proceeds from the hospital setting to the home. Wearable sensors allow for assessing and measuring a stroke patient's motor behaviors, which can then be used to forecast clinical results [41].

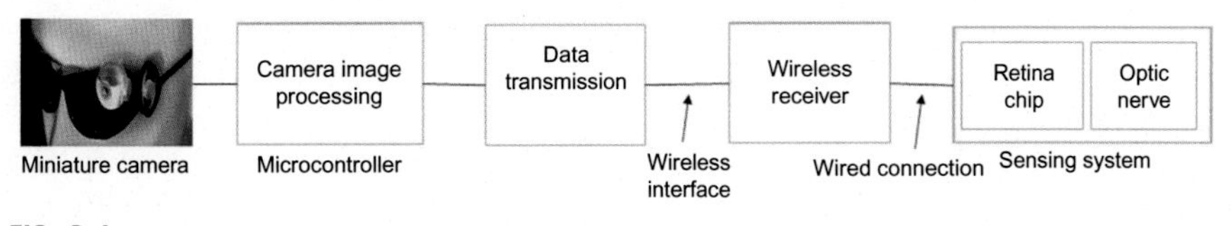

FIG. 2.4

Processing system of the artificial retina.

2.5 Challenges in wireless sensor networks and wearable sensors

Sensitivity of sensors: Sensor sensitivity is important, especially in turbulent environments. Transducers of sensory appliances are affected by sweat, resulting in a reduction in sensor sensitivity.

Deployment: Node deployment is a fundamental issue to be solved in WSNs. A proper node deployment scheme can reduce the complexities of problems. Managing and deploying a large number of nodes requires some special techniques and skills.

Design constraints: The most basic aim of WSN design is to create cheaper, smaller, and more efficient devices. The design of WSN and sensor nodes is affected by a variety of additional changes.

Security: One of the major challenges in WSN is providing high security requirements with constrained resources. Security issues include data confidentiality and node authentication. The deployment sensor must pass a node authentication examination by its corresponding manager nodes. In fact, a sensor network needs a new solution for better establishment, distribution, node authentication, and secrecy.

Limited bandwidth: Bandwidth limitation directly affects message exchange among sensors. Presently, wireless communication is limited to a data rate in the order of 10–100kbps. As a result, much less power is consumed in processing the data than transmitting it. These wireless communication links operate in the radio infrared or optical range.

Node cost: A WSN consist of a large set of sensor nodes. Overall, the cost of an individual node is very high. The cost of each sensor node should be kept low for global acceptance. If the overall cost of the sensors are appropriate then users will be more likely to accept it.

Energy: One of the important design challenges for a WSN is energy efficiency. Power consumption is very important to three functional domains: (1) sensing, (2) communication, and (3) data processing, each of which requires optimization. The life span of a sensor node depends upon the lifetime of the battery. Sensor nodes operate with a limited energy budget. Batteries are capable of powering sensors and must be recharged when depleted. For nonrechargeable batteries, a sensor node should be able to operate until either its mission time has ended or the battery can be removed. Mission time depends upon the type of application [42].

Unobtrusiveness: Unobtrusiveness is still an issue related to the layout and advancement of wearable sensor appliances. Unobtrusiveness is a significant problem when sensors are being carried by patients [43, 44]. A combination of various types of sensors is required to get one solution that gives together the units related to LiveNet [45] and tracks. These wearable sensor devices are weighty and invasive.

Compatibility problems: Compatibility issues arise when trying to unify multiple sensory devices at dissimilar frequencies. Devices communicate with each other and use a variety of bands and protocols that can result in intervention between devices.

2.6 Discussion and conclusion

Nowadays there is a growing enthusiasm for creating specialized solutions to various problems. The aging population presents difficulties for the medical services industry to convey quality services. This chapter is an attempt to show various aspects of healthcare techniques and presents the architecture of a WSN platform that supports and encourages medical information transmission and gathering. In addition, the chapter provides a deep insight into health care and chronic disease supervision. It offers a broad view of recent advances in sensor devices and internet applications. The WSN has emerged as one of the most promising technologies. This has been enabled by advances in technology and the availability of inexpensive smart sensors, which are easily deployable and cost-effective. A WSN has the ability to minimize these future difficulties by improving the utilization of therapeutic equipment and machinery. The design of an advanced medical sensor network seems to be a good solution; minimal cost advances are required to help in the conveyance of quality services while at same time reducing cost. As a result, WSNs are becoming important for patient monitoring both in the clinical setting and in the home. This chapter studied the application of WSNs in the healthcare industry. The application of WSN technology in health care can be done by monitoring patients from any location, whether it be the office or the home. Sensors link the physical world with the digital world by capturing and revealing real-world phenomena and converting these into a concrete form that can be stored, processed, and used to make decisions about the state of patient health and wellbeing. Continuous monitoring enhances data quality as well as allows data to be analyzed in an attempt to solve problems in the healthcare industry. Moreover, it is shown that a WSN provides that kind of unique technology to build wireless sensing and create a giant infrastructure for multiple data gathering. WSN application in the healthcare industry is being researched and deployed all over the world. With the rise of this application, implications will also increase rapidly. However, improvement is still needed in some areas, including further upgrades to WSN communication.

References

[1] B. Krishnamachari, Networking Wireless Sensors, Cambridge University Press, 2006.

[2] J. Yoo, L. Yan, S. Lee, Y. Kim, H. Yoo, A 5.2 mw self-configured wearable body sensor network controller and a 12 w wirelessly powered sensor for a continuous health monitoring system, IEEE J. Solid-State Circ. 45 (2010) 178–188.

[3] M. Renaud, K. Karakaya, T. Sterken, P. Fiorini, C. Hoof, R. Puers, Fabrication, modelingandcharacterization of MEMS piezoelectric vibration harvesters, Sens. Actuator A 145–146 (2008) 380–386.

[4] C. Lauterbach, M. Strasser, S. Jung, W. Weber, Smart clothes self-powered by body heat, in: Proceedings of Avantex Symposium, Frankfurt, Germany, 2002, pp. 5259–5263.

[5] T. Gao, D. Greenspan, M. Welsh, R. Juang, A. Alm, Vital signs monitoring and patient tracking over a wireless network, in: Proceedings of the 27th Annual International Conference of the IEEE EMBS, Shanghai, China, 1–4 September, 2005, pp. 102–105.

[6] T.B. Tang, E.A. Johannessen, L. Wang, et al., Toward a miniature wireless integrated multisensor microsystem for industrial and biomedical applications, IEEE Sensors J. 2 (6) (2002) 628–635.
[7] L. Machaya, S. Tembo, A study on memory management in wireless sensor nodes during key agreement generation, Int. J. Netw. Commun. 6 (2) (2016) 19–23, https://doi.org/10.5923/j.ijnc.20160602.01.
[8] H. Abelson, et al., Amorphus computing, CACM 43 (5) (2000) 74–82.
[9] E 802.11 Working Group, IEEE 802.11-2007: Wireless LAN Medium Access Control (MAC) and Physical Layer (PHY) Specifications, 2007. ISBN 0-7381-5656-5659.
[10] http://www.bluetooth.com/bluetooth/.
[11] www.zigbee.com.
[12] A. Minaie, A. Sanati-Mehrizy, P. Sanati-Mehrizy, R. Sanati-Mehrizy, Application of wireless sensor networks in health care system, Age 23 (1) (2013).
[13] https://www.researchgate.net/publication/266618899.
[14] Y. Yin, Y. Zeng, X. Chen, Y. Fan, The internet of things in healthcare: an overview, J. Ind. Inf. Integr. 1 (2016) 3–13, https://doi.org/10.1016/j.jii.2016.03.004.
[15] O.O. Olaniyi, N.M. Nadine, Epidemiology of dementia among the elderly in Sub-Sahara Africa, Int. J. Alzheimers Dis. 2014 (195750) (2014).
[16] R. Varatharajan, G. Manogaran, M.K. Priyan, R. Sundarasekar, Wearable sensor devices for early detection of Alzheimer disease using dynamic time warping algorithm, Clust. Comput. 21 (1) (2017) 681–690, https://doi.org/10.1007/s10586-017-0977-2.
[17] Z. Zhang, R. Tavenard, A. Bailly, X. Tang, P. Tang, T. Corpetti, Dynamic time warping under limited warping path length, Inf. Sci. 31 (393) (2017) 91–107.
[18] Y. Wan, X.L. Chen, Y. Shi, Adaptive cost dynamic time warping distance in time series analysis for classification, J. Comput. Appl. Math. 1 (319) (2017) 514–520.
[19] M. Masoli, D. Fabian, S. Holt, R. Beasley, The global burden of asthma: executive summary of the GINA Dissemination Committee Report, Allergy 59 (5) (2004) 469–478, https://doi.org/10.1111/j.1398-9995.2004.00526.x.
[20] A. Hosseini, C.M. Buonocore, S. Hashemzadeh, H. Hojaiji, H. Kalantarian, C. Sideris, et al., HIPAA compliant wireless sensing smartwatch application for the self-management of pediatric asthma, in: 2016 IEEE 13th International Conference on Wearable and Implantable Body Sensor Networks (BSN), 2016. https://doi.org/10.1109/bsn.2016.7516231.
[21] H.-K. Ra, J.A. Stankovic, A. Salekin, H.J. Yoon, J. Kim, S. Nirjon, D. Stone, S. Kim, J.-M. Lee, S.H. Son, Demo: AsthmaGuide: An Ecosystem for Asthma Monitoring and Advice, (2015) pp. 451–452, https://doi.org/10.1145/2809695.2817849.
[22] G. Appelboom, E. Camacho, M.E. Abraham, S.S. Bruce, E.L. Dumont, B.E. Zacharia, et al., Smart wearable body sensors for patient self-assessment and monitoring, Archives Public Health 72 (1) (2014). https://doi.org/10.1186/2049-3258-72-28.
[23] Z. Pang, L. Zheng, J. Tian, S. Kao-Walter, E. Dubrova, Q. Chen, Design of a terminal solution for integration of in-home health care devices and services towards the Internet-of-Things, Enterprise Inform. Syst. 9 (1) (2015) 86–116, https://doi.org/10.1080/17517575.2013.776118.
[24] Y. Zhang, M. Qiu, C.-W. Tsai, M.M. Hassan, A. Alamri, Health-CPS: healthcare cyber-physical system assisted by cloud and big data, IEEE Syst. J. 11 (1) (2017) 88–95, https://doi.org/10.1109/jsyst.2015.2460747.
[25] J. Ferlay, M. Colombet, I. Soerjomataram, C. Mathers, D.M. Parkin, M. Piñeros, et al., Estimating the global cancer incidence and mortality in 2018: GLOBOCAN sources and methods, Int. J. Cancer (2018), https://doi.org/10.1002/ijc.31937.

[26] T. Guy-Anne, W. Andrew, A.A. Azad, S. Benjamin, S. Siva, Radiotherapy and immunotherapy: a synergistic effect in cancer care, MJA 210 (1) (2019) 47–53, https://doi.org/10.5694/mja2.12046.
[27] C.K. Yoon, M. Ochoa, A. Kim, R. Rahimi, J. Zhou, B. Ziaie, Yeast metabolic response as an indicator of radiation damage in biological tissue, Adv. Biosyst. (2018), 1800126. https://doi.org/10.1002/adbi.201800126.
[28] V.R. Singh, K. Singh, Wireless sensor networks for biomedical applications in cancer hyperthermia, in: 2008 30th Annual International Conference of the IEEE Engineering in Medicine and Biology Society, Vancouver, BC, 2008, pp. 5160–5163. https://doi.org/10.1109/IEMBS.2008.4650376.
[29] K. Raza, Improving the prediction accuracy of heart disease with ensemble learning and majority voting rule, in: U-Healthcare Monitoring Systems, Academic Press, 2019, pp. 179–196.
[30] U. Anliker, J.A. Ward, P. Lukowicz, G. Troster, F. Dolveck, M. Baer, M. Vuskovic, AMON: a wearable multiparameter medical monitoring and alert system, IEEE Trans. Inf. Technol. Biomed. 8 (4) (2004) 415–427, https://doi.org/10.1109/titb.2004.837888.
[31] J.M. Cano-Garcia, E. Gonzalez-Parada, V. Alarcon-Collantes, E. Casilari-Perez, A PDA-based portable wireless ECG monitor for medical personal area networks, in: MELECON 2006, 2006 IEEE Mediterranean Electrotechnical Conference, 2006. https://doi.org/10.1109/melcon.2006.1653199.
[32] K.J. Kappiarukudil, M.V. Ramesh, Real-time monitoring and detection of "heart attack" using wireless sensor networks, in: 2010 Fourth International Conference on Sensor Technologies and Applications, Venice, 2010, pp. 632–636.
[33] P. Kakria, N.K. Tripathi, P. Kitipawang, A real-time health monitoring system for remote cardiac patients using smartphone and wearable sensors, Int. J. Telemed. Appl. 2015 (2015) 1–11, https://doi.org/10.1155/2015/373474.
[34] H.C. Ahn, B. Delshad, J. Baranowski, An implantable pressure sensor for long-term wireless monitoring of cardiac function-first study in man, J. Cardiovasc. Dis. Diagn. 4 (2016), 252. https://doi.org/10.4172/2329-9517.1000252.
[35] N. Dey, A.S. Ashour, F. Shi, S.J. Fong, R.S. Sherratt, Developing residential wireless sensor networks for ECG healthcare monitoring, IEEE Trans. Consum. Electron. 63 (4) (2017) 442–449, https://doi.org/10.1109/tce.2017.015063.
[36] C. Veraart, F. Duret, M. Brelen, J. Delbeke, Vision rehabilitation with the optic nerve visual prosthesis, in: Proceedings of the 26th Annual International Conference of the IEEE EMBS, San Francisco, USA, 2004, pp. 4163–4165.
[37] M.K. Watfa, Practical applications and connectivity algorithms in future wireless sensor networks, Int. J. Inform. Commun. Eng. 4 (2008) 5.
[38] L. Schwiebert, S.K.S. Gupta, P.S.G. Auner, G. Abrams, R. Iezzi, P. McAllister, A biomedical smart sensor for the visually impaired, Proc. IEEE Sensors (2002), https://doi.org/10.1109/icsens.2002.1037187.
[39] Y. Miura, T. Hachida, M. Kimura, Artificial retina using thin-film transistors driven by wireless power supply, IEEE Sensors J. 11 (7) (2011) 1564–1567, https://doi.org/10.1109/jsen.2010.2096807.
[40] D. Malan, T. Fulford-Jones, M. Welsh, S. Moulton, Code Blue: an ad hoc sensor network infrastructure for emergency medical care, in: Proceedings of the MobiSys/Workshop on Applications for Mobile Embedded Systems, 2004, pp. 12–14.
[41] P. Neves, M. Stachyra, J. Rodrigues, Application of Wireless Sensor Networks to Healthcare Promotion, 2008.

[42] S. Karthik, A.A. Kumar, Challenges of wireless sensor networks and issues associated with time synchronization, in: Proceedings of the UGC Sponsored National Conference on Advanced Networking and Applications, 2015.

[43] A. Purwar, D.U. Jeong, W.Y. Chung, Activity monitoring from realtime triaxial accelerometer data using sensor network, in: Proceedings of International Conference on Control, Automation and Systems, Hong Kong, 21–23 March, 2007, pp. 2402–2406.

[44] C. Baker, K. Armijo, S. Belka, M. Benhabib, V. Bhargava, N. Burkhart, A. Der Minassians, G. Dervisoglu, L. Gutnik, M. Haick, C. Ho, M. Koplow, J. Mangold, S. Robinson, M. Rosa, M. Schwartz, C. Sims, H. Stoffregen, A. Waterbury, E. Leland, T. Pering, P. Wright, Wireless sensor networks for home health care, in: Proceedings of the 21st International Conference on Advanced Information Networking and Applications Workshops, Niagara Falls, Canada, 21–23 May, 2007, , pp. 832–837.

[45] M. Sung, C. Marci, A. Pentland, Wearable feedback systems for rehabilitation, J. Neuro Eng. Rehabil. 1 (2005) 2–17.

CHAPTER

A comprehensive dialogue for U-body sensor network (UBSN) with experimental case study

3

S.P. Sonavane[1], S.G. Tamhankar[1], M.G. Rathi[1], F.S. Kazi[2]

[1]*Department of Information Technology, Electronics, WCE, Sangli, India* [2]*Department of Information Technology, Electronics and Electrical, VJTI, Mumbai, India*

3.1 Introduction

Wearable body sensor network is a highly recognized e-healthcare monitoring system for well-being of people and hence the society. By practice, home health monitoring system (HMS) is globally recommended with remote access for limited range of coverage [1]. The sophisticated way to visualize this real-time HMS is framed using embedded body sensors with Wireless Sensor Network (WSN).

Nowadays, smart wearable devices can sense human heart rate (beats per minute), body temperature, step count, calories burned, and similar information in the form of data streams. Home healthcare is an affordable option to conceive the status of the body. HMS provides continuous data access from multichannel embedded device and smart wearable that are orchestrated with mobile applications [2, 3]. These phenomenal systems can reduce the health risks of a patient/user for physical weakening in human body. Such systems are becoming popular as wearable sensor network provides health monitoring without disturbing the routine work of the user. The overall setup for a typical system is shown in Fig. 3.1.

However due to physiological changes, the environmental issues are proved as a barrier or a matter of challenge for setting WSN in remote healthcare conditions. These challenges may be related to hardware used, software implemented, or communication level depending on the circumstances of the field application. These issues encourage researchers to concentrate their studies toward optimization of the parameters considering computational platform, network delay, sensor response time, synchronization, mutual exclusion, etc.

The chapter aims to elaborate UBSN that provides the multichannel polling through single gateway protocol for fetching the activities of the user with wireless sensor nodes placed at significant positions in the body sensor network [3]. Spatial distribution of independent nodes increases the coverage area of the body to detect the activities performed by the user in multiple epochs. To improve the quality of data communication, IEEE 802.15.4 channel along with shared gateway is used with

Sensors for Health Monitoring. https://doi.org/10.1016/B978-0-12-819361-7.00003-8

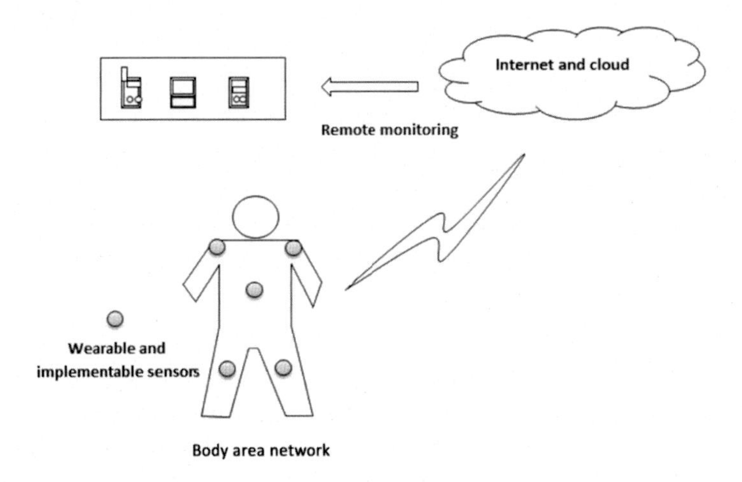

FIG. 3.1

Body area network in health monitoring system (HMS).

cooperative transmission on the link. Data polling with multiple channels and streaming of heart rate from multiple UBSN networks are synchronized at the gateway. The algorithms are developed with multichannel gateway that implement sliding window protocol to upgrade the window at cloud database server periodically with time stamp of 250 ms. Further, the received data is processed at database server for individual occupant using Artificial Neural Network (ANN) to infer the activities and notify back to the user's mobile phone through android interfaces [4]. This process arrives with a certain recommendation system providing health profile of the user with activities concerned to be performed defining his/her regular exercise patterns.

3.2 Related work

In view of deciding the methodology of the work, the literature survey is carried out in relevance to the major functional blocks in similar systems. ANN techniques are compared for adopting suitable algorithm for handling the structured data and implementing supervised learning model to enhance the experimental results. This exercise has been proved to be helpful to come up with near-optimal solution space in healthcare problem.

In general, the primary focus of the researchers is on wearable body sensors to statically stream data from body area network. The data are sampled across suitable time intervals (few seconds to a day) to define and represent user's activity chart [5, 6]. Efforts also are carried out on vital part of the body to sense temperature, heart rate, ECG monitor, giving the systematic representation of patient's profile in embedded system [7]. A web-based diabetic support management system is proposed to detect sugar level of the patient helpful for further precautions [8].

Internet of thing (IOT)-based decision support system is recommended for e-healthcare providing personnel medical assistance to the patient [3, 9]. The device-based IOT system comprises programming logic, configured sensor and network interface to integrate the sensor network in HMS. This makes significant impact in e-healthcare system to connect the device, node, and people together. Existing technologies such as machine learning, big data analytics, WSN, and cloud loudly support the IOT-based paradigm in intelligent U-healthcare system. Smart healthcare system with wearable devices qualifies the IOT-based UBSN that provides the updated user profile for period of time [10].

Home ECG monitoring system is useful to monitor the patient's appropriate heart functioning periodically and allows patient to avoid hospitalization and enjoy at home place. This system continuously observes the beats per minutes (bpm) and data interpretation and analysis [11].

Foundation to data analytics in clinical support system has gained the valuable attention in recent years. Few attempts are made to combine smart home system with ubiquitous healthcare networking through suitable communication protocols with collective data processing [12]. Biological attributes of pregnant women are extracted with WSN to diagnose health status of a mother and the fetus [13]. Multipatient body network is proposed to store, process, and analyze the data collected in a group with a single gateway [3, 14]. Frame relay and base station are the coordinators in mobile healthcare systems that evidently define the communication channel with battery backup, data aggregation at sink node [5, 15]. A typical WSN model is extended toward intelligent therapy management to diagnose, monitor, and cure in biomedical application [16].

Handling noisy data in real-time application is a crucial aspect to investigate the health attributes but can be solved using regression and clustering model in data mining [3, 17]. The specific attention of study is continued further to learn supervised as well as unsupervised model training data in HMS. Such ANN back propagation model is used to detect the medical impairment of the patient. Wireless health monitoring has numerous applications with personal SMS or email notifications that can communicate either in textual or verbal form providing more user-friendly system [18, 19]. Data cleaning is required as preprocessing step in case of sensitive applications such as anomaly detection in received signal responses [20, 21]. On the contrary, few applications require robust data processing techniques with removal of fault tolerance in the communication channel [22]. Thus, the methodology to be decided is dependent on the domain scope and the experts involved in the system.

3.3 Motivation and background

Emerging trends in HMS provide the opportunity to the researchers to find better options to inculcate wider solutions. Wearable devices can monitor the logs of the user profile on activity performed. The prime solution for such a pivotal region can be possible with multidisciplinary approach. In today's modern era, due to fast changing life style of the individual, it is necessary to take due precautions of the health.

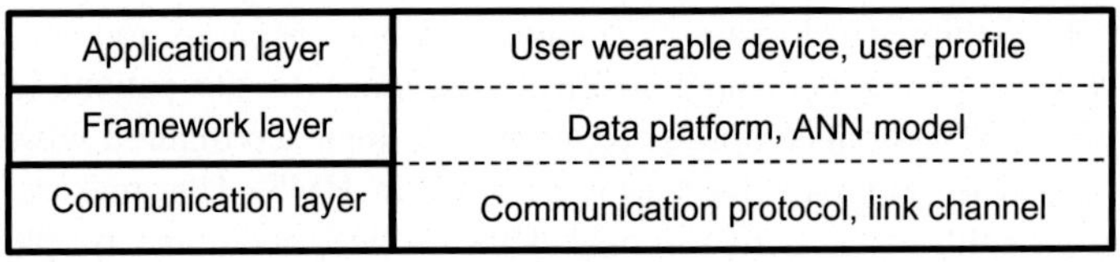

Application layer	User wearable device, user profile
Framework layer	Data platform, ANN model
Communication layer	Communication protocol, link channel

FIG. 3.2

Layered architecture for UBSN.

Hence, availability of the modern devices and sensors is becoming modest and handy for simplified use. The requirement of the user is that he/she expects a priori alarming information of routine actions at the end of the day.

Multichannel data can be streamed or polled with body sensors to observe the routine actions made by the user. Embedded U-healthcare carries the effective approach to generate the user profile in order to recommend the notifications toward health improvement [23]. Smart wearable working with WSN along with data analytics is able to fulfill the very need. Thus, activity detection, prevention, and recommendation can improve the the patient's life [24].

This background suggests some intelligent integrated solution with ever-increasing scope for developing further with comprehensive and précised solutions.

The UBSN architecture can broadly be layered, based on the functionality as shown in Fig. 3.2.

- **Communication layer:** It accomplishes deployed nodes to communicate the signal response between sensors in line with gateway. Time synchronization and link utilization is succeeded by bit delay at this layer. It addresses various communication protocols and link transmission over the topology with collision avoidance.
- **Framework layer:** It comprises the layout of data fetching from sensor nodes placed at several positions acting as a data collection and processing platform. Structured data are trained in ANN to pursue the activity of user.
- **Application layer:** It embraces the smart wearable device to monitor the heart rate, links interface for generation of user profile and communication agent with system interpretation.

3.4 System design

Overall system comprises component for sensing, communication, data storage, computing, and decision making.

3.4.1 Sensor

A sensor producing an electrical output while combined with interfacing electronic circuits is called as "Smart Sensor." It simply converts physical, biological, or chemical inputs into the measurable values in a digital format.

Conventionally the typical sensor selection criteria include sensing range, connection type, and electrical output. However, in order to get the better results, users need to concentrate on additional parameter and its critical analysis.

3.4.2 Sensing, communication, and computing

Sensors are usually used to deploy in the system to capture the desired data. There are varieties of sensors available for sensing the typical parametric data. Sensor selection is majorly application specific based on the parameters like range, accuracy, response time, resolution, and type of required output etc. Preprocessing like amplification and filtering is essentially required to process the data further. Nowadays smart sensors are available with all these capabilities with local storage capabilities additional communication and limited processing.

Gateway is integral part of such system, which carries data to server/cloud for further processing and decision making. Gateways play an import role in transferring data from physical system to server/Cloud. It is available with multiprotocol functionality; on one side of gateway short-distance protocols like BLE, Zigbee, Bluetooth are functioning, and other side TCP/IP is required to carry that data to server/ cloud. It is essential to take into account how data need to be transfer like polled data, streamed data, or regular interval data.

To store, monitor, analyze, and control the data; it is obvious to connect this physical system with internet. The appropriate selection of algorithm helps to process data and content into desired format. The integration of wireless protocol with IOT devices along with sensor gateway is the basic structure towards system design.

In the coming age of IoT, internet connectivity to wireless non-IP protocol will play important role. Hence, sensor Gateway and protocol plays important role in system design.

3.4.3 Gateway design

As shown in figure, gateway is a device is used to connect two different networks, especially a connection to the Internet (Fig. 3.3).

Many proprietary ZigBee gateways are available that have a single channel. These gateways can be implemented individually considering various features such as low-power consumption, optimized data rate, and low cost [25].

ZigBee gateway is designed using two radio modules for increasing data transmission rate of ZigBee and serves low latency. In this gateway, two ZigBee modules are embedded for remote patient monitoring. The ZigBee transceivers are used for polling data transfer in the medical applications. Another transceiver is dedicated for ECG/EMG sensor data monitoring. A 32-bit controller is used for scheduling the data received at the gateway [26].

3.4.3.1 Gateway limitations

Many gateways have finite local storage capacity with limited processing power and RAM. This limits the expandability and becomes very complex. The gateway must be evaluated in the context of its requirements that has a limited ability to translate

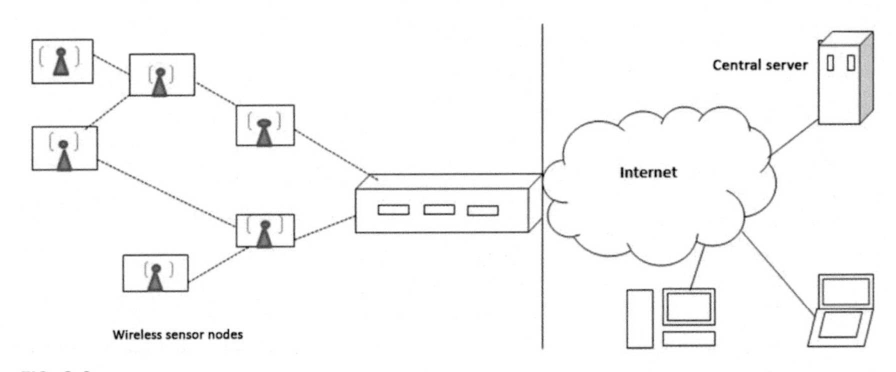

FIG. 3.3

System overview in UBSN.

dissimilar concepts with the capability of protocol conversion from short-distance wired or wireless protocols to standard TCP/IP packets. For the device configuration from a remote location, it is difficult to pass the messages through the gateway; hence, bypassing the gateway is the only alternative solution. So specifications of the gateway must be enough clear to state about the collected information. As most of the gateways are application specific only, information access is limited through the gateway that leads to difficulty in troubleshooting the integrated applications. Implementing prioritization of data is complex because of speed, processing overhead, and irregular delay. Thus, these are the important issues that are to be necessarily addressed.

3.4.3.2 SMART gateway implementation

To overcome the limitations of conventional gateway, following are the design considerations for SMART gateway:

- **Real-time monitoring:** A closer look at embedded gateway will almost always reveal that the data are delayed or otherwise not presented in a useful way. System should have very small delay, high data transmission rate, and low latency.
- **Security:** All communication between sensors and server needs to be highly secure. Slight interference with data can cause a lot of damage.
- **Adaptive nature:** By managing the buffer, jitter can be controlled.
- **Size and cost:** Real-time gateways that are available for large-scale automation have high cost and are bulky in nature.
- **Power consumption:** As gateway runs continuously, power consumption over a time becomes decisive factor.

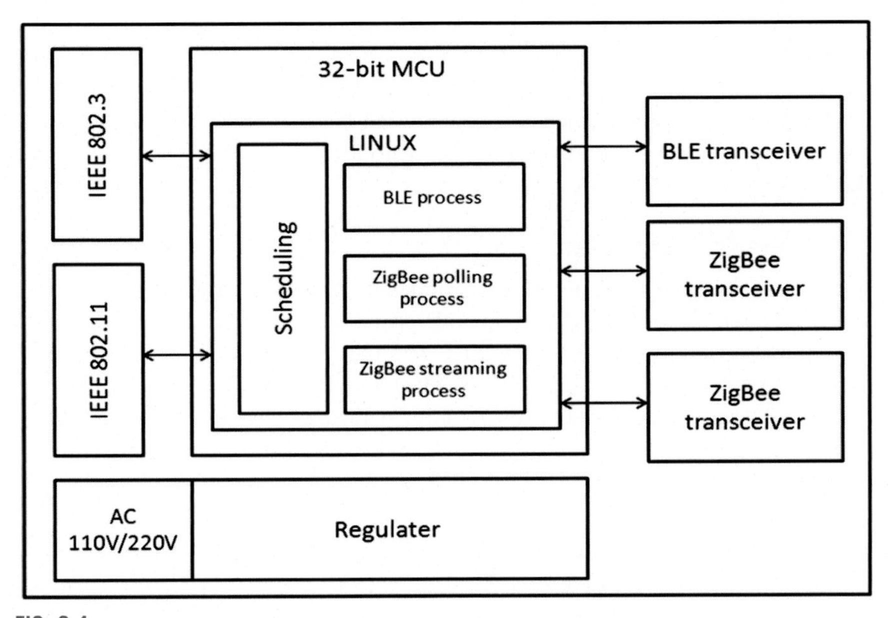

FIG. 3.4

Block diagram for gateway.

The main task of gateway is to acquire the data received from physical system and pass them to internet. On one side of the gateway, wireless short-distance protocols like Zigbee, Bluetooth, and Bluetooth Low Energy (BLE) are active, and, on the other side, TCP/IP protocol is active. Whole gateway is modeled using ARM Cortex-A8-based TI sitara AM3358 and supported with 512 MB RAM and 4GB flash memory. The gateway is built on top of Linux kernel (Fig. 3.4).

In ZigBee polling process, data are collected in periodic intervals or on demand, then transferred to central server using socket. In ZigBee streaming process, ZigBee is optimized and the received data are transferred to central server.

3.5 Proposed methodology

The proposed model has been used with WSN. The experimental model has been proposed with IEEE 802.15.4 for radio wave communication with Zigbee network IEEE 802.5 Token Ring protocol is realized for communication between the sensor nodes. Received frame from sensor node and wearable blood pressure monitor have been processed with ANN to detect activity of user. The Recommendation system has been proposed to suggest the exercise to the user [4].

UBSN system comprises three parts Sensor system, Communication system, Data storage, Computing and Decision-making system.

Proposed system with UBSN is realized with four wireless sensors placed at Chest, Right Ankle, Left Ankle, and Wrist Blood Pressure Monitor. Token ring protocol has been implemented to communicate the sensor node with the gateway.

Token Ring has been formed between sensor nodes A, B, and C programmed with TinyOs. Node A is a gateway programmed with the public key of the cloud to send the data.

3.5.1 Sensor parameter

Four sensor nodes A, B, C, and D are parameterized with minimal response time as stated in Table 3.1 and Fig. 3.5. Range of sensor is about 10 m with gateway node configured as Single-Chip 2.4 GHz IEEE 802.15.4 [6]. Table 3.1 reflects the comparative study of gateways used in WSN. The proposed gateway to be used in system design is disclosed with respect to the required parameters and accordingly supported with justifying remarks.

3.5.2 Network setup

Ubiquitous Sensor Network is commonly implemented with IEEE 802.15.4 Wireless Personal Area Network (WPAN) with nodes A, B, C, and D. It is largely utilized for MAC and Physical layer operation in UBSN [15]. 802.15.4 is a wireless communication standard configured with Zigbee and Bluetooth talks. It uploads the data (250 kb) at low-power cost and is reasonable to embed the system with this standard. It operates on data link layer with all motes in WSN. Though the bandwidth is low, it emphasizes the small distance and low-cost communication with frame transmission at physical layer.

Table 3.1 Comparative analysis of gateways in WSN

Existing gateway	Proposed gateway	Parameter	Remark
Multiprotocol functionality but no multichannel functionality	Multiprotocol and multichannel functionality	Only one gateway will serve purpose	Cost saving
Fixed buffer	Adaptive buffer	Performance improvement	Minimize packet loss by 30% in typical scenario
No security at gateway level	Minimal security feature implemented	Helps in detection of intrusion and Man in middle attack	Overheads of algorithm implementation in optimized way
Static and fixed	Programmable, reconfiguration as per user requirement	Application-specific configuration possible	Time saving in case of reconfiguration
Bulky in size	Tiny in nature	Placement in any environment	Suitable for IoT

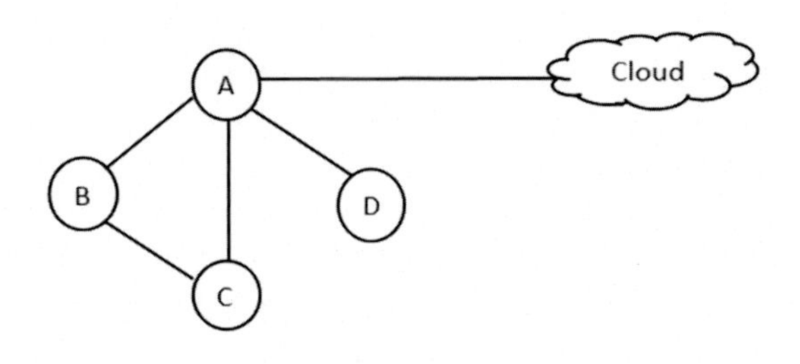

FIG. 3.5

Data polling with sensor nodes A, B, C, and D.

Table 3.2 Network parameters for system design

Network parameter	Value
Number of nodes	4
Accelerometer sensor response time	0.001 (s)
Gyroscope sensor response time	0.002 (s)
Sensor range	10 m (approx.)
Bit rate	250 kbits/s
Channel frequency	2.4 GHz
Node RAM	128 kB
Battery life	1–10 days
Transmission delay	4 ms

Gateway router converts the sense signal to the received response time (RRT) for every 250 ms of epoch in WSN (Table 3.2).

3.5.2.1 Communication protocol

IEEE 802.5 token ring protocol has been followed to communicate the sensors A, B, and C in WSN. Node having the token frame can transmit the beacon frame over the link. The entire beacon frame is collected by gateway with RRT. A token is under the custody for 50 ms on respective node to transmit beacon frame in ring. In order to avoid the collision, if node A is transmitting, nodes B and C should be in listen mode. Node A should transmit the beacon frame after the 50 ms. to the next node and so on. Estimated ring latency is 250 ms; hence, data frame should get collected in 250 ms. Heart beat for a person is monitored with wearable blood pressure monitor.

3.5.2.2 Data polling

A client program is configured with motes A, B, and C to fetch the data at different intervals of time as shown in Fig. 3.5. Beacon frame is uploaded to the gateway mote

A in 250 ms for a single epoch. Busy waiting is implemented with token ring protocol to avoid the conflicts and frame collision. It synchronizes the data fetch operation with software and hardware I/O mechanism. Mote A is configured with private key generated with cloud storage to upload the data simultaneously over the cloud. All the motes are polled within the 250 ms to get back the RRT of wearable embedded sensor user. Data are continuously polled from the motes A, B, C, and D for 120 s in the form of beacon frame with RRT as a payload and stored in the database at cloud.

Data are polled with RRT between nodes AB, BC, and AC along with the heart beat rate for 120 s for a user. Data are streamed to storage platform in relational database for data processing and analytics.

Fig. 3.6 illustrates the sequence of polling data transfer over the Zigbee network to coordinate the system with gateway. UBSN is configured with four nodes in body sensor network to sense the response time of the nodes.

1. Sensor node is configured with user ID and Age, generate the profile.
2. User profile is sent to gateway with IEEE 802.15.4
3. According to user profile, Zigbee sends data over gateway at specified interval
4. Gateway buffers the upgraded frame to validate the user and connection *connect()* is established to cloud server
5. Cloud server sends the acknowledgement response to data *write()* request

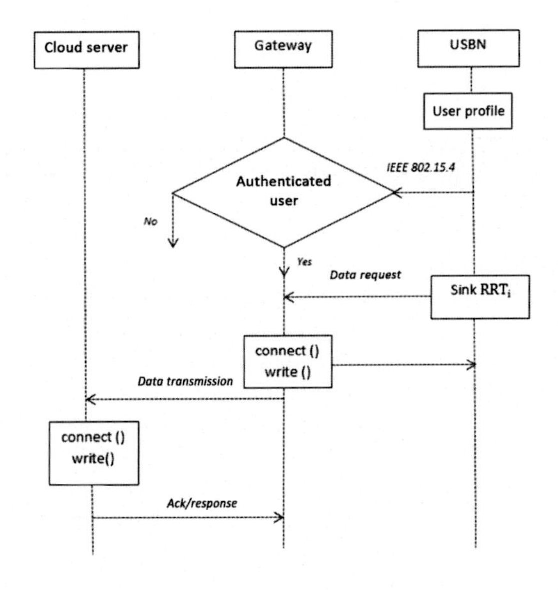

FIG. 3.6

Sequence of data polling in Zigbee.

3.5.2.3 Token ring protocol management

IEEE 802.5 Token Ring provides the Ethernet communication in WSN with synchronization of token management. Operational modes of token ring are Listen Mode, Transmission Mode, and Bypass Mode. In UBSN, these modes have been followed to provide the mutual exclusion of the beacon transmission on link in Ethernet.

a. Listen mode

It comprises the one-bit delay at the sensor to process the beacon frame sent by other nodes. Read-only operation is performed with payload and transmits the data to next node in a ring (Fig. 3.7).

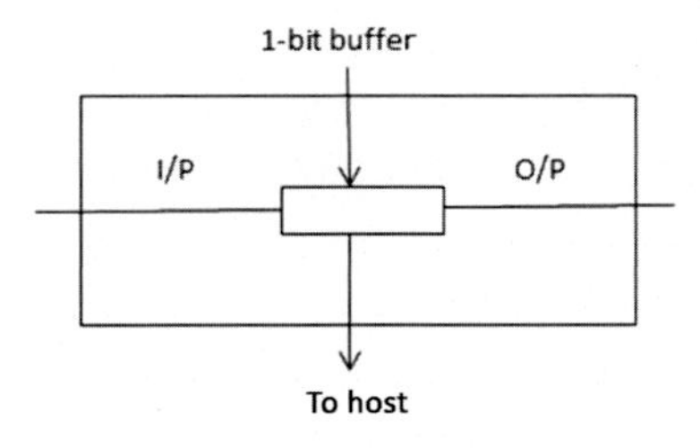

FIG. 3.7

Token ring management in listen mode.

b. Transmission mode

It allows the node to transmit the frame on link when other nodes are in listening or bypass mode (Fig. 3.8).

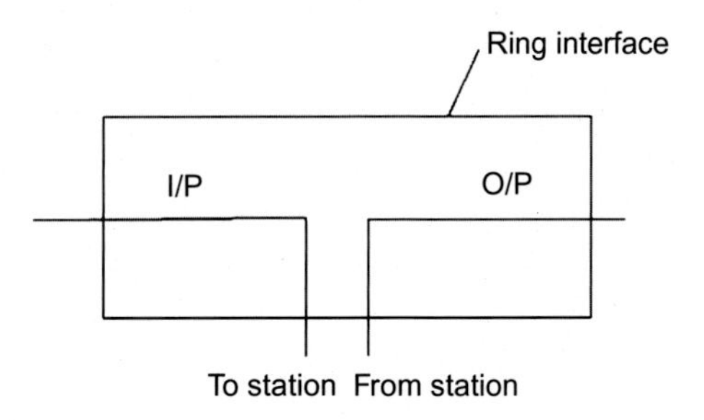

FIG. 3.8

Token ring management in transmission mode.

c. By-pass mode

When one of the sensor nodes goes down, By-Pass mode side steps the data without any processing and bit delay (Fig. 3.9).

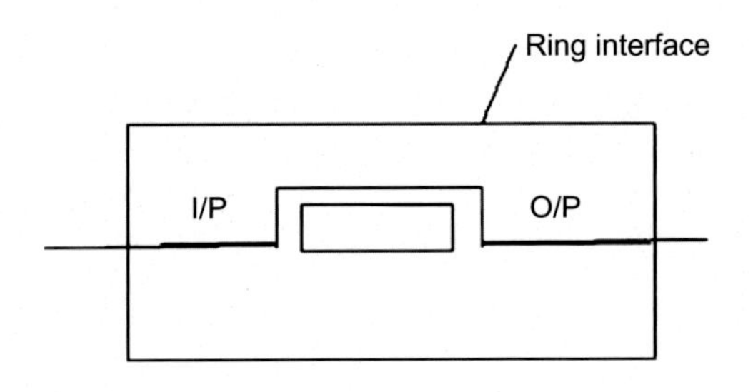

FIG. 3.9

Token ring management in by-pass mode.

3.6 Working model

The experimental set up elaborated in Section 3.4.2 is always utilized to perform the communication in WSN with wearable body sensor for HMS. UBSN describes the types of the motes/sensor nodes with actuation, communication, and sensation from hardware to application.

3.6.1 System architecture

The schematic representation of architectural view of UBSN comprises IEEE 802.15.4 and IEEE 802.5 is shown in Fig. 3.10:

Tiny-OS firmware is operational software middleware, configured with the IP address of gateway node to sink the frames. Token Ring provides the UBSN, a communication link to measure the response time for user performing activities like *Bending*, *Cycling*, *Walking*, *Lying*, and *Standing*. The featured series of response time is the principal key to detect user activities for iterative epoch at single gateway in a network [6].

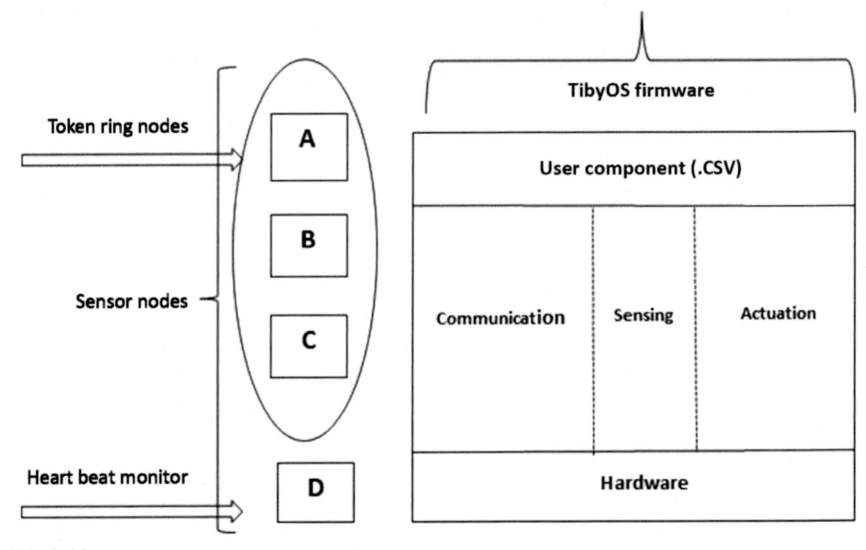

FIG. 3.10

Architectural view of UBSN.

Following algorithm executes the working model to fetch and process the data:

Algorithm 3.1. First level of algorithm for token ring communication

```
Initialize Token t for k Nodes in Token Ring
1. Let i is counter for Token
   for (i=1, i ≤ k, i++)
   if (t==i)
   {
      Transmission Mode();
   }
   else
   {
      Listening Mode();
   }
2. Pass Token, i←k+1, Recurring step 1
3. Sink frame, upload RRTi
```

Algorithm 3.1 is first level of algorithm to link channel communication with IEEE 802.5 and is configured with public cloud over the internet. If token t is under the custody of node, it executes the transmission of packet in privileged *Transmission Mode()* else *Listening Mode()*.

Algorithm 3.2. Second level of algorithm for activity detection

```
Set RRTi as RRTAB, RRTBC, RRTAC
1. Input RRTi to ANN
2. Learn RRTi for Target
3. if(RRTi==Target)
{
   Forward Pass();
}
   else
{
   Backward Pass();
}
end if;
```

Algorithm 3.2 elaborates on a RRT RRT_{AB} between nodes A and B, RRT_{BC} between nodes B and C. The scenario is explained in detail in Section 3.6.3. These response times RRT_i are aggregated at public cloud for further data processing. While performing the activity, RRT_i are received for series of time intervals from UBSN regulated on body of user. Supervised ANN learning is followed to compare the average response time with target value. After training, if RRT_i matches with

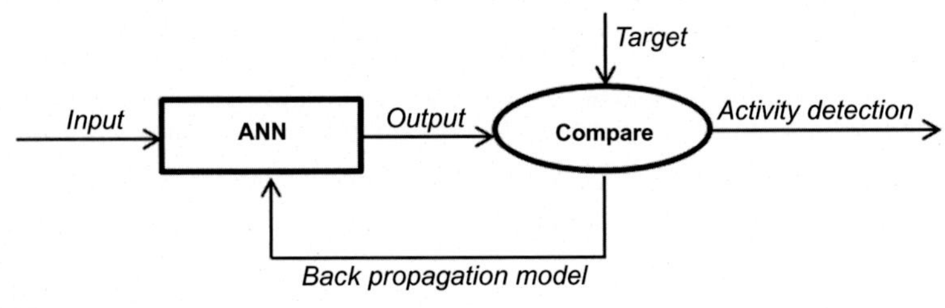

FIG. 3.11

Artificial neural network with supervised learning.

target dataset *Forward Pass()* function is executed to detect the activity of user else *Backward Pass()* to recur the training in ANN as shown in Fig. 3.11.

3.6.2 Data classification

The received multichannel data are classified as per age group for further detailing. Table 3.3 gives the three classes of age, viz., Young, Middle, and Old.

Classified values are imported in PHP MYSQL database for processing. Heart rate fetched for a user can be differentiated with respect to epoch period of 120 s. The average value of heart rate is stored against the single user to avoid the complexity of the data. Relational PHP MYSQL database consists of age wise activity to be performed by user along with the standard heart rate for a given age prescribed by medical professionals. The SQL query fired from the console retrieves the heart rate for the user that helps to identify cardiac and noncardiac health in the user. The basic disorder information is effectively utilized to recommend the exercise for the person.

Table 3.4 indicates the database consisting of a log file of users distinguished by their user ID, class, activity performed, and heart rate.

Table 3.5 is updated accordingly with cardiac functionality into low, high, and normal beats per minute (bpm). This valued information is utilized further to recommend the exercise to the user in consensus with the physiotherapist.

Table 3.3 Age groups

Class	Age window
"Y" Young age	21–35
"M" Middle age	36–50
"O" Old age	51–65

Table 3.4 User data with heart rate (beats/min)

User ID	Class	Activity performed	Heart rate (beats/min)	Ideal heart rate (beats/min)
101	"Y"	Standing, walking, bending	70	60–100
102	"Y"	Standing, walking, bending, cycling	73	60–100
103	"O"	Standing, walking	55	70–100
104	"Y"	Standing, walking, cycling	120	60–100
115	"M"	Standing, walking, bending	115	60–100
116	"Y"	Standing, walking	75	60–100
117	"M"	Standing, walking, bending	75	60–100

Table 3.5 User data with cardiac functionality (beats/min)

User ID	Class	Activity performed	Heart rate (beats/min)	Ideal heart rate (beats/min)	Cardiac functionality
101	"Y"	Standing, walking, bending	70	60–100	Normal bpm
102	"Y"	Standing, walking, bending, cycling	73	60–100	Norma1 bpm
103	"O"	Standing, walking	55	70–100	Low bpm
104	"Y"	Standing, walking, cycling	120	60–100	High bpm
115	"M"	Standing, walking, bending	115	60–100	High bpm
116	"Y"	Standing, walking	75	60–100	Norma1 bpm
117	"M"	Standing, walking, bending	75	60–100	Normal bpm

3.6.3 Experimental case study

For the verification of the proposed system, an experimental bed is set up with "Y" Young user with wearable body sensors as shown in Fig. 3.12. Here, A, B, and C indicate three body sensor nodes involved in token ring set up. Node A is a gateway in Zigbee network with IEEE 802.15.4 standard for multiple communication channels. The wrist node D is connected to gateway A monitoring heart rate in UBSN (Figs. 3.13 and 3.14).

Similar observations are carried out for cycling, walking, and lying positions as shown in Fig. 3.15:

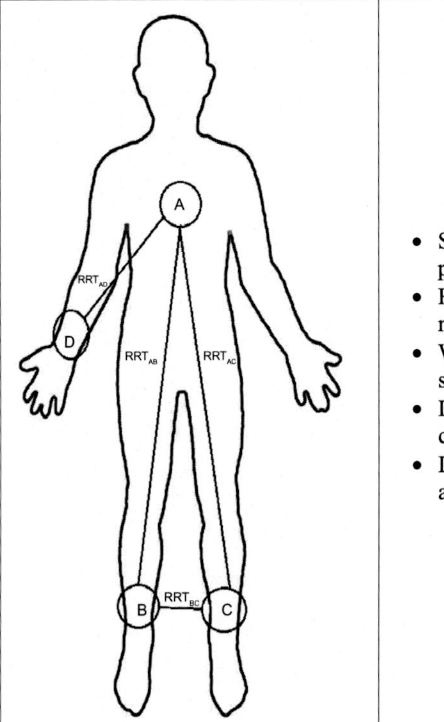

- Sensor node placement at various positions on body from class 'Y'
- RRT_{AB}, RRT_{BC} and RRT_{AC} are sink at node A as a gateway for 250ms
- Wrist band with heart beat monitor D is streamed for 120s
- Data from gateway A is imported to cloud with GSM Wi-Fi modem
- Data is segregated in the relational form as discussed in table IV and V

FIG. 3.12

Test bed for UBSN.

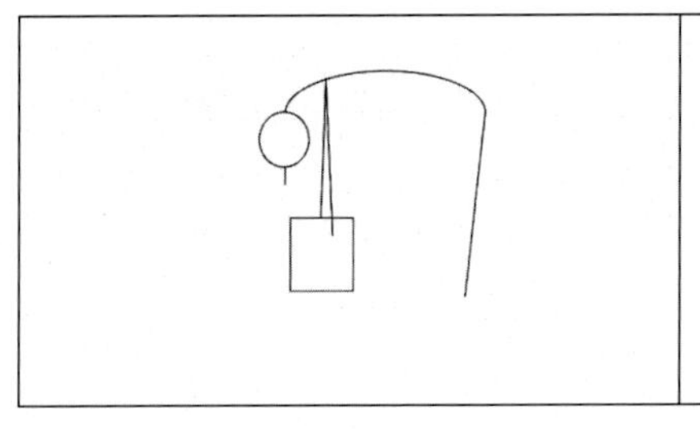

- When user is under bending position, these RRT values are reduced between all the nodes as the distance between nodes is reduced with respect to referential position
- These RRT statistics are the features to detect the activity in ANN

FIG. 3.13

Bending 1 activity.

Graph shown in Figs. 3.16–3.21 detects the activity model with RRT fetched from UBSN. Using these values, activities like walking, cycling, bending, and lying are detected with supervised learning model.

These activities are considered in suggestion model to notify the user with sport actions for health improvement.

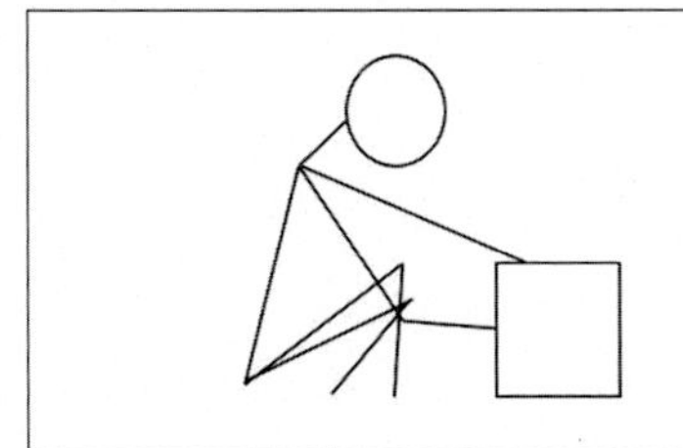	• Bending2 RRT (with minimum node distances) is differentiable than Bending1 RRT • Reading variance is calculated to observe the movement fluctuation for every 250ms

FIG. 3.14

Bending 2 activity.

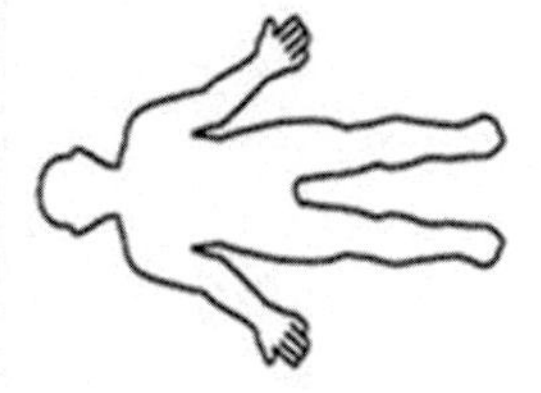

FIG. 3.15

Activity detection for cycling, walking, and lying position in UBSN.

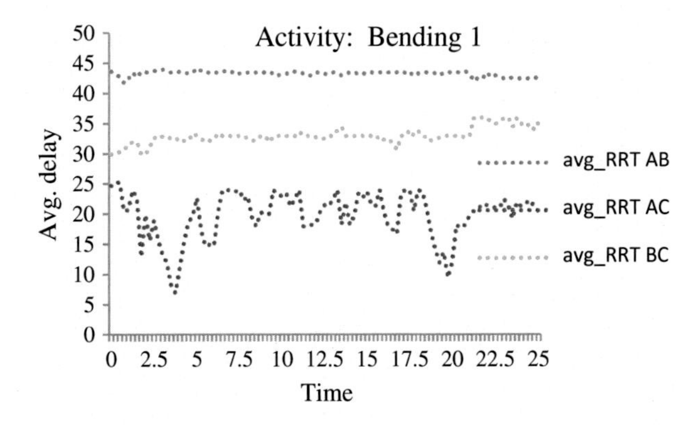

FIG. 3.16

Bending1 activity response.

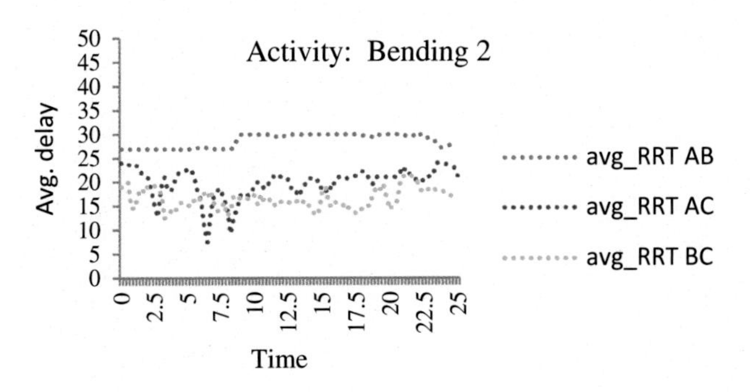

FIG. 3.17

Bending2 activity response.

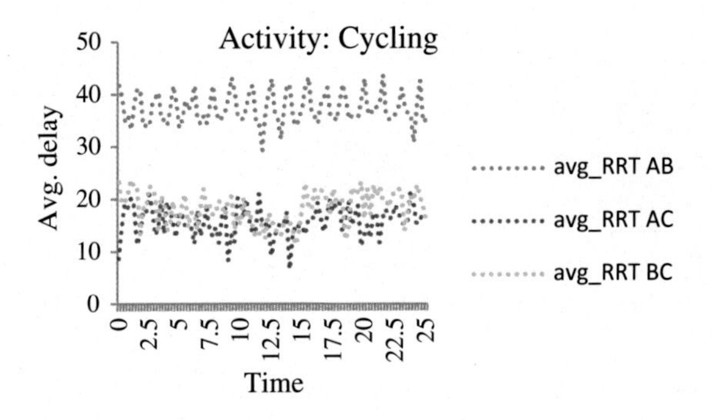

FIG. 3.18

Cycling activity response.

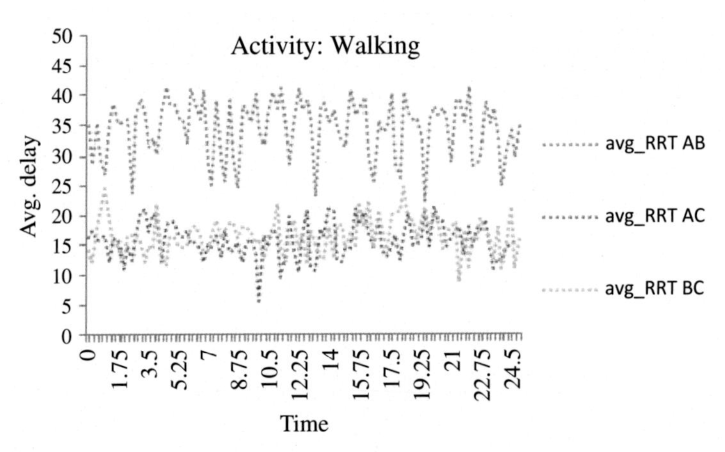

FIG. 3.19

Walking activity response.

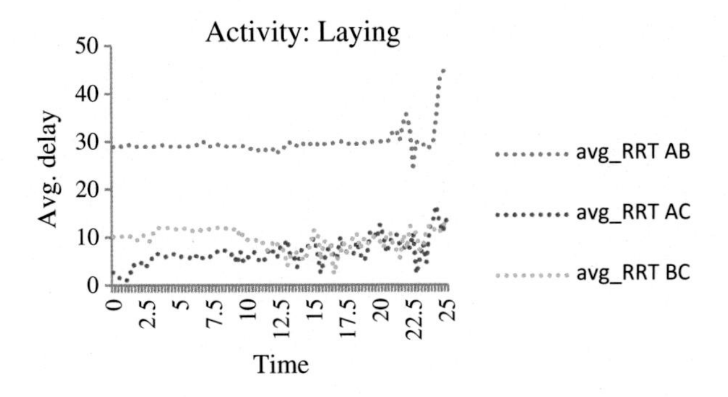

FIG. 3.20

Laying activity response.

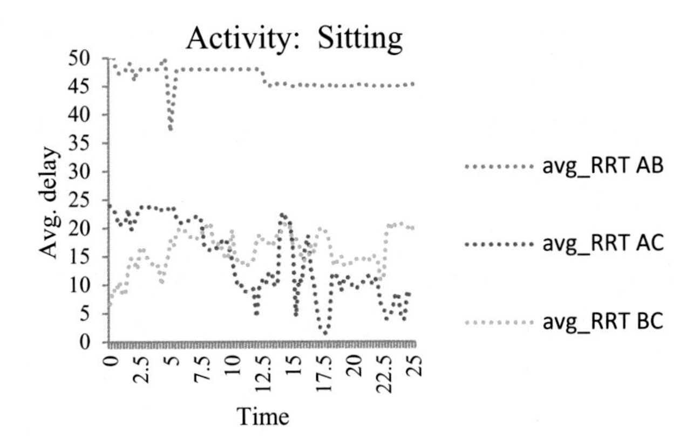

FIG. 3.21

Sitting activity response.

3.7 Recommendation system

Database consists of activities performed by user along with heart beat rate as shown in Table 3.5. Recommendation model suggests the activities to be performed in concern with cardiac information of the user. If the person is suffering from cardiac disorder, then mild activities have to be suggested. Standard dataset in concern with Physiotherapist is taken under consideration to define the ideal heart rate for each class in database. Activity-Activity collaborative learning has been followed between performed and standard activities for a class[].

Let us consider for user ID 101, "Y" class user with normal blood pressure is limited to perform type 2 bending, and cycling is missing in day-to-day activities, similar kind of exercise like cycling, running, and treadmill has been recommended by the system.

Table 3.6 Activity recommendation in database

User ID	Class	Activity performed	Heart rate (beats/min)	Ideal heart rate (beats/min)	Cardiac functionality	Activity recommended
101	"Y"	Standing, walking, bending 1	70	60–100	Normal bpm	Cycling, treadmill, running
102	"Y"	Standing, walking, bending 2, cycling	73	60–100	Normal bpm	Cycling, treadmill, running
103	"O"	Standing, walking, cycling	55	70–100	Low bpm	Walking, slow cycling
104	"Y"	Standing, walking, cycling	120	60–100	High bpm	Cycling, treadmill, running
115	"M"	Standing walking, bending 2	115	60–100	High bpm	Walking

Similarly, consider user ID 103, "O" class user, low blood pressure, and performed standing, walking, and cycling. As user is suffering from cardiac disorder (low bpm), walking and slow cycling has been recommended by system (Table 3.6).

3.8 Remark

This chapter proposes a UBSN inspired toward a healthcare system with embedded WSN. Requirement of HMS is justified with proper solution and relevant background with available solutions. In order to implement body sensor network, standard data link layer protocols are followed. This unified network is polled at various interval of time to generate the user profile for further reference in data computation. Cross-validation of RRT is carried out between the nodes, which is an effective measure to detect the activity the user in concern with data science.

Formulation of multichannel gateway in the framework coordinates the single window to communicate and retrieve the information from computational platform. More importantly, UBSN is synchronized with heart rate monitor to classify the cardiac information of the user in relational database. Age wise classification of the user is very informative toward suggesting the exercise to the health consumer.

The experimental case study is elaborated with ANN toward the bending1 and bending 2 activity performed by the user with wearable body sensors. It shows that change in response time with supervised learning makes system able to differentiate the user activities. Recommendation system is based on Activity-Activity statistical comparison of user profile with endorsed action for specific age class.

References

[1] N.G. Bourbakis, Prognosis—a wearable health-monitoring system for people at risk: methodology and modeling, IEEE Trans. Inf. Technol. Biomed. 14 (3) (2010) 613–621.
[2] M. Pavel, H.B. Jimison, I. Korhonen, C.M. Gordon, N. Saranummi, Behavioral informatics and computational modeling in support of proactive health management and care, IEEE Trans. Biomed. Eng. 62 (12) (2015) 2763–2775.
[3] P. Chatterjee, R.L. Armentano, L.J. Cymberknop, Internet of Things and decision support system for eHealth—applied to cardiometabolic diseases, in: International Conference on Machine Learning and Data Science (MLDS), 2017, pp. 75–79.
[4] D. Estrin, R. Govindan, J. Heidemann, Embedding the Internet, Commun. ACM 43 (5) (2000) 39–41.
[5] M. Aminian, H.R. Naji, A hospital healthcare monitoring system using wireless sensor networks, J. Health Med. Inform. 2157-7420, 4 (2013).
[6] H. Huo, Y. Xu, An elderly health care system using wireless sensor networks at home, in: 3rd International Conference on Sensor Technologies and Applications. Sensor comm, 2009, pp. 158–163.
[7] Y.-J. Chang, C.-H. Chen, L.-F. Lin, R.-P. Han, W.-T. Huang, G.-C. Lee, Wireless sensor networks for vital signs monitoring: application in a nursing home, Int. J. Distrib. Sens. Netw. 4 (2012) 342–356.
[8] S. Misra, S. Sarkar, Priority-based time-slot allocation in wireless body area networks during medical emergency situations: an evolutionary game-theoretic perspective, IEEE J. Biomed. Health Inform. 19 (2) (2015) 541–548.
[9] A.M.M. Celestina, K.A. Kumar, An auction based health monitoring scheme using group management techniques in wireless sensor network, in: International Conference on Communication and Signal Processing (ICCSP), ISBN 978-1-5386-3522-3. 2018.
[10] C. Bhatt, N. Dey, A.S. Ashour, Internet of Things and Big Data Technologies for Next Generation Healthcare, Springer, 2017.
[11] N. Dey, A.S. Ashour, F. Shi, S.J. Fong, R.S. Sherratt, Developing residential wireless sensor networks for ECG healthcare monitoring, IEEE Trans. Consum. Electron. (2017) 442–449.
[12] A. Marcos, J. Simplicio, H.I. Leonardo, M.B. Bruno, C.M.B.C. Tereza, M. Näslund, Secure health: a delay-tolerant security framework for mobile health data collection, IEEE J. Biomed. Health Inform. 19 (2) (2015) 761–772 (IEEE Trans. INF TECHNOL B).
[13] R.X. Lu, X.D. Lin, X.M. (Sherman) Shen, SPOC: a secure and privacy-preserving opportunistic computing framework for mobile-healthcare emergency, IEEE Trans. Parallel Distrib. Syst. 24 (3) (2013) 614–624.
[14] P.T. Sivasankar, M. Ramakrishnan, Active key management scheme to avoid clone attack in wireless sensor network, in: Proc. of 4th Int. Conf. on Computing Communications and Networking Technologies (ICCCNT'13), 2013, pp. 1–4.

[15] L.K. Guo, C. Zhang, J.Y. Sun, Y.G. Fang, A privacy-preserving attribute-based authentication system for mobile health networks, IEEE Trans. Mob. Comput. 13 (9) (2014) 1927–1941.

[16] M.U.H. Al Rasyid, F.A. Saputra, M.H.R. Ismar, Performance of multi-hop networks using beacon and non-beacon scheduling in wireless sensor network (WSN), in: IEEE International Electronics Symposium (IES), ISBN 978-1-4673-9344-7. 2015.

[17] N.A. Khan, N. Javaid, Z.A. Khan, M. Jaffer, U. Rafiq, A. Bibi, Ubiquitous healthcare in wireless body area networks, in: 11th IEEE International Conference on Trust Security and Privacy in Computing and Communications, 2012, pp. 1960–1967.

[18] R.T. Hameed, O.A. Mohamad, O.T. Hamid, N. Tapus, Design of e-healthcare management system based on cloud and service oriented architecture, in: E-Health and Bioengineering Conference (EHB), ISBN 978-1-4673-7545-0. 2015.

[19] S.S. Chandeep, S.M. Maha, S.K. Manpreet, S. Altamash, et al., Proposal of a cloud computing system for management of health data, Int. J. Comput. Appl. 114 (15) (2015) 0975–8887.

[20] N.S. Abual Karim, M. Ahmed, An overview of electronic health record (HER) implementation framework and impact on health care organizations in Malaysia: a case study, in: IEEE International Conference on Management of Innovation and Technology (ICMIT), 2010.

[21] R. Wu, G.-J. Ahn, H. Hu, Secure sharing of electronic health records in clouds, in: 8th International Conference on Collaborative Computing: Networking Applications and Worksharing (CollaborateCom), 2012, pp. 711–718.

[22] D. Sobhy, Y. El-Sonbaty, M. Abou Elnasr, MedCloud: healthcare cloud computing system, in: Internet Technology and Secured Transactions, 2013, pp. 161–166.

[23] M. Mahmoud, A.M. Zeki, Security issues with health care information technology, Inter. J. Sci. Res. (2015) 1021–1024.

[24] I. Chiuchisan, D.-G. Balan, O. Geman, I. Chiuchisan, I. Gordin, A security approach for health care information systems, in: 6th IEEE International Conference on E-Health and Bioengineering—EHB, ISBN 978-1-5386-0358-1. 2017.

[25] M. Hawelikar, S. Tamhankar, A design of Linux based ZigBee and Bluetooth low energy wireless gateway for remote parameter monitoring. in: International Conference on Circuits, Power and Computing Technologies, ICCPCT-2015, 2015, pp. 1–4, https://doi.org/10.1109/ICCPCT.2015.7159375.

[26] H. Costin, C. Rotariu, A. Pasarica, Mental stress detection using heart rate variability and morphologic variability of ECG signals, in: Proc. of the International Conference and Exposition on Electrical and Power Engineering, 2012, pp. 591–596.

CHAPTER

Compressive sensing in medical signal processing and imaging systems

4

Thales Wulfert Cabral*, Mahdi Khosravy*, Felipe Meneguitti Dias*, Henrique Luis Moreira Monteiro*, Marcelo Antônio Alves Lima*, Leandro Rodrigues Manso Silva*, Rayen Naji†, Carlos Augusto Duque*

**Department of Electrical Engineering, Federal University of Juiz de Fora, Juiz de Fora, Brazil*
†Medical School, Federal University of Juiz de Fora, Juiz de Fora, Brazil

4.1 Introduction

Electrocardiogram (ECG), magnetic resonance imaging (MRI), and computed tomography (CT) were developed as auxiliary tools for physicians in order to aid them in the diagnosis of various diseases. These tools use data processing for the purpose of assessing the condition of the patient. However, such data must follow an acceptable quality threshold so that clinical judgment is not jeopardized. Thus compressive sensing (CS) acts as a new path that benefits both the speed and quality of the diagnosis. The term "health monitoring" refers to the technologies that monitor the vital signals of the human body. There are several possible approaches: body monitoring networks, wearable, and nonmobile equipment in hospitals.

When it comes to the ECG technique, improvement of data acquisition together with proper analysis can assist clinicians in making more accurate diagnoses. ECG is a biosignal that captures the electrical activities of the heart. Thus any movement of contraction or relaxation of this organ is preceded by an electrical signal. The capture of these signals enables the physician to verify that the heart is functioning normally. MRI is a technique that allows visualization of certain living materials and tissues [1]. Therefore, it becomes possible to collect information from parts of a patient's body without the need for invasive techniques that may cause harm. Another important tool is CT, which is an X-ray technique wherein various body structures can be reproduced in a single plane.

4.1.1 Compressive sensing in health monitoring

The modern world demands more and more of the population. Social pressures for results have made people's habits less healthy. Levels of sedentarism are alarming and the diet of the population is getting worse. This panorama, along with the increasing number of people in advanced age, contributed to the increased development of chronic diseases and cancer cases [2]. Early diagnosis of diseases facilitates

Sensors for Health Monitoring. https://doi.org/10.1016/B978-0-12-819361-7.00004-X

treatments and increases the chances of survival. Technological development enables increasingly early diagnosis of diseases, and modern medicine is increasingly linked to the use of technology.

The invention of the X-ray machine in 1895 was the precursor for imaging-based diagnostics. Using this machine, it was possible to see the inside of the human body, making possible the diagnosis of tuberculosis and breast cancer and to verify bone fractures. An advance of the X-ray techniques was the CT scan that through the emission of X-rays in several angles was able to make the volumetric reconstruction of the body with higher quality than the X-ray machine. This improvement allowed for the earlier diagnosis of cancer, heart disease, and infectious diseases. The invention of the MRI enabled even more detailed imaging than tomography without the use of X-ray waves since this equipment is based on the use of large magnets to generate a magnetic field. The major disadvantage of the MRI is the longer acquisition time. Keeping the quality of the images obtained on CT and MRI scans with the least possible exposure time to the effects of the equipment is one of the main challenges in medicine today.

In addition to image-based examinations, there are others of paramount importance for diagnosis. Among them, the ECG examination is one of the most important. The ECG allowed the mapping of the electrical activity of the heart in order to diagnose if the heart is malfunctioning. The new trend for ECG technologies is toward the development of portable equipment that can take continuous measurements of the electrical activity of the heart. Battery capacity is one of the main problems of this approach. CT, MRI, and ECG are three important tests for health monitoring. In this section, we introduce applications of CS in health monitoring to tackle the main problem of high data volume in each of these biomedical equipment. CS directly samples the signal in a compressed manner with fewer samples per time unit required by the Nyquist-Shanon theory of sampling. In short, the resultant lower sampling rate requirement brings the advantage of shortening the examination time for CT and MRI and reducing battery consumption for ECG. This chapter invites the reader to a journey through the most varied techniques of health monitoring with a comprehensive view of CS. The intentions behind the use of CS and the importance of the relationship between health monitoring and CS will be clear in this chapter.

4.2 A brief review of compressive sensing and its role in health monitoring

CS was introduced by Donoho, Candès, Romberg, and Tao in 2004 [3–5]. This theory challenges the Nyquist-Shannon theorem by claiming that a signal with Ω bandwidth can be sampled at a sampling frequency $f \ll \Omega$, that is, a sampling frequency much smaller than the Nyquist frequency. This section introduces the main concepts of CS theory. Initially, the concepts of signal sampling and then the recovery of that signal will be introduced.

4.2.1 Compressive sensing signal acquisition

The acquisition using the CS theory is mathematically defined as measuring a length of n-sampled signal vector $\boldsymbol{x}$ by a CS matrix Φ of the size $m \times n$ as much shorter vector signal $\mathbf{y}$ of size m samples:

$$y_{m\times 1} = \Phi_{m\times n} x_{n\times m} \quad m \ll n \tag{4.1}$$

The choice of Φ is the subject of CS research in a considerable number of literatures. It is mainly taken as a random matrix.

4.2.2 Compressive sensed signal reconstruction

Most of the signals are naturally sparse or have a sparse representation in some other domain. Let $\boldsymbol{s}$ be the sparse representation of $\boldsymbol{x}$ and Ψ the sparcifying matrix. You can write this mapping to a sparse domain mathematically as:

$$\boldsymbol{x} = \Psi \boldsymbol{s} \tag{4.2}$$

Let $\boldsymbol{s}$ be the sparse representation of $\boldsymbol{x}$. Eq. (4.3) is obtained by combining Eqs. (4.1), (4.2):

$$\boldsymbol{y} = \Phi \Psi \boldsymbol{s} \tag{4.3}$$

Given $\boldsymbol{y}$, Φ, and Ψ, it is possible to recover $\boldsymbol{s}$ solving an optimization problem. This can be solved using the norms L_0, L_1, or L_2. The use of the L_0-norm results in the exact result, but it is an nonprogrammable (NP)-hard problem and therefore it is not normally used. The L_2-norm results in large errors and therefore its use is not recommended. The norm L_1 has smaller errors and it is the most used norm for this optimization problem.

Eq. (4.4) shows the optimization problem to be solved:

$$\hat{\boldsymbol{s}} = \min_{s} \| \boldsymbol{s} \|_1 \text{ subject to } \boldsymbol{y} = \Phi \Psi \boldsymbol{s} \tag{4.4}$$

In order to guarantee that the signal is recovered, it is necessary that the restricted isometry property (RIP) is satisfied. This property is defined in Eq. (4.5):

$$1 - \delta \leq \frac{\| \Phi \Psi u \|_2}{\| u \|_2} \leq 1 + \delta \tag{4.5}$$

where $\delta > 0$ is known as the restricted isometry constant and u has the same k-nonzero entries as $\mathbf{x}$.

However, RIP is difficult to verify. Another property that ensures signal recovery is the property of inconsistency. This is defined in Eq. (4.6).

$$\mu(\Phi, \Psi) = \sqrt{n} \max_{1 \leq i, j \leq N} |\langle \Phi_i, \Psi_j \rangle| \tag{4.6}$$

4.2.3 Compressive sensing metrics

There are some metrics to measure the quality of signal reconstruction and the amount of compression that has been obtained. The percentage root-mean-squared difference (PRD) and the compression ratio (CR) are two well-known metrics widely used in the CS literature. The mathematical definition of them is shown in Eqs. (4.7), (4.8):

$$\mathrm{CR} = \frac{n}{m} \tag{4.7}$$

$$\mathrm{PRD}\,(\%) = \frac{\| \boldsymbol{x} - \hat{\boldsymbol{x}} \|}{\| \boldsymbol{x} \|} \times 100 \tag{4.8}$$

where $\hat{\boldsymbol{x}}$ is the reconstructed $\boldsymbol{x}$ signal.

In the reconstruction of images, such as tomography or MRI images, the literature generally measures the quality of reconstruction in visual manners.

4.2.4 Compressive sensing and health monitoring

While every moment requirement of comprehensive monitoring of all vital signals/images in the medical system becomes increasingly essential, managing the high volume of medical data becomes a big challenge. CS with theoretical strong background in reconstruction of data by multiple less sampled signal vectors brings several advantages, especially in health monitoring. It also reduces the side effects of unnecessary redundant exposure of patients to the radio waves and electromagnetic fields used in medical equipment, such as CT and MRI. First, the low data samples make it possible to perform MRI and CT imaging much faster, in a matter of seconds, instead of the current several minutes. Apart from the economic advantages of increasing the lifespan of this very expensive equipment, and apart from lower power consumption, another benefit is that the patient is less exposed to radiation. In addition, in cases where a patient suffers from claustrophobia, the process is faster and more acceptable. Also, the data will require lower transmission rates and less processing time for noise removal. Fig. 4.1 shows the general process of implementing CS in health monitoring.

4.3 Compressive sensing of electrocardiogram

The heart is the organ responsible for pumping blood rich in oxygen and nutrients to the whole body and collecting blood poor in oxygen and sending it to the lungs. All functioning of this organ is electrically controlled. A region of the heart called the sinoatrial node spontaneously generates electrical impulses that initiate the cycle of contraction (systole) and relaxation (diastole) of the four different chambers present inside the heart: left and right atria and left and right ventricles. Correct coordination of this cycle ensures that the heart functions properly.

The ECG works by capturing the electrical activities of the heart. Every movement of contraction or relaxation of this organ is preceded by an electrical signal. The correct capture of these signals enables the physician to check if the heart is functioning properly.

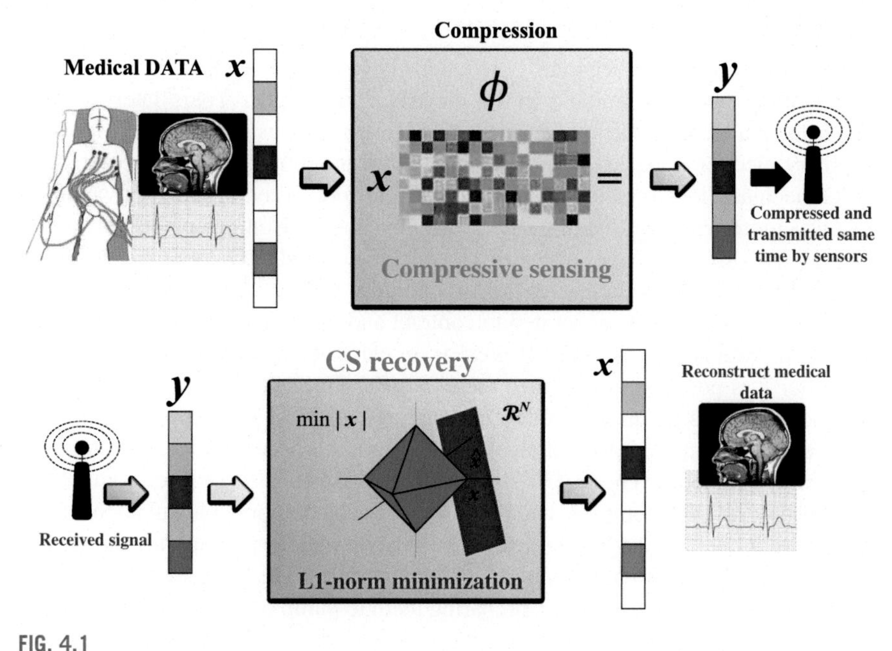

FIG. 4.1

General process of implementation of CS in health monitoring.

The number of people suffering from cardiovascular problems around the world is increasing and heart disease is already the leading cause of death worldwide [6]. This situation requires people to seek healthier habits and medical attention if an anomaly is noticed. There are different types of tests that help doctors diagnose heart problems. The ECG is one of the main tests that aids in the diagnosis of cardiac problems. It is a simple, quick, cheap, and safe examination.

The new trend in the field of health technological applications is residential wireless sensor networks for ECG healthcare monitoring [7], and even the use of wearable devices that monitor vital signals, like the ECG signal. Some of the main challenges of these devices are battery capacity and data transferring, which is a major energy consumer. The CS technique has emerged to reduce the number of samples a signal must have to be reconstructed, thereby reducing the amount of battery being used. As ECG records are very susceptible to noise, ECG preprocessing [8–10] is a conventional approach for having a higher quality of ECG for computer-aided processing [11] or clinical observation. A strong nonlinear approach suggested for noise cancelation and baseline correction of ECG in the literature is morphological filtering [12–14]. Besides, other multiarray signal processing techniques like blind source separation [15–18] or blind component processing [19] can be used for noise cancelation and detection.

The historical context of the ECG appearance, the theoretical basis of its acquisition, and some applications of CS in ECG signals will be shown in this section.

Historical background of electrocardiogram

The history of the ECG began in 1781 when Galvani observed that the muscles of a frog contract when stimulated by an electric current [20]. Years later, in 1887, Augustus Waller was able to capture the first human ECG using a Gabriel Lippman's capillary electrometer. In 1901, Willem Einthoven, considered the creator of the ECG machine, developed an ECG measurement tool called a string galvanometer that was more sensitive than the device used by Waller. Einthoven was able to capture the five different waves of the ECG, which he named P, Q, R, S, and T. He won the Nobel Prize for Medicine for this invention in 1924.

The electrical signals generated to contract and relax the atria and the ventricles propagate throughout the body. If there is an electrode placed on the body and an amplifier it is possible to visualize these signals. Since the generated signal has a vector characteristic, an electrode only captures an ECG projection at that point. In order to obtain more information about the heart, it is necessary to position a larger number of electrodes. The ECG's standard waveform consists of three components: the P wave, the QRS complex, and the T wave. Each component is linked to an electrical activity performed by the heart. The P wave indicates the polarization of the atria, preparing them to pump blood into the ventricles. The QRS complex indicates the polarization of the ventricles, preparing them to pump oxygenated blood into the body and deoxygenated blood into the lungs. The T wave indicates the repolarization of the ventricles, that is, when they return to the relaxed state. A physician examining a waveform of an ECG signal observes changes in each of these components in the search for signs of a heart problem.

The ECG examination is easy to perform. Usually, a trained healthcare professional will attach multiple electrodes to specific regions of the patient's body. These electrodes are connected to an ECG machine and the examination lasts only a few minutes. The examination is safe and offers no risk to the patient. Fig. 4.2 presents a standard waveform [21].

4.3.1 Compressive sensing applications to ECG

Several solutions have already been presented aiming to decrease the number of samples required for the reconstruction of ECG signals using CS. However, some factors must be addressed before showing the results obtained in the literature. The most important factor to be checked on ECG data compression is the maximum permissible reconstruction error that would still guarantee good diagnostic quality for the signal. Zigel et al. [22] addressed this problem. In this work, compression with different reconstruction errors is delivered to medical specialists who must evaluate the quality of the CS. A standard limit value was obtained for ECG signal compression in this paper: signals with PRD $< 9\%$ were considered good enough for a diagnosis.

In the evaluation of CS techniques for ECG signals, it is usual to use a public database. A widely used database in the literature is the MIT-BIH database [23]. This database was generated by the Massachusetts Institute of Technology and

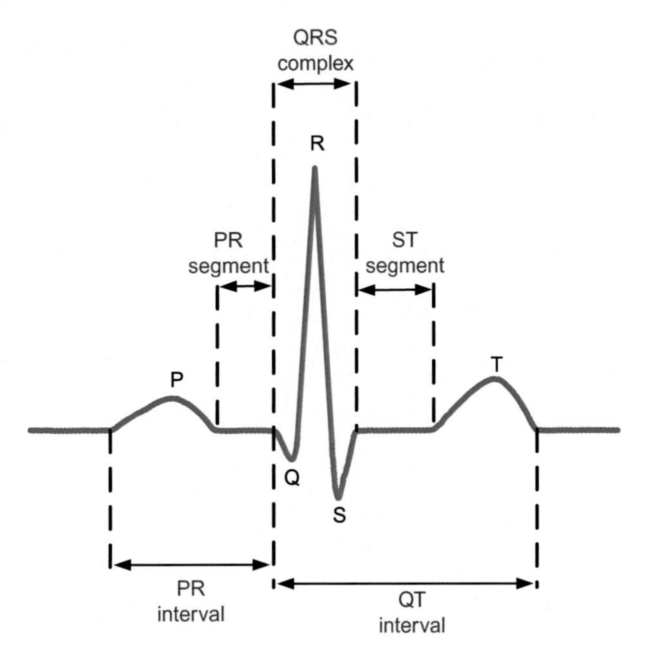

FIG. 4.2

ECG standard waveform.

Boston Beth Israel Hospital. In this database, there are 48 30-min recordings of 47 patients obtained between 1975 and 1979. The recordings were scanned using a sampling rate of 360 samples/s using an 11-bit resolution in signal quantization. Several approaches have been made in the literature to apply CS to ECG signals. Some of these approaches will be addressed in the sections that follow.

4.3.1.1 Reconstruction algorithms

Dixon et al. compared [24] reconstruction signals in order to determine what provides the signal with the least possible reconstruction error. Among the observed algorithms are Basis Pursuit Denoising, Orthogonal Matching Pursuit, Compressive Matching Pursuit, and Normalizes Interactive Hard Thresholding.

4.3.1.2 Body area network applications

In embedded devices that aim to save as much energy as possible, it is necessary to look for techniques for an efficient use of energy. CS decreases power consumption by data transmission because it samples less data, therefore, there is less data to be transmitted. Some papers focused on verifying CS's ability to save energy. Among them, we can cite work by Balouchestani et al. [25], which achieved a reduction of energy consumption by about 40% without signal degradation. Another work that focused on the reduction of energy consumption compared CS to [26] discrete wavelet transform (DWT) compression. It was observed that an increase in battery life of 37.1% was achieved using CS as compared to using the DWT technique.

4.3.1.3 Compressive sensing vs. discrete wavelet transform

Data compression using the DWT technique was considered the state of the art for ECG signals, as it allows high compression rates with small reconstruction error rates. Different studies involving the application of CS on ECG made this comparison between CS and DWT. Chae et al. [27] compared the CS and DWT techniques in noise situations. This work showed an unquestionable superiority of DWT in relation to compression levels with the same reconstruction error. However, the authors of this work emphasize that although DWT has a superior result, the CS technique should not be discarded because its great advantage is in the simplification of the encoder, transferring all the complexity to the decoding part.

4.3.1.4 Compressive sensing using the ECG structure

Most of the approaches to ECG ignore the peculiar structure of this signal when applying the technique. Some the authors opted for an approach that explored this structure [28]. Initially, a preprocessing stage was implemented to detect the QRS complex. Then the heart rate was fixed, and a window was applied to include a few beats. The analysis was done in this window. This approach increased compression rates but also increased the encoder complexity.

Another work also explored the structure of the ECG signal [29]. In this work, a specific binary uniform sensing matrix that focuses specifically on the region where most of the ECG signal information is located (the QRS complex) was used. An increase in compression rates has also been noted; however, the sensing matrices must be transmitted along with the data.

4.4 Compressive sensing in magnetic resonance imaging

The history of MRI began in 1937, when Isidor Rabi developed a method to measure the movement of the nucleus of atoms [30]. He noticed that the nucleus of the atom shows its existence by absorbing or emitting radio waves when exposed to a strong magnetic field. This method is called nuclear magnetic resonance (NMR) and is the basis for the full development of MRI technology. In 1946, Edward Purcell and Felix Bloch simultaneously extended Rabi's work by using NMR techniques in solids and liquids, not just isolated atoms. Initially, NMR was used for analysis of chemical structures, but a doctor named Raymond Damadian proposed that this method could be used to detect cancerous tissue because it was known that these tissues have a greater amount of water than healthy tissues and, consequently, they have more atoms of hydrogen. Therefore this region would absorb more radio waves than the others.

After reading Damadian's work, Paul Lauterbur, who had already used NMR techniques to create images of water glasses and peppers, thought he could generate a three-dimensional (3D) model of an object by grouping two-dimensional (2D) images using the NMR technique. Peter Mansfield developed a technique that would generate NMR images faster. He was able to generate the first image of a human body part using the NMR technology.

Theoretical basis: The MRI test is based on the properties of the nucleus of certain atoms. The examination shows that they have a different number of protons and neutrons. As the human body is formed largely by water, hydrogen atoms are one of the major atoms for generating MRI images [31].

Hydrogen atoms, because they have only one proton and no neutrons, have a spin, that is, they have a small magnetic moment. These atoms align with the magnetic field after a strong magnetic field is applied. Then, radio-frequency waves are inserted, which modify the inclination of the atoms in relation to the previously applied magnetic field. After removing the source of radio-frequency waves, the atoms realign with the magnetic field and produce radio-frequency waves during their return to their former condition. The resultant RF waves are picked up by antennas in the MRI device. Fig. 4.3 shows this theoretical process.

From the patient's point of view, the MRI scan is painless and does not use X-rays. The radio waves used are in the frequency spectrum, which does not cause harm to human beings.

Nowadays, MRI examinations are performed in specialized locations. In the examination, the patient lies on a bed that moves into a tubular-shaped device. The appliance is operated by a trained professional who will control the appliance in a separate room using a computer. During the examination, the patient will be able to communicate with the operator if they experience any malaise. The examination requires the patient to remain as still as possible for the duration of the examination, which may vary from 15 to 90 min depending on the region being scanned. Although an examination considered safe, MRI is not recommended in cases where the patient has implanted metals, such as a pacemaker, or during pregnancy [32].

MRI importance: MRI is a technique that allows the visualization of certain living materials and tissues [1]. Because of its noninvasive nature, modern medicine has

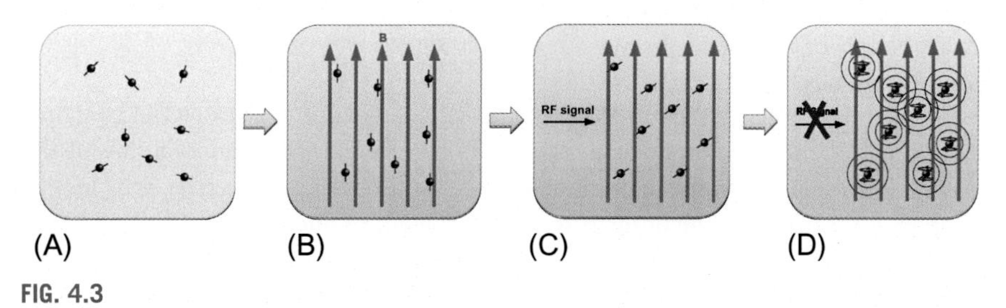

FIG. 4.3

Magnetic resonance imaging theoretical process in making atoms radiating RF waves. (A) Normal condition of atoms, (B) alignment of the atoms with strong magnetic field, (C) inclination of the atoms by applying RF, (D) realignment of atoms with magnetic field after removing RF as it produces RF.

adopted it as an important analytical tool in the fight against diseases. It is possible to collect information from parts of the body of a patient without the need to produce lesions in it. A physician is able to use it for the most varied diagnoses such as damage to the blood vessels, heart problems, brain injury, tumors, and other diseases. Additionally, the area of oncology was provided with productive results using the MRI, such as the diagnosis of prostate cancer. It is possible to obtain structural information of the prostate and verify the carcinogenic condition of the gland using signs of abnormality [33].

Taylor and Reddick demonstrated the relevance of the MRI in the treatment of tumors [34]. It contributed to the reduction of mutilations inherited from the surgeries, and consequently, improved the postoperative period of patients. This article also emphasized the contribution of MRI to the demarcation of the cancerous region, supporting better diagnoses.

Important advances were promoted from the MRI, one of which is dynamic contrast-enhanced MRI (DCE-MRI). DCE-MRI is an MRI use from another perspective and it has the ability to produce morphological data sequences, collecting "MRI frames," or simply "movies" that provide quick images for diagnoses [35]. DCE-MRI provides information on tumor microcirculation and angiogenesis (the formation of new blood vessels) [36]. Thus this examination provides conditions for observing the therapeutic course in cervical cancer.

MRI is very present in modern medicine, especially in the treatment of tumors and related diseases where a visual analysis is often necessary. Therefore, MRI provides new hope for the treatment and diagnosis of these serious diseases.

This technique became implementable through NRM signal [1]. The NRM is produced through some elementary nuclei, such as 1H and 13C, if struck by a strong magnetic field and radio waves [1]. This is physically possible because of the alignment between the applied magnetic field and the nuclear magnetic moment. MRI can be used to verify the way the water is spatially arranged, so the signal amplitude will reveal the distribution of the water at a specific location within a sample. MRI uses the knowledge of magnetic field gradients to encode the NRM with spatial information, and the magnetic fields are linearly variant with tridirectional orthogonal positions applied by gradient coils. The technique also uses a set of frames for the construction of an image, becoming directly dependent on the way of acquisition of these frames.

Researchers always seek to improve the MRI technique in various fields in order to offer more diagnostic power to the MRI. The DCE-MRI is an example of this, however, an analysis from the point of view of signal processing can be done in order to apply state-of-the-art techniques to improve MRI. In this context, CS is an important tool for MRI. After all, in the field of data processing, especially image processing, any improvement or intelligent way to handle the information is always welcome and may even make the process less expensive.

As previously mentioned, MRI uses a frame sequence for image construction. Thus it is necessary to deal with the acquisition of the data for the formation of the image, which are constructed from this sequence. Each acquisition deals with the collection of a set of frames. MRI images, the same as all other medical images, go under various image processing techniques where image preprocessing [37], enhancement [38], and adaptation [39, 40] prior to any operation can assist more successful results. For all the processing techniques, there are always numerical indexes to validate the processing technique's efficiency [41].

4.4.1 Compressive sensing applications to MRI

A transverse magnetization originating from a radio-frequency excitation is generated for a specific acquisition [42]. Then, it is sampled during a trajectory in k-space. By default, this sampling is designed in accordance with Nyquist's theorem, which when breached can produce anomalies in the reconstruction of the image. It is also possible to choose the k-space trajectory according to each acquisition; however, it adopts straight lines of a Cartesian grid, since this configuration presents robustness to the anomalies that may arise in the system. However, there are radial and spiral trajectories in some systems. The reference points out that the acquisition is the act of crossing curves in multidimensional k-space and the crossing speed has a restricted range, determined by the amplitudes of the gradients. If these amplitudes were high they could stimulate peripheral nerves, which would not be desirable to the patient. Thus the system has a speed limit. This limitation was an incentive to seek better alternatives that could use less data and still preserve the performance of the system. Fortunately, at this stage, science was already aware that the data from the MRI contained repeated information. Thus Lustig et al. [42] highlight the possibility of taking advantage of this fact and design systems in which sampling could be intelligent and more effective. This situation was aligned with the CS concepts. CS is used in MRI for the purpose of improving the speed of acquisition and in the context of imaging, a set of different diagnostic techniques based on images of parts of the human body is relevant for the reduction of acquisition time [43]. However, it is necessary to provide conditions for CS to be used in this type of application. For this, the literature lists important points that corroborate and justify, based on CS theory, its use in MRI. The authors expose that the images generated by MRI are sparse in a certain domain. The reference alludes to the fact that this concept of sparsity is applicable to compressibility in the treatment of reconstruction of the original signal, that is, it is perfectly acceptable that some coefficients can affect repetitions, which is applicable to CS, since the technique uses only the prominent coefficients for reconstruction. As a result, after image acquisition, compression is usually done to reduce data storage costs. By the previous fact, only the most important coefficients are saved, preserving the "constitutive inheritance" of the original signal.

4.4.2 MRI compressive sensing requirements

According to Geethanath et al. [43], for the actual implementation of CS in MRI, there is a need to fulfill three requirements: (1) transform sparsity, (2) pseudo-random undersampling, and (3) nonlinear reconstruction. Transform sparsity can report the number of samples required to achieve a true reconstruction of the original signal. In general, wavelets are used because of their good compression ratio. Pseudo-random undersampling is required since the scanners collect the spatial frequency (or k-space) data. So the acquired k-space needs to undergo a process of subsampling so that aliasing does not occur. It should be noted that more dispersed data results in fewer samples needed for recovery. However, the reference warns that there are two limitations: the difficulty of obtaining a pure k-space random sampling result and most of the energy from the collected data is contained in the low-frequency components, so a uniform weighting of the samples would be detrimental the SNR. Thus implementations of pseudo-random undersampling are used. Finally, there is nonlinear reconstruction, which is an image reconstruction process for CS [43, 44]. According to Lustig et al. [45], the latter can be described by a restricted optimization problem according to the following equation:

$$\text{Minimize} \quad \|\Psi m\|_1 \tag{4.9}$$

$$\text{s.t.} \quad \|F_u m - y\|_2 < \epsilon \tag{4.10}$$

where Ψ is a linear operator that transforms the pixel representation to a sparse representation, m is a vector representing the image of interest (reconstructed), F_u is the subsampled Fourier transform, y is the acquired data in k-space and controls the accuracy of reconstruction fidelity. The equation is an objective function described by $L1$-norm, defined as $\|x\|_1 = \sum_i |x_i|$. The minimization proposed by Eq. (4.9) will promote sparseness as desirable.

There is a vast field of applications that can employ CS for aid in clinical medicine techniques. In this step, CS applications in MRI will be listed. Each variant of MRI has its own characteristics and specific purposes. However, the contribution of CS is necessary for the development of the applications.

4.5 Compressive sensing in angiograms

Angiograms are examinations that express the state of vascularization: like veins and arteries [43]. According to Lustig et al. [45], sparse information in the images conveys the conclusion that there are few pixels with relevance. The reference points out that, in this case, the angiograms are very sparse when represented by pixels, and may be sparse to the naked eye. This fact is not an absolute rule for more complex medical images; however, they may exhibit sparsity if they are represented in terms of finite differences, through wavelet transformations or other suitable transformations. The authors describe the application of contrast-enhanced 3D angiography whose application with the aid of CS presents promising results. In these results, blood vessel information is preserved using only 5% of the transform coefficients, which translates into significant importance to justify the use of this method anchored in CS. According to Lustig et al. [42], CS also has the ability to reduce anomalies/artifacts generated from subsampling, in particular in the reconstruction stage of the data that were extracted from situations of subsampling. In work by Çukur et al. [46], magnetic resonance angiography (MRA) is evidenced. It is a breakdown of the MRI technique that can be applied in the area of vascular disease treatments. In this aspect, CS is also able to remove interferences produced in reconstructed vascular images. In work by Trzasko et al. [47], a nonconcave CS model is proposed for the reconstruction of subsamples angiograms. Finally, some authors present a technique capable of promoting the correction for high-frequency signal loss, with the purpose of mitigating the loss of resolution [48].

4.6 Compressive sensing in neuroimaging

Neuroimaging or brain imaging is an area of neurological medicine whose purpose is to aid in brain diagnosis. This area has great relevance since it works with the collection of information from the brain, which has unquestionable importance to the good functioning of the human body. Thus brain-related diagnoses need to be accurate so that the clinician has realistic conditions to obtain a prognosis consistent with the real conditions of the patient.

Among the techniques of brain imaging, there are some branches such as functional MRI, which measures brain activity [49] and magnetoencephalography, which is a technique used to measure magnetic fields produced by brain activity [50].

In the context of CS, some authors were able to obtain progress and present satisfactory results when applying this technique in the area related to MRI. In this case, CS has the utility of reducing collection time and improve the resolution of the image [42]. In work by Zhang et al. [51], the authors alluded to the use of a new rapid CS technique on MRI; however, such implementation was not mature enough for clinical use. Thus they presented an alternative to it. The application of CS produces a superior result to the methods based on regularization [52]. Therefore this area of research is always looking for improvements in its noninvasive method of diagnosis, seeking to produce benefits for the sick population.

4.7 Compressive sensing in cardiac magnetic resonance

Cardiac imaging is a form of diagnosis present in modern medicine. Cardiac magnetic resonance (Cardiac MR) is able to demonstrate the structure, function, and myocardial state of the heart with great capacity [53]. This type of resonance is widely used in cardiovascular science, although its clinical applicability is not yet widespread [53]. However, this popularization depends on improvements, for example, the improvement of the pulse sequences.

At that moment, some scientific works will be approached that, in a propositive way, contributed to the applications of CS related to the area of Cardiac MR in MRI. In this case, CS presents a more incisive performance characteristic in signal processing, always contributing to data processing.

In the context of applications, the work developed by Otazo et al. [54] deals with first-pass cardiac perfusion MRI, a technique capable of evaluating coronary artery disease in a noninvasive way. The authors point out that this application is an appropriate candidate for acceleration CS, once it presents a sparse representation in the temporal and spatial Fourier domain. Liang et al. [55] presented a variant of Cardiac MR, named Dynamic Cardiac MRI, which captures a series of images of cardiac motion. In this case, CS is used to shorten the data acquisition time. The work proposes a support detection method for the purpose of improving CS reconstruction for Dynamic Cardiac MRI with an iterative procedure. Usman et al. [56] deal with a structure that promotes motion-corrected CS, using CS for reconstruction. According to the authors, the importance in this context is due to the fact that the movement when the magnetic resonance is in operation promotes anomalies in the reconstructed images.

4.8 Compressive sensed MRSI

Magnetic resonance spectroscopic imaging (MRSI) is an advent of clinical imaging that assists in the characterization of prostatic, mammographic, and neurological diseases [57]. Another modern tool that presents itself as a propositive solution for clinical diagnoses.

In the context of this kind of application, CS can be used to reduce acquisition time [43]. Some specific applications developed with CS support will be presented.

Hu et al. [58] suggested a high-resolution MRSI; however, the number of phase coding fails at the time of acquisition. Therefore, CS is applied in order to promote an improvement in resolution without harming the acquisition time. In another application [59], a 3D MR spectroscopic imaging method is developed with the ability to acquire hyperpolarized 13C data via CS. Askin et al. [60] propose the evaluation of the CS method for faster phosphorus MRSI using the brain and its proposal of use in it.

4.9 Compressive sensing in musculoskeletal system

The musculoskeletal system, or locomotor system [43], comprises a class of muscles and can be defined as a complex structure endowed with a hierarchy of tissues and muscle fibers, in which each muscle fiber consists of "contractile machines" of skeletal muscle [61].

There are several models of this type of structure according to the purpose of the study, models that deal with the cellular properties and the mechanical and kinematic function of this musculoskeletal system [61]. The importance of studying this type of structure is so great that it is possible to divide the General Anatomy and the Musculoskeletal System into four "regions": general anatomy, trunk wall, upper limb, and lower limb [62, 63]. In this context, MRI is used to extract information from the musculoskeletal system. There are works in MRI that present an approach for automatic multiorgan registration [64]. In other cases, the state of the art is highlighted providing an overview in this context of the application [65]. However, science has a point of convergence. There is undeniable clinical importance of the use of MRI in musculoskeletal systems. Any other technique that can help or contribute to these applications will be well seen. Thus CS can be used to reduce the time to acquire muscle images via MRI [43]. In work by Geethanath et al. [66], for example, CS was used to increase the speed of acquisition of magnetic resonance, especially in Concurrent Dephasing and Excitation (CODE) used for human knee analysis.

4.10 Compressive sensing of computed tomography

Tomography is an X-ray photography technique where you get a single plane containing information from various structures of other planes. The X-ray was discovered by Wilheim Conrad Röntgen in an accidental way wherein he first tried it over a book and then on his wife's hand [67].

Because the structural information of object planes appear in one plane, radiography has a visual limitation due to overlapping structures, as the volume of the three dimensions of the human body is compressed into a 2D X-ray image. In the case of the X-ray procedure of some part of the body, such as the lung, other structures such as bones may become evident. Also, the X-ray image has a reduced ability to differentiate a low-contrast object from the background [68]. Due to these reasons, conventional tomography was developed.

One of the pioneers of tomography was A.E.M. Bacage. He developed an apparatus around 1921. Bocage's invention consisted of an X-ray tube, an X-ray film, and a mechanical connection to move the tube and the film. Due to some issues, such as the movement of the tube and X-ray film, the position of the point to be focused, and the structures in close planes, conventional tomography has some limitations. It does not have good resolution to identify certain structures, and it delivers a high dose of X-ray to the patient. Conventional tomography was limited in clinical applications because of these reasons.

Due to these issues, along with the development of the technology, there is an interest in developing strategies to reproduce images in depth, through the acquisition of some data and improving the image processing. The tomosynthesis technique was defined to achieve this goal, where a smaller dose of X-ray can be used in certain clinical applications, such as breast and lung cancer [68]. The use of images obtained from projections began in 1940. The first images were not obtained with the use of computers. They were acquired using some pieces of equipment and techniques of optical rear projection. Since then, several studies have been carried out, aiming to improve the application of tomography, through methodologies and equipment.

The first reports of a computerized scanner machine are from 1963, by Allan M. Cormack. The development of a clinical CT began in 1967 with Godfrey N. Hounsfield. Since then, the devices have evolved more and more, seeking to reproduce the image in a clear way, using low doses of X-ray, and reducing the time of data acquisition. The advances are notorious, being applied in people, animals, and forests.

Computed tomography importance: The application of CT has become essential in the analysis and detection of several anomalies and assisting other techniques. Using CT, we can analyze various conditions in the foot such as tarsal collision, subtalar arthritis, tarsal tumors, soft tissue abnormalities of the heel and peroneal tendons, preoperative evaluation of the hallucinium helix valgus, crooked foot in elderly patients, and fractures of tarsus [69].

Also, CT can assist in determining the correct location for the pedicle screw [70], staging the carcinoma of the cervix [71], interpreting the peritoneal cancer index in colorectal carcinoma [72], and the diagnosis of pulmonary embolism [73]. It can also be used as a noninvasive analysis in patients after vascular events in the posterior circulation [74]. However, radiation exposure in the application of CT must be considered [75, 76].

Guyer et al. [69] discussed the use of CT to analyze many foot conditions. A good analysis using CT is important because there are a large number of cases of fractures in the calcaneus. Using CT, the degree of comminution and the main fragments of the fracture, the presence of subtalar joint incongruence, and the status of the soft tissue structures that need to be repaired can be identified.

Another application of CT is in determining the correct place to insert the pedicle screw [70]. Considering only conventional tomography, it is difficult to define the correct location of the screw, but with CT the location and the diameter of the pedicle screw can be determined. Thus, the application of pedicular screws became efficient. This work shows that irritation was found in only 0.5% of the patients. It is also reported that experienced physicians have 80% accuracy in the implantation of pedicle screws and it may be improved using image-guided insertion equipment.

Although MRI has better accuracy in the staging of cervical carcinoma CT is also used [71]. In this work, a summary was made on the available evidence, and the estimates of the performance of each methodology were obtained. Tomography had a lower percentage of efficiency in all of the scenarios. Regarding the accuracy and clinical relevance of CT in the interpretation of peritoneal metastases, Esquivel et al. [72] present results in which they are not accurate. The study was conducted with 52 patients from 16 different institutions. The precision of the computerized tomography was worse than the precision obtained from MRI.

Carrier et al. [73] defined that computerized pulmonary angiotomography with multiple detectors has greater sensitivity for pulmonary embolism. This study was carried out analyzing 22 patients with suspected pulmonary embolism, and it was obtained that the proportion of patients diagnosed seems to increase without reducing the risk of thromboembolism.

A study was carried out on the diagnosis of computerized angiotomography in the posterior circulation as a noninvasive substitute for intraarterial digital subtraction angiography [74]. Although the tomography may have several applications, attention should be paid to the X-ray exposure it provides. Some analyses are presented regarding the exposure of the patients to the radiation [75, 76]. Although there is little data on the risk of cancer that can be caused, an estimate can be made [75]. Also, the radiation dose associated with common studies can be estimated [76]. It was perceived that the risk varies markedly and it is considerably higher for younger women and for combined cardiac and aortic tests [75]. Smith-Bindman et al. [76] defined that the estimated number of CT scans that will lead to the development of cancer varies greatly, depending on the type of examination, and the age and gender of the patient.

Thus, it is noticeable that a reduction in doses is necessary when CT is used. For this reason, techniques that provide a more accurate detection are necessary, considering less clear images, then would happen with the reduction of the doses in the applications of CT. Thus, CS techniques can help in a very expressive way, obtaining more accurate results in the diagnoses and also helping to reduce the doses to which patients are submitted. The following are several CS techniques applied in the processing of the images obtained by CT.

4.10.1 Compressed sensing implementation on computed tomography

CT has great relevance in the clinical area. Modern medicine needed this advent for possible advances in medical diagnoses. However, it is necessary to keep a fixed vision in the future considering the new possibilities that science offers to improve the implementation of this technology. A possible solution is handled by the signal processing area, and again, CS can be used as an efficient way to manipulate the data and preserve the fidelity of the data when you want to rebuild it.

Through the knowledge acquired in this section, some applications involving CS and CT will be highlighted so that the reader can obtain the awareness of the capillarity that CS can acquire in health monitoring applications, especially in CT. Yu et al. [77] applied an internal reconstruction method via CS in practice, obtaining relevant results compared to which can be reconstructed according to global projections. In this case, the method implements an iterative interior reconstruction algorithm based on CS. Thus, a higher computational cost is required when compared with analytical algorithms.

van Sloun et al. [78] used a CT unfolding, so-called ultrasound computed tomography (UCT), for reconstruction of measurable tissue characteristics (e.g., mass density). For this purpose, CS is used to assist UCT with reduced sensing. Thus, the acquisitions become more compact. The authors further conclude that the proposed method was able to reduce acquisition time without impairing the high resolution of the image.

In the paper presented by Szczykutowicz and Chen [79], double-CT is used. Such equipment is characterized by a CT that uses a dual-source scanner. In this context, the reconstruction technique allows the reduction of the slew rate. The reference emphasizes the use of the prior image constrained compressed sensing (PICCS), which is a reconstruction algorithm based on CS. Another consistent application with CT is cone-beam computed tomography (CBCT), whose purpose is to verify the patient's position and identify the target of the treatment when there is a need to perform image-guided radiation therapy in treatments [80]. However, there are several construction limitations. Artifacts in the images coming from the movement of the patient due to breathing, such as in the thorax, for example. Thus, the authors propose the use of PICCS for image reconstruction, allowing the reconstruction of the image with cone-beam projections and also previous images, fully sampled.

Park et al. [81] point out that compressed sensing was able to assist CBCT in accurate reconstruction. Even so, the time of reconstruction is costly, making this method infeasible if referred to the CBCT. Therefore, the authors suggest an alternative method to CS, based on a new gradient projection algorithm.

The work presented by Lauzier and Chen [82] exposes the need to promote patient care on CT scans since the patient is exposed to ionizing radiation. Thus, the authors propose a study regarding the performance of a scheme known as reduction using prior image constrained compressed sensing (DR-PICCS) in order to reduce this dose. In the reconstruction stage, two algorithms were used: the filtered backprojection (FBP) and the PICCS. In this case, the authors concluded that DR-PICCS allows the reconstruction of images with lower noise than the use of FBP.

In the paper presented by Choi and Brady [83], the function of the role of coding apertures in X-ray transform measurement systems is studied in order to observe the data efficiency and the fidelity of reconstruction of the same via CS. In Choi et al. [84], the paradigm of CBCT image reconstruction is considered to be about potentially noisy samples. The authors propose compressed sensing in a conventional way, inserted in a sparsity context and based on the minimization of the norm $L1$. Finally, they conclude that CS exceeds the traditional algorithm.

According to the work of Sidky and Pan [85], the authors propose an iterative algorithm based on CS for the reconstruction of the volume image from the scanning of a cone-beam circular beam. The work emphasizes that the algorithm can also be applied in scanning geometries for other X-ray emitting trajectories. Throughout this application the wide range of CT applications and their variants is evident. Nevertheless, computed tomography in 4D (4D CT) is yet another unfolding of this. Such a tool is capable of analyzing the movement of the patient's anatomy during the respiratory phases [86], to the detriment of the reconstruction for each respiratory phase. Thus a 4D CT reconstruction algorithm was proposed [86].

Jia et al. developed [86], a reconstruction algorithm that can fit in most cases, according to the authors. It benefits from principles of CS theory. This method is based on the vector of difference, whose proposal is to offer the difference between the reconstructed image and the original in which this result is sparse. The method also uses the minimization of the $L1$-norm to reconstruct the difference vector.

In the work developed by Gross et al. [87], methods are established for quantum state tomography using the compressed sensing theory, presenting a new modality.

Meng et al. [88] deal with optic resolution ocular acoustic microscopy, which is a research tool for microcirculation. Thus a CT bifurcation, the optical-resolution photoacoustic computed tomography (OR-PACT), is inserted in this context of microvascularization. The reference proposes a strategy of photoacoustic reconstruction via CS.

4.11 Conclusion

Technological developments have emerged as aids to medicine. Since the introduction of the first medical technological equipment, such as the X-ray, it has been possible to perform earlier diagnoses more accurately. The result of this partnership between medicine and technology can be verified by the increased life expectancy around the world.

With the advancement of technology, some examinations have become standard and, in some cases, indispensable for physicians. ECG, MRI, and CT scans can be cited as essential. Cardiac diseases, such as atrial arrhythmia, can be previously diagnosed using portable ECG devices (Holter device). Cancers and tumors can be combated in the early stages with the help of these examinations in the diagnosis stages. There are complex parts of the human body, such as the musculoskeletal system, that can be treated based on information extracted from examinations performed by these technologies. Health monitoring gained a great prominence because it is able to promote better monitoring of the human body in situations in which speed and fidelity of diagnosis are necessary. For such reasons, CS becomes a natural option for these applications. It carries the necessary conditions to promote speed, simplicity, and fidelity in the treatment of the data. In other words, it helps health monitoring becomes more feasible.

Finally, this chapter looked at the applications of CS in health monitoring in order to inform the reader of the importance of this technique and the extension of the possible applications. We have shown applications with different approaches, and after reading this review, the reader should be able to understand the intricacies related to the theme and perceive the vastness of applications.

References

[1] J.C. Richardson, R.W. Bowtell, K. Mäder, C.D. Melia, Pharmaceutical applications of magnetic resonance imaging (MRI), Adv. Drug Deliv. Rev. 57 (8) (2005) 1191–1209.

[2] F.T. Denton, B.G. Spencer, Chronic health conditions: changing prevalence in an aging population and some implications for the delivery of health care services, Can. J. Aging 29 (1) (2010) 11–21.

[3] E.J. Candès, J. Romberg, T. Tao, Robust uncertainty principles: exact signal reconstruction from highly incomplete frequency information, IEEE Trans. Inf. Theory 52 (2) (2006) 489–509.

[4] D.L. Donoho, Compressed sensing, IEEE Trans. Inf. Theory 52 (4) (2006) 1289–1306.

[5] E.J. Candes, T. Tao, Near-optimal signal recovery from random projections: universal encoding strategies? IEEE Trans. Inf. Theory 52 (12) (2006) 5406–5425.

[6] N.J. Pagidipati, T.A. Gaziano, Estimating deaths from cardiovascular disease: a review of global methodologies of mortality measurement, Circulation 127 (6) (2013) 749–756.

[7] N. Dey, A.S. Ashour, F. Shi, S.J. Fong, R.S. Sherratt, Developing residential wireless sensor networks for ECG healthcare monitoring, IEEE Trans. Consum. Electron. 63 (4) (2017) 442–449.

[8] M.H. Sedaaghi, M. Khosravi, Morphological ECG signal preprocessing with more efficient baseline drift removal, in: 7th IASTED International Conference, ASC, 2003, pp. 205–209.

[9] M.R.A. Mahdi Khosravy, M.H. Sedaaghi, Morphological adult and fetal ECG preprocessing: employing mediated morphology, in: IEICE Tech. Rep., vol. 107, IEICE, 2008, pp. 363–369.

[10] M. Khosravi, M.H. Sedaaghi, Impulsive noise suppression of electrocardiogram signals with mediated morphological filters, in: 11th Iranian Conference on Biomedical Engineering, ICBME, 2004, pp. 207–212.

[11] S. Acharjee, N. Dey, S. Samanta, D. Das, R. Roy, S. Chakraborty, S.S. Chaudhuri, Electrocardiograph signal compression using ant weight lifting algorithm for tele-monitoring, J. Med. Imaging Health Inf. 6 (1) (2016) 244–251.

[12] M.H. Sedaaghi, R. Daj, M. Khosravi, Mediated morphological filters, in: 2001 International Conference on Image Processing, vol. 3, IEEE, 2001, pp. 692–695.

[13] M. Khosravy, N. Gupta, N. Marina, I.K. Sethi, M.R. Asharif, Morphological filters: an inspiration from natural geometrical erosion and dilation, in: Nature-Inspired Computing and Optimization, Springer, Cham, 2017, pp. 349–379.

[14] M. Khosravy, M.R. Asharif, M.H. Sedaaghi, Medical image noise suppression using mediated morphology, in: IEICE Tech. Rep., IEICE, 2008, pp. 265–270.

[15] M. Khosravy, M.R. Asharif, K. Yamashita, A PDF-matched short-term linear predictability approach to blind source separation, Int. J. Innov. Comput. Inf. Control 5 (11) (2009) 3677–3690.

[16] M. Khosravy, M.R. Asharif, K. Yamashita, A theoretical discussion on the foundation of Stone's blind source separation, Signal Image Video Process. 5 (3) (2011) 379–388.

[17] M. Khosravy, M. Asharif, K. Yamashita, A probabilistic short-length linear predictability approach to blind source separation, in: 23rd International Technical Conference on Circuits/Systems, Computers and Communications (ITC-CSCC 2008), Yamaguchi, Japan, 2008, pp. 381–384.

[18] M. Khosravy, M.R. Alsharif, K. Yamashita, A PDF-matched modification to Stone's measure of predictability for blind source separation, in: International Symposium on Neural Networks, Springer, Berlin, Heidelberg, 2009, pp. 219–222.

[19] M. Khosravy, M. Gupta, M. Marina, M.R. Asharif, F. Asharif, I.K. Sethi, Blind components processing a novel approach to array signal processing: a research orientation, in: 2015 International Conference on Intelligent Informatics and Biomedical Sciences, ICIIBMS, 2015, pp. 20–26.

[20] M. AlGhatrif, J. Lindsay, A brief review: history to understand fundamentals of electrocardiography, J. Commun. Hosp. Intern. Med. Perspect. 2 (1) (2012) 14383.

[21] Agateller, Schematic diagram of normal sinus rhythm for a human heart as seen on ECG, 2007. https://commons.wikimedia.org/wiki/File:SinusRhythmLabels.svg. Accessed 31 January 2019.

[22] Y. Zigel, A. Cohen, A. Katz, The weighted diagnostic distortion (WDD) measure for ECG signal compression, IEEE Trans. Biomed. Eng. 47 (11) (2000) 1422–1430.

[23] G.B. Moody, R.G. Mark, The MIT-BIH arrhythmia database on CD-ROM and software for use with it, in: Computers in Cardiology 1990, Proceedings, IEEE, 1990, pp. 185–188.

[24] A.M.R. Dixon, E.G. Allstot, D. Gangopadhyay, D.J. Allstot, Compressed sensing system considerations for ECG and EMG wireless biosensors, IEEE Trans. Biomed. Circuits Syst. 6 (2) (2012) 156–166.

[25] M. Balouchestani, K. Raahemifar, S. Krishnan, Wireless body area networks with compressed sensing theory, in: 2012 ICME International Conference on Complex Medical Engineering (CME), IEEE, 2012, pp. 364–369.
[26] H. Mamaghanian, N. Khaled, D. Atienza, P. Vandergheynst, Compressed sensing for real-time energy-efficient ECG compression on wireless body sensor nodes, IEEE Trans. Biomed. Eng. 58 (9) (2011) 2456–2466.
[27] D.H. Chae, Y.F. Alem, S. Durrani, R.A. Kennedy, et al., Performance study of compressive sampling for ECG signal compression in noisy and varying sparsity acquisition, in: -ICASSP, 2013, pp. 1306–1309.
[28] L.F. Polania, R.E. Carrillo, M. Blanco-Velasco, K.E. Barner, Compressed sensing based method for ECG compression, in: 2011 IEEE International Conference on Acoustics, Speech and Signal Processing (ICASSP), IEEE, 2011, pp. 761–764.
[29] F. Ansari-Ram, S. Hosseini-Khayat, ECG signal compression using compressed sensing with nonuniform binary matrices, in: 2012 16th CSI International Symposium on Artificial Intelligence and Signal Processing (AISP), IEEE, 2012, pp. 305–309.
[30] R.R. Edelman, The history of MR imaging as seen through the pages of radiology, Radiology 273 (2S) (2014) S181–S200.
[31] E.M. Haacke, R.W. Brown, M.R. Thompson, R. Venkatesan, et al., Magnetic Resonance Imaging: Physical Principles and Sequence Design, vol. 82, Wiley-Liss, New York, NY, 1999.
[32] N.H. Service, Overview—MRI scan, 2019. https://www.nhs.uk/conditions/mri-scan/ Accessed 31 January 2019.
[33] A.R. Padhani, Dynamic contrast-enhanced MRI of prostate cancer, in: Dynamic Contrast-Enhanced Magnetic Resonance Imaging in Oncology, Springer, New York, NY, 2005, pp. 191–213.
[34] J.S. Taylor, W.E. Reddick, Dynamic contrast-enhanced MR imaging in musculoskeletal tumors, in: Dynamic Contrast-Enhanced Magnetic Resonance Imaging in Oncology, Springer, New York, NY, 2005, pp. 215–237.
[35] I.S. Gribbestad, K.I. Gjesdal, G. Nilsen, S. Lundgren, M.H.B. Hjelstuen, A. Jackson, An introduction to dynamic contrast-enhanced MRI in oncology, in: Dynamic Contrast-Enhanced Magnetic Resonance Imaging in Oncology, Springer, New York, NY, 2005, pp. 1–22.
[36] J.F. Montebello, N.A. Mayr, W.T.C. Yuh, D.S. McMeekin, D.H. Wu, M.W. Knopp, Dynamic contrast-enhanced MR imaging for predicting tumor control in patients with cervical cancer, in: Dynamic Contrast-Enhanced Magnetic Resonance Imaging in Oncology, Springer, New York, NY, 2005, pp. 175–189.
[37] Z. Tian, N. Dey, A.S. Ashour, P. McCauley, F. Shi, Morphological segmenting and neighborhood pixel-based locality preserving projection on brain fMRI dataset for semantic feature extraction: an affective computing study, Neural Comput. Appl. 30 (12) (2018) 3733–3748.
[38] N. Dey, A.S. Ashour, S. Beagum, D.S. Pistola, M. Gospodinov, E.P. Gospodinova, J.M. R.S. Tavares, Parameter optimization for local polynomial approximation based intersection confidence interval filter using genetic algorithm: an application for brain MRI image de-noising, J. Imaging 1 (1) (2015) 60–84.
[39] M. Khosravy, N. Gupta, N. Marina, I.K. Sethi, M.R. Asharif, Brain action inspired morphological image enhancement, in: Nature-Inspired Computing and Optimization, Springer, Cham, 2017, pp. 381–407.

[40] M. Khosravy, N. Gupta, N. Marina, I.K. Sethi, M.R. Asharifa, Perceptual adaptation of image based on Chevreul-Mach bands visual phenomenon, IEEE Signal Process. Lett. 24 (5) (2017) 594–598.
[41] M. Khosravy, N. Patel, N. Gupta, I.K. Sethi, Image quality assessment: a review to full reference indexes, in: Recent Trends in Communication, Computing, and Electronics, Springer, New York, NY, 2019, pp. 279–288.
[42] M. Lustig, D.L. Donoho, J.M. Santos, J.M. Pauly, Compressed sensing MRI: a look at how CS can improve on current imaging techniques, IEEE Signal Process. Mag. 25 (2008) 72–82.
[43] S. Geethanath, R. Reddy, A.S. Konar, S. Imam, R. Sundaresan, R. Venkatesan, Compressed sensing MRI: a review, Crit. Rev. Biomed. Eng. 41 (3) (2013).
[44] A.R. Farias, D. de Castro Medeiros, H.A. Magalhães, M.F.D. Moraes, E.M.A. M. Mendes, A novel approach for accelerating mouse abdominal MRI by combining respiratory gating and compressed sensing, Magn. Reson. Imaging 50 (2018) 45–53.
[45] M. Lustig, D. Donoho, J.M. Pauly, Sparse MRI: the application of compressed sensing for rapid MR imaging, Magn. Reson. Med. 58 (6) (2007) 1182–1195.
[46] T. Çukur, M. Lustig, D.G. Nishimura, Improving non-contrast-enhanced steady-state free precession angiography with compressed sensing, Magn. Reson. Med. 61 (5) (2009) 1122–1131.
[47] J. Trzasko, C. Haider, A. Manduca, Practical nonconvex compressive sensing reconstruction of highly-accelerated 3D parallel MR angiograms, in: IEEE International Symposium on Biomedical Imaging: From Nano to Macro, 2009, ISBI'09, IEEE, 2009, pp. 274–277.
[48] T. Cukur, M. Lustig, E.U. Saritas, D.G. Nishimura, Signal compensation and compressed sensing for magnetization-prepared MR angiography, IEEE Trans. Med. Imaging 30 (5) (2011) 1017.
[49] R.B. Buxton, Introduction to Functional Magnetic Resonance Imaging: Principles and Techniques, Cambridge University Press, Cambridge, UK, 2009.
[50] M. Hämäläinen, R. Hari, R.J. Ilmoniemi, J. Knuutila, O.V. Lounasmaa, Magnetoencephalography—theory, instrumentation, and applications to noninvasive studies of the working human brain, Rev. Mod. Phys. 65 (2) (1993) 413.
[51] Y. Zhang, Z. Dong, P. Phillips, S. Wang, G. Ji, J. Yang, Exponential wavelet iterative shrinkage thresholding algorithm for compressed sensing magnetic resonance imaging, Inf. Sci. 322 (2015) 115–132.
[52] B. Wu, W. Li, A. Guidon, C. Liu, Whole brain susceptibility mapping using compressed sensing, Magn. Reson. Med. 67 (1) (2012) 137–147.
[53] J.P. Earls, V.B. Ho, T.K. Foo, E. Castillo, S.D. Flamm, Cardiac MRI: recent progress and continued challenges, J. Magn. Reson. Imaging 16 (2) (2002) 111–127.
[54] R. Otazo, D. Kim, L. Axel, D.K. Sodickson, Combination of compressed sensing and parallel imaging for highly accelerated first-pass cardiac perfusion MRI, Magn. Reson. Med. 64 (3) (2010) 767–776.
[55] D. Liang, E.V.R. DiBella, R.-R. Chen, L. Ying, K-T ISD: dynamic cardiac MR imaging using compressed sensing with iterative support detection, Magn. Reson. Med. 68 (1) (2012) 41–53.
[56] M. Usman, D. Atkinson, F. Odille, C. Kolbitsch, G. Vaillant, T. Schaeffter, P.G. Batchelor, C. Prieto, Motion corrected compressed sensing for free-breathing dynamic cardiac MRI, Magn. Reson. Med. 70 (2) (2013) 504–516.
[57] S. Posse, R. Otazo, S.R. Dager, J. Alger, MR spectroscopic imaging: principles and recent advances, J. Magn. Reson. Imaging 37 (6) (2013) 1301–1325.

[58] S. Hu, M. Lustig, A.P. Chen, J. Crane, A. Kerr, D.A. Kelley, R. Hurd, J. Kurhanewicz, S.J. Nelson, J.M. Pauly, et al., Compressed sensing for resolution enhancement of hyperpolarized 13C flyback 3D-MRSI, J. Magn. Reson. 192 (2) (2008) 258–264.
[59] P.E.Z. Larson, S. Hu, M. Lustig, A.B. Kerr, S.J. Nelson, J. Kurhanewicz, J.M. Pauly, D.B. Vigneron, Fast dynamic 3D MR spectroscopic imaging with compressed sensing and multiband excitation pulses for hyperpolarized 13C studies, Magn. Reson. Med. 65 (3) (2011) 610–619.
[60] N.C. Askin, B. Atis, E. Ozturk-Isik, Accelerated phosphorus magnetic resonance spectroscopic imaging using compressed sensing, in: Engineering in Medicine and Biology Society (EMBC), 2012 Annual International Conference of the IEEE, IEEE, 2012, pp. 1106–1109.
[61] K.M.-P.M.F. Nielsen, A. Wittek, K. Miller, B. Doyle, G.R. Joldes, M.P. Nash, Computational Biomechanics for Medicine, Springer, New York, NY, 2010.
[62] M. Schuenke, E. Schulte, U. Schumacher, General Anatomy and Musculoskeletal System (THIEME Atlas of Anatomy), Thieme, Germany, 2006.
[63] M. Schünke, E. Schulte, U. Schumacher, L.M. Ross, E.D. Lamperti, Thieme Atlas of Anatomy: General Anatomy and Musculoskeletal System, vol. 1, Thieme Stuttgart, New York, NY, 2006.
[64] B. Gilles, L. Moccozet, N. Magnenat-Thalmann, Anatomical modelling of the musculoskeletal system from MRI, in: International Conference on Medical Image Computing and Computer-Assisted Intervention, Springer, 2006, pp. 289–296.
[65] G.P. Schmidt, M.F. Reiser, A. Baur-Melnyk, Whole-body imaging of the musculoskeletal system: the value of MR imaging, Skelet. Radiol. 36 (12) (2007) 1109–1119.
[66] S. Geethanath, S. Moeller, V.D. Kodibagkar, Accelerated 3D radial short echo-time MRI of the knee using compressed sensing, in: Proceedings of ISMRM Annual Meeting, 2012, p. 3286.
[67] I. Zenger, The history of computed tomography at Siemens: a retrospective, 2019. https://www.siemens.com/history/pool/newsarchiv/downloads/20151201_medhistory_milestones_history_of_ct_at_siemens_english.pdf. Accessed 31 January 2019.
[68] J. Hsieh, et al., in: Computed Tomography: Principles, Design, Artifacts, and Recent Advances, in: SPIE, Bellingham, WA, 2009.
[69] B.H. Guyer, E.M. Levinsohn, B.E. Fredrickson, G.L. Bailey, M. Formikell, Computed tomography of calcaneal fractures: anatomy, pathology, dosimetry, and clinical relevance, AJR Am. J. Roentgenol. 145 (5) (1985) 911–919.
[70] C.J. Schulze, E. Munzinger, U. Weber, Clinical relevance of accuracy of pedicle screw placement: a computed tomographic-supported analysis, Spine 23 (20) (1998) 2215–2220.
[71] S. Bipat, A.S. Glas, J. van der Velden, A.H. Zwinderman, P.M. Bossuyt, J. Stoker, Computed tomography and magnetic resonance imaging in staging of uterine cervical carcinoma: a systematic review, Gynecol. Oncol. 91 (1) (2003) 59–66.
[72] J. Esquivel, T.C. Chua, A. Stojadinovic, J.T. Melero, E.A. Levine, M. Gutman, R. Howard, P. Piso, A. Nissan, A. Gomez-Portilla, et al., Accuracy and clinical relevance of computed tomography scan interpretation of peritoneal cancer index in colorectal cancer peritoneal carcinomatosis: a multi-institutional study, J. Surg. Oncol. 102 (6) (2010) 565–570.
[73] M. Carrier, M. Righini, P.S. Wells, A. Perrier, D.R. Anderson, M.A. Rodger, S. Pleasance, G. Le Gal, Subsegmental pulmonary embolism diagnosed by computed tomography: incidence and clinical implications. A systematic review and meta-analysis of the management outcome studies, J. Thromb. Haemost. 8 (8) (2010) 1716–1722.

[74] J. Graf, B. Skutta, F.-P. Kuhn, A. Ferbert, Computed tomographic angiography findings in 103 patients following vascular events in the posterior circulation; potential and clinical relevance, J. Neurol. 247 (10) (2000) 760–766.
[75] A.J. Einstein, M.J. Henzlova, S. Rajagopalan, Estimating risk of cancer associated with radiation exposure from 64-slice computed tomography coronary angiography, JAMA 298 (3) (2007) 317–323.
[76] R. Smith-Bindman, J. Lipson, R. Marcus, K.-P. Kim, M. Mahesh, R. Gould, A.B. De González, D.L. Miglioretti, Radiation dose associated with common computed tomography examinations and the associated lifetime attributable risk of cancer, Arch. Intern. Med. 169 (22) (2009) 2078–2086.
[77] H. Yu, G. Wang, J. Hsieh, D.W. Entrikin, S. Ellis, B. Liu, J.J. Carr, Compressive sensing-based interior tomography: preliminary clinical application, J. Comput. Assist. Tomogr. 35 (6) (2011) 762.
[78] R. van Sloun, A. Pandharipande, M. Mischi, L. Demi, Compressed sensing for ultrasound computed tomography, IEEE Trans. Biomed. Eng. 62 (6) (2015) 1660–1664.
[79] T.P. Szczykutowicz, G.-H. Chen, Dual energy CT using slow kVp switching acquisition and prior image constrained compressed sensing, Phys. Med. Biol. 55 (21) (2010) 6411.
[80] S. Leng, J. Tang, J. Zambelli, B. Nett, R. Tolakanahalli, G.-H. Chen, High temporal resolution and streak-free four-dimensional cone-beam computed tomography, Phys. Med. Biol. 53 (20) (2008) 5653.
[81] J.C. Park, B. Song, J.S. Kim, S.H. Park, H.K. Kim, Z. Liu, T.S. Suh, W.Y. Song, Fast compressed sensing-based CBCT reconstruction using Barzilai-Borwein formulation for application to on-line IGRT, Med. Phys. 39 (3) (2012) 1207–1217.
[82] P.T. Lauzier, G.-H. Chen, Characterization of statistical prior image constrained compressed sensing (PICCS): II. Application to dose reduction, Med. Phys. 40 (2) (2013). 021902, https://doi.org/10.1118/1.4773866.
[83] K. Choi, D.J. Brady, Coded aperture computed tomography, in: Adaptive Coded Aperture Imaging, Non-Imaging, and Unconventional Imaging Sensor Systems, vol. 7468, International Society for Optics and Photonics, Bellingham, WA, USA, 2009, p. 74680B.
[84] K. Choi, J. Wang, L. Zhu, T.-S. Suh, S. Boyd, L. Xing, Compressed sensing based cone-beam computed tomography reconstruction with a first-order method, Med. Phys. 37 (9) (2010) 5113–5125.
[85] E.Y. Sidky, X. Pan, Image reconstruction in circular cone-beam computed tomography by constrained, total-variation minimization, Phys. Med. Biol. 53 (17) (2008) 4777–4807.
[86] X. Jia, Y. Lou, B. Dong, Z. Tian, S. Jiang, 4D computed tomography reconstruction from few-projection data via temporal non-local regularization, in: International Conference on Medical Image Computing and Computer-Assisted Intervention, Springer, 2010, pp. 143–150.
[87] D. Gross, Y.-K. Liu, S.T. Flammia, S. Becker, J. Eisert, Quantum state tomography via compressed sensing, Phys. Rev. Lett. 105 (15) (2010) 150401.
[88] J. Meng, L.V. Wang, D. Liang, L. Song, In vivo optical-resolution photoacoustic computed tomography with compressed sensing, Opt. Lett. 37 (22) (2012) 4573–4575.

PART 2

Internet of things for U-healthcare

CHAPTER

5 Nanopore sequencing technology and Internet of living things: A big hope for U-healthcare

Khalid Raza, Sahar Qazi

Department of Computer Science, Jamia Millia Islamia, New Delhi, India

5.1 Introduction

MinION, the nanopore sequencing device developed by Oxford Nanopore Technologies (ONT), is a portable pocket-sized device that can be connected to a common laptop through a USB connector [1]. Its small size and cost of less than $1000 makes it a game changer in the sequencing market. Other variants of nanopore sequencers such as PromethION, GridION, and SmidhION are also becoming popular [2]. Theses sequencers are not only being used to sequence DNA and RNA, but also to build an "Internet of Living Things" (IoLT). Sequencing devices can be miniaturized and sensors placed on the body to monitor human health and vital signs. Advancement in sensor technologies gave rise to DNA-reading sensors that can be used for almost everything, from food production equipment to water bodies to farms. For instance, MinION is capable of sensing an individual's DNA strands passing through their pores due to changes in ion, and interpreting the data into nucleotides (A, T, G, and C). These nucleotide sequences are streamed onto the cloud server where they are available to researchers, doctors, and scientists for further processing and analysis.

ONT has recently developed an even smaller device called SmidgION that uses the same sensing technology as MinION and PromethION, but that is designed to be used with smartphones or low-power devices. In other words, real-time mobile DNA sequencing and analysis is possible through SmidgION, which may have several potential applications, including remote monitoring of viruses, pathogens, or infections in an outbreak [3].

Under the Internet of things (IoT) framework, it is possible to connect portable MinION, PromethION, and SmidgION sequencers to other technical systems such as mobile phones, hospitals, airports, insurance companies, and so on. As a result, scientists, doctors, patients, employees—everyone —will be able to monitor the DNA of their own bodies on shared cloud computing labs. In addition, equipped DNA-reading sensors can identify the nature, transmission paths, and mutations of deadly viruses, lethal pathogens, and engineered bacteria, and alert to outbreaks

Sensors for Health Monitoring. https://doi.org/10.1016/B978-0-12-819361-7.00005-1

at a very early stage. Further, they can be used by individuals to detect the most vital biomarkers of early-stage cancers or viral agents. Hence, millions of individuals streaming these data to the cloud will build a powerful system for predictive, preventive, personalized, and participatory (P4) medicine. In a nutshell, genetic identity of any living things will take on new life on the Internet and the world will enter into the age of the IoLT.

The use of wearable sensors connected to the Internet has wide applications in healthcare. It is also the need of the hour for elderly people. Sensors can be used for the early detection of ailments, which not only helps to control disease but may also prevent disease from occurring in the first place. Early detection and treatment of an ailment before it becomes critical will bring about a profound breakthrough in the healthcare sector. The IoLT model can collect, analyze, and infer knowledge and give health recommendations to patients in real time, allowing for the use of current technologies such as IoT, cloud computing, big data analytics, and complex computational and artificial intelligence algorithms. In other words, the IoLT model is bringing a new revolution by promoting general well-being through continuous monitoring and analysis of biological systems. Some of the current wider applications of IoLT are monitoring an aging family member, and scalable, continuous heart rate monitoring [4, 5].

This chapter presents an introduction to nanopore sequencing technology, its applications to U-healthcare, IoLT concepts and their applications in U-healthcare, and the convergence of nanopore technology with IoLT concepts for U-healthcare. We show how far we have come, and discuss future promises, opportunities, and challenges.

5.2 Nanopore sequencing technology

5.2.1 Evolution of sequencing technologies

The field of genomics has been revolutionized over the last four decades, and sequencing has advanced from first- to third-generation technologies (Fig. 5.1). The first generation of sequencing was developed by two different research groups in parallel: Frederick Sanger in 1975 (chain-termination method), and Maxam and

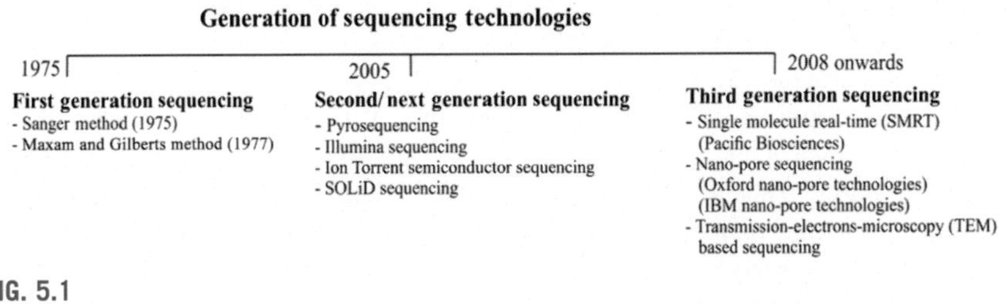

FIG. 5.1

Evolution of sequencing technologies [6].

Gilbert in 1977 (chemical sequencing method) [3]. Sanger's chain-termination method was extensively used as first-generation sequencing due to its less complex protocol and higher scalability compared to the chemical sequencing method. The high cost and low throughput of first-generation sequencing led to the development of new sequencing methodology, shifting to second-generation sequencing (SGS), also known as next-generation sequencing (NGS), massively parallel sequencing, or deep sequencing. SGS started with the launch of Roche 454's pyrosequencing in 2005, followed by sequence-by-synthesis-based Illumina Genome Analyzer in 2007, and so on [6]. SGS produces a large volume of short reads cheaply and therefore it has dominated the sequencing market.

The drawback of SGS is short read lengths and nonportability of the devices, which makes it applications difficult to real-time surveillance and onsite sequencing. The need for longer reads and shorter sequencing times led to the advent of third-generation sequencing (TGS). TGS directly sequences single DNA molecules, enabling real-time sequencing and reducing sequencing time from a few days to a few hours. The single-molecule real-time sequencing (SMRT) technology, introduced by Helicos and Pacific Biosciences, suffers from higher error rate (10%–15%). In 2014, ONT launched a new TGS platform, the MinION, as previously discussed. Currently, the error rate of MinION ranges between 5% and 15% [4, 7]. We discuss the details of ONT and its various sequencers in the next section.

5.2.2 Nanopore sequencing technologies

Nanopore-based sequencers are small size with low equipment costs, capable of sequencing the entire human genome quickly (within a few hours) and reliably for less than $1000. The nonnanopore sequencers require a lot of sample preparation and complex algorithms for processing data, which limits their capabilities; they have low throughputs, high costs, and short read lengths. Some of the advantages of nanopore sequencers are that they are high-throughput, label-free, capable of ultra-long reads, and have low material requirements [8]. Recently, nanopore-based detection of single molecules has appeared as one of the most powerful sequencing technologies that allows the study of DNA-protein interaction as well as protein-protein interaction. In other words, nanopore-based sensing technologies of single molecules have opened up a new avenue to investigate at the single-molecule scale and have several potential applications including analysis of ions, DNA, RNA, polymers, peptides, proteins, drugs, and macromolecules.

We discuss some of the nanopore sequencers launched by ONT in the following sections.

5.2.2.1 MinION (2015)

MinION is a pocket-sized, portable sequencing USB device that is attracting the interest of the genomics community, especially for pathogen surveillance, environmental monitoring, and clinical diagnostic applications. It is a real-time sequencing device that weighs less than 100 g, can be plugged into a laptop or PC via a USB cable, and is usable in a nonlaboratory environment (such as mountains, jungles, or the international space station). The single consumable flow cell may generate 10–30 Gb of DNA sequences (www.nanoporetech.com/products/minion).

5.2.2.2 PromethION

This device follows the same workflow and uses similar technology as MinION and GridION but at a much larger scale (e.g., population-scale sequencing). PromethION is developed for core centralized sequencing labs and genome centers. PromethION has two modules: (1) sequencing module and (2) computer module. The computer module has integrated base-call accelerators and real-time data analysis in the device. Since January 2019, PromethION X24 and X48 have offered up to 24 and 48 flow cells, respectively; each of them can be started and stopped independently (https://nanoporetech.com/products/promethion).

5.2.2.3 SmidgION

The SmidgION is a very small nanopore sequencer designed to be run on smartphones. It uses a similar sensing technology as MinION and PromethION. In fact, it is a smaller version of MinION. The DNA samples are loaded into the tiny instrument that can be directly plugged into a smartphone. The SmidgION device uses the phone's battery power and is operated by mobile app. The generated data are either stored to the phone's memory or uploaded to the cloud. This little device is a big attempt to simplify the sequencing process and bring it to the consumer market. In other words, both SmidgION and MinION devices are an attempt to bring us closer to the "lab on a chip," that is, facilitating complex laboratory analysis with the help of a simple-to-use, tiny instrument in a nonlaboratory setting. As a result, medical clinics can screen for new viruses in seconds, and researchers can get DNA sequences in real time.

5.2.2.4 GridION

This is a benchtop sequencer capable of running and analyzing up to five MinION flow cells either concurrently or individually. It is ideal for laboratories with multiple projects. The GridION X5 is a compact device allowing laboratories to offer nanopore sequencing as a service to the users.

Table 5.1 presents a comparison of different nanopore sequencers. Fig. 5.2 depicts these sequencers and their sizes and weights.

5.3 Role of nanopore sequencing in U-healthcare

Ubiquitous healthcare (U-healthcare), that is, a combination of electronic and mobile healthcare, is more concerned with person-centric therapy rather than traditional hospital healthcare. With the rise of technology, people today are aware of acute and severe diseases, as well as specific remedial treatment. People cannot be deceived or misled by the healthcare support system as everything today is just a click away!

Table 5.1 Comparison of different sequencers from ONT

ONT devices	Launching year	Purpose	Average read length	Run time	Speed (48 h)	Instrument starter pack cost	Reagent cost per run
MinION	2015	Sequencing outside the lab for small projects	Variable (up to 2 Mb)	1 min–72 h	10–50 Gb	~$1k	$99
PromethION	2015	Similar to MinION, but it offers large-scale sequencing	Variable (up to 2 Mb)	1 min–64 h	Up to 15 Tb	$165k–$285k	$99
SmidgION	2016	Smallest sequencing device to date, designed to be run on smartphones	N.A.	N.A.	N.A.	N.A.	N.A.
GridION	2017	A compact benchtop sequencer designed to run and analyze multiple flow cells and ideal for offering sequencing as a service	Variable (up to 2 Mb)	1 min–72 h	50–100 Gb	$50k	$99

Source: https://nanoporetech.com/products/comparison.

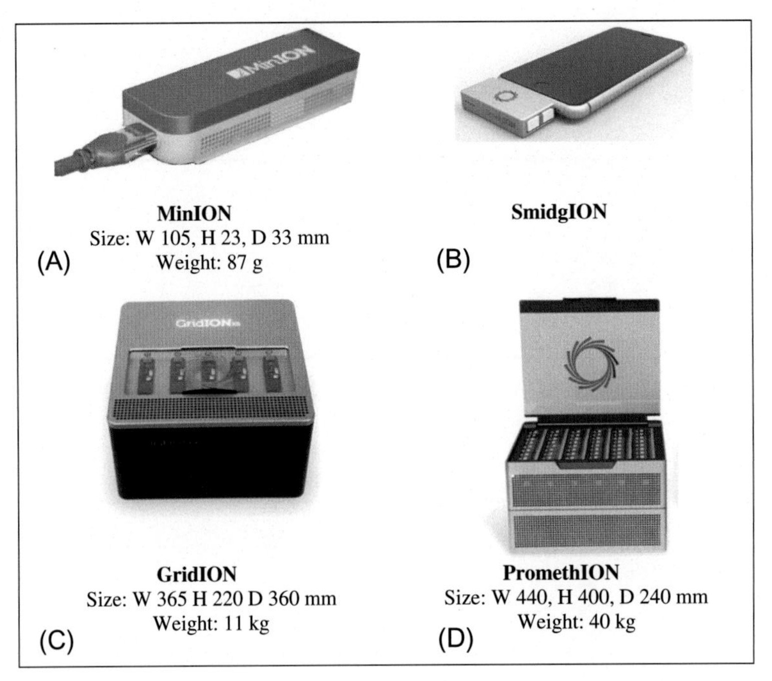

FIG. 5.2

Nanopore sequencers, their sizes, and weights (image sizes are not to scale).

Image credit/source: https://nanoporetech.com/.

All thanks go to the Internet, which revolutionized healthcare by making it mobile, creating a new field of study called *tele-based healthcare systems*. Tele-based healthcare units are independent of direct consultant-patient interactions. They have helped patients as they no longer have to wait in long queues for appointments or medicines; it is the most convenient and user-friendly way of dealing with health-based issues irrespective of distance or boundaries. Scientific research prophesizes that mobile and ubiquitous tele-healthcare combined with Wireless Body Area Networks (WBANs) is optimal for paving the way to next-generation U-health [9]. For this purpose, nanopore sequencing techniques have been appreciated and accepted worldwide. The essential component of nanopore sequencing is its miniature-sized pores (holes), which easily puncture the biological membranes. ONT provides a wide range of options to its users such as direct DNA/RNA sequencing and real-time sequencing. Their sequencers incur no extra capital expenditure, provide extremely long reads (~2 Mb), and are easily portable. They also automated simple usage, library preparation, and increased yields for humongous genomes. Nanopore sequencing has myriad applications in many biological domains such as microbiology, microbiome, environment-based research, human genetics research, cancer and epigenetics research, population-wide research, transcriptomics analysis, and plant and animal research to name a few [10].

5.3.1 Applications of nanopore technologies in U-healthcare

5.3.1.1 Virus control and surveillance

Poultry workers and breeders find it hard to cope with outbreaks of Newcastle disease (ND), which can be managed only if the viral agent causing the disease is rapidly identified. The evolutionary mechanisms of the Newcastle disease virus (NDV) hinder effective diagnosis of the same, and so diagnosis is a tedious and tough process. Amplicon sequencing (AmpSeq) is a sensitive mechanistic technique for predicting virulence and genotype that is useful for NDV samples; it is a third-generation, real-time (rt) DNA sequencing platform. In a study by Butt et al. [11], the authors executed 1D MinION sequencing of barcoded virus amplicons carried out on 33 egg-grown isolates that were inclusive of 23 unique lineages and 15 different NDV genotypes and from 15 clinical swab samples from field outbreaks. The assembly-based analysis was done by Galaxy-based AmpSeq. For all egg-grown samples, NDV was identified and virulence and genotype were also determined. For 15 swab clinical samples, NDV was rectified to be in 10 of 11 NDV samples. Six of the clinical samples were composed of two mixed genotypes, of which the MinION identified both genotypes in four of those samples. Furthermore examination of a dilution series of one NDV sample turned out to be helpful in identifying NDV with around 50% egg infectious dose (EID50) as low as 101 EID50/mL, which was successfully gained in 7 min of sequencing time along with a 98.37% sequence identity as compared to the expected consensus outcome. Their study clearly discerns that this technique can be useful in the identification of other viral agents as well. In another study conducted by Votintseva et al. [12], researchers used MinION for antibiotic susceptibility predictions in *Mycobacterium tuberculosis*, which had the potential to continue sequencing until threshold coverage is gained.

The salmonella epidemic in 2014 at Heartlands Hospital (Birmingham) was a shock and proved to be the optimal platform for examining ONT's miniature portable DNA sequencer. A small group of researchers from Birmingham University identified the origins of epidemics using the MinION device [13].

5.3.1.2 Real-time monitoring of body fluids

Healthcare has become useful because of nanopore sensors that directly measure multiple metabolites in the body, allowing healthcare providers to keep an eye on their patients' health via blood, sweat, and saliva samples. Even though markets are full of gadgets where biosensors are integrated into electronic devices, these gadgets are limited to measuring only a few metabolites like glucose. For a personalized U-healthcare, nanopore-based devices are indeed required for direct sampling of body fluids where the biosensors are miniaturized, making them flexible for the user [14]. Identification of structural alterations in a protein is a simpler way to quantify metabolites. ONT's biosensor also works for single molecules by neglecting the need for calibration, allowing many proteins to be read by a single nanopore simultaneously. What's in my Pot? (WIMP) is an analysis workflow combined with ONT's MinION. Beginning with a raw sample, WIMP develops the sample's sequence data and categorizes microorganisms such as bacteria, viruses, and fungi present in the sample into sub-species and strain level in a meticulous manner in around 3.5 h [15].

5.3.1.3 Molecular level of understanding disease mechanism

ONT's MinION has been successful in unbiased diagnosis of many diseases. For instance, differential diagnosis of acute febrile illness on a MinION sequencer can yield results within 6 h of the experimental run, which can then be analyzed in real time by using sequence-based ultrarapid pathogen identification, real-time (SURPIrt), which is an easy version of SURPI used for medical diagnosis in hospitals. Clinical and lethal diseases for which MinION has successfully provided a clear understanding of molecular dynamics are Zika virus (ZikV), Ebola virus (EboV), Lassa virus, chikungunya virus, dengue virus, influenza virus, and the malarial parasite *Plasmodium falciparum* [16, 17].

5.3.1.4 Wastewater management

Mismanaged sewer systems can cause chaos in urban areas. Sewer units are composed of wastewater and stormwater sewers. Wastewater is only processed before discharging. Damage of pipelines or connections in networks can be catastrophic as untreated water can then enter into natural flowing water bodies via the stormwater system. The traditional tracing mechanism used is *Escherichia coli*, which is a very simple, cheap method but also time-consuming and nonspecific. Instead of using this method, two sequencing-based methods (IlluminaMiSeq 16S rRNA gene amplicon sequencing and MinION shotgun metagenomic sequencing) were been employed for 73 stormwater samples. The results were compared with the ones obtained by the *E. coli* method of tracing wastewater and it was found that the amplicon data holds information on the source of contamination and how much time has passed since the filth matter has entered the system [18]. Another finding where ONT's technology has played a vital role is in bulking [19].

5.3.1.5 Inferring evolutionary relationship

A combination of Illumina and ONT's sequencing methodology was employed to deduce an entire metagenomic sequence of 5.7 Mbp for *Candidatus Amarolineaceae* phylotype, which gave a lucid understanding of the phylogenetic and evolutionary relationship of the same. The ONT-based annotation discerned that the phylotype had the potential of aerobic respiration, fermentation, and conversion of nitrate to ammonia [19].

5.3.2 Role of ONT in revolutionizing U-healthcare

Individualized genomic medicine (IGM), or personalization, focuses on revolutionizing healthcare approaches by aggregating and analyzing a patient's unique genomic data. The basic requirements of personalization are initial diagnosis, robust and efficient prognosis, and treatment exclusive of medication side effects [20]. Personalized U-healthcare is largely dependent on the development of novel techniques that are portable and can sequence the entire DNA sequence. In order to reduce instrumentation costs, nanopore technology has been employed as the backbone for NGS, which can perform simple and serene sequencing of genomic material and is better than the traditionally used sequencing techniques available in the markets. Nanopore sensors are mainly electrical, having the potential to determine greater DNA concentrations/volumes from a patient's blood or saliva [21]. Moreover, ONT's devices have defined a dramatic increment of DNA reads, from 450 bases to more than 10kb bases.

In 2012, ONT unveiled a USB-based, single-use device costing about $900 that can sequence an entire bacteriophage genome [22]. Impressed by its successful sequencing, many other companies, including IBM, Electronic BioSciences, Genia, and others, also have invested in developing such nanopore sequencers. As far as academia and research fraternities are concerned, nanopore sequencing approaches have been adapted and used for the betterment of nucleotide sequencing and its myriad components, which are studied using an inclusive patch-clamp amplifier technology accordingly.

Nanopores are single-molecule sequencers that can be better understood as biological drilling machines that form a natural hole/etch, called a *biological pore* [23, 24], and/or an opening in the solid-state substrate (solid-state pore) [25]. Utilizing a sensitive patch-clamp amplifier, nanopore devices validate the ionic movement caused by application of the voltage across the membranes by a single pore, which differentiates into two domains: *cis* and *trans*. It is then filtered, sampled, and analyzed. This is very fruitful for studying the complex structures of nucleic acids (DNA/RNA) [26–28], for the identification of various affiliations between DNA-DNA binding proteins [29, 30].

ONT has been highly successful in bringing up nanopore sequencing to the commercial industry, which proved to be the optimal ultrasensitive platform for sequencing biological macromolecules. The future is bright for ONT in U-healthcare, as the nanopore devices have started to work on the multidimensional challenges by balancing the fluctuating signal-noise ratios (SNR), slowing down DNA mobility via nanopores, decreasing the background noise, and generating automated systems for parallel arrays of nanopore sequencing for gigantic data [31]. There are high hopes for ONT's devices that they can sequence a full phage genome where long DNA/RNA reads are rapidly possible. Many challenges still exist, however, for instance, the requirement of an algorithm that is capable of gleaning the sequence from raw data, optimizing the device in such a way that it has integrated circuitry to functionally charge every biological pore in an array simultaneously, and a high demand for microfluidics to guard the array. ONT has paved the way for healthy and user-friendly sequencing of nucleic acids rapidly and with longer reads. However, many more modifications are required to make sequencing even more robust in the future [32].

5.3.3 Applications of nanopore technologies in space stations

ONT was lucky enough to get their MinION sequencer used in a special NASA mission in 2016. It was the first time DNA was successfully sequenced in microgravity as part of the Biomolecule Sequencer experiment performed aboard the International Space Station [33]. This inclusion of the device pointed at the regulation of alterations in microorganisms in the space environment. A potential future application of the device may be in identifying and determining the existence of DNA-based life in the universe (if any) [34]. Researchers on Earth performed simultaneous evaluations in order to deduce efficiency of MinION. International Space Station crew members successfully accomplished sequencing of DNA from microorganisms such as bacteria, bacteriophage, and rodent samples that were retrieved from Earth [35].

5.3.4 Maintenance and issues of ONT in U-healthcare

ONT developed a high-yield, cheap alternative to NGS technologies [1]. MinION, a versatile DNA/RNA sequencing ONT product, is a simple-to-use, USB-based device that is minimal in its requirements and is therefore easily connected to a computer [36]. However, the humongous increase in MinION throughput has highlighted the urgent need for decreased per-base data handling and storage demands in present and future scenarios.

A major concern today revolves around data storage issues and data management and maintenance. State-of-the-art strategies have aggregated eukaryotic genome assembly data, which was generated by multiple flow cells so as to derive satisfactory indemnity of the genome [37]. To cope with this situation, ONT initiated the precommercial launch of PromethION, a workbench nanopore sequencing device composed of 3000 channels and 48 flow cells. In comparison, the MinION has only 512 channels. PromethION generates an enormous amount of data, almost six terabases per day [2].

Myriad methodologies have been put into existence for the excellent analysis of humongous nanopore datasets generated by sequencing devices. Unfortunately, only a few data storage methods are available to store and maintain this data. European Nucleotide Archive (ENA) is one way that a nanopore sequencer can upload a single data file; however, the data cannot be compressed. In order to decrease the file size, so that data can be compressed and stored, Picopore was developed. It is a tool for decreasing the storage footprint of the sequencing device without hindering users from using their preferred analysis tools. It employs an amalgamation of storage compression along with efficient memory allocation, nonredundancy of data, and elimination of unnecessary data created by base-calling by the end user [38].

Both MinION and PromethION pledge to provide apex datasets as the technology drifts toward industrial launch. The disk space needed to successfully run and store the datasets from the two devices usually causes a problem for both service providers and users, thus Picopore comes in handy.

It is evident that less computational time, data retention, and disk space are some issues that cannot be completely solved. However, Picopore has provided some solutions in order to reduce these issues, for instance, real-time compression, lower bandwidth utilization for transfer of datasets between laboratories, and decreasing the storage on data servers [38].

5.4 Internet of living things: Concepts and applications in U-healthcare

The IoT is focused on the overall connection of physical things (devices) with the Internet in order to make an interchangeable communication network enabling more efficiency and easy usage [39, 40]. It can be interpreted as a connected set of six ANYs: (1) anyone, (2) anything, (3) anywhere, (4) anytime, and (5) any network (see Fig. 5.3) [41]. The IoT provides umbrella-wide solutions to trending problems such as smart city developments, traffic management, waste management, medical healthcare systems, industrialization, logistics, and so on [42–46].

The IoT is not just restricted to nonliving beings, hence it is now called the Internet of living things (IoLT). It has the capacity to pave the way for many medical applications, including healthcare monitoring, fitness management, severe diseases maintenance, and many others. Two major components of medical healthcare are the diagnosis and prognosis of a particular disease, and so IoLT has the potential to provide treatment and medication at home [41]. Henceforth many medical devices, bodily sensors, and diagnostic and imaging instruments can now be as smart as smartphones, or any intelligent devices for that matter. The IoLT-based healthcare devices and services are helpful in reducing healthcare expenditure and increasing the quality of healthcare management. IoLT encapsulates all the instrumentations required for an efficient healthcare system. It uses smart serene devices for robust healthcare that are both pocket-friendly and efficient. The IoLT is an apex trend in next-generation techniques, which can be significant for the entire industrial, commercial, academia, medical fraternities [42] (Fig. 5.4).

Efficient diagnosis, prognosis, and treatment at home by healthcare providers is an important application of IoLT-based healthcare systems. Many devices, applications, and prototypes have been designed and developed. The main components of IoLT include: (a) network architectures, (b) network platforms, and (c) network

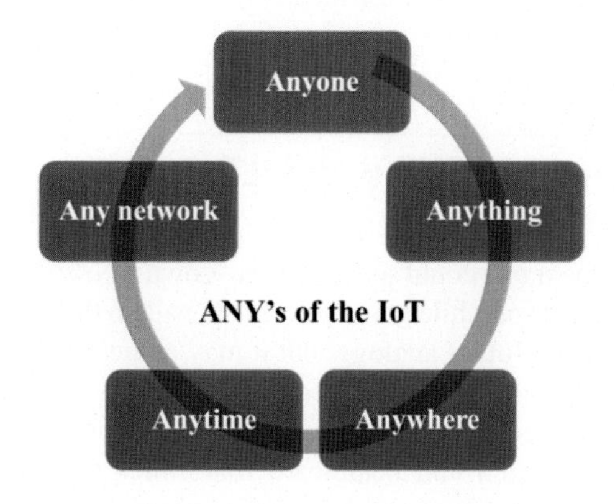

FIG. 5.3

The five ANYs of the internet of things (IoT).

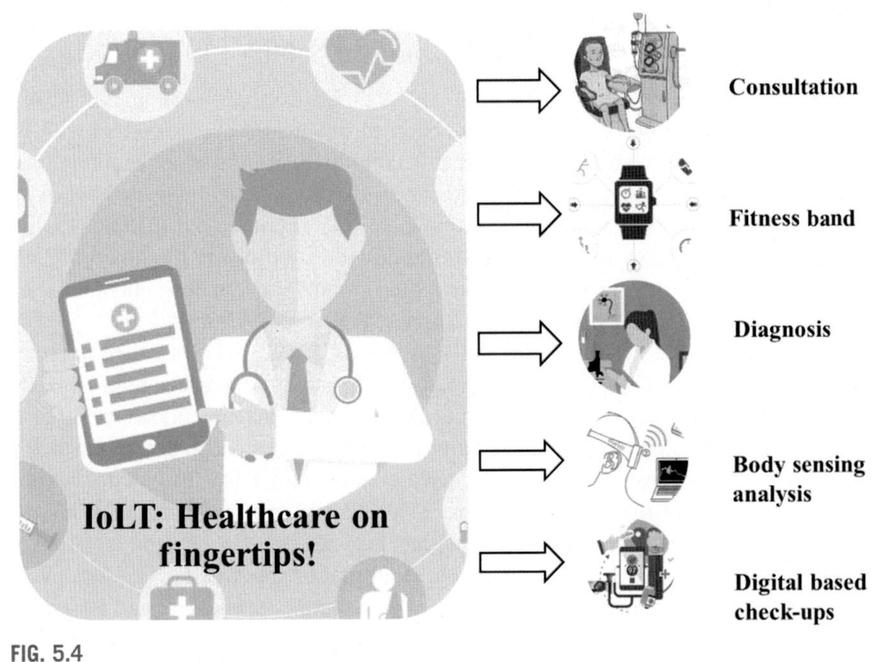

FIG. 5.4

IoLT-based technology in medical healthcare.

topology inclusive of novel services, products, and platforms. Since many people are skeptical of using this newer technology in the medical field, patient safety and privacy policies and guidelines have been proposed for the excellence of IoLT around the globe. However, IoLT is still growing and expanding and remains a novel field of research. Table 5.2 summarizes the services and applications provided by IoLT that have been successfully employed for medical healthcare by people worldwide.

5.5 Convergence of ONT with IoLT and other technologies for U-healthcare

How great would it be if one could sequence one's own DNA on mobile phones while commuting on the metro or while going for a brisk walk in the evening? The idea may sound like a far-fetched future analogy, but it may come into existence sooner than you think. The benefits of IoLT have already been discussed in the previous section, however, it has again been reiterated that the IoLT is a prosperous outlook for scientific biomedical research and healthcare fraternities as it gets everything and everyone on one major platform where specific subunits can interact and associate with one another to bring about better outcomes.

Table 5.2 Services and applications of IoLT in healthcare

IoLT service	Description	IoLT application	Description
Ambient assisted living (AAL)	Artificial intelligence-based IoLT platform for elderly people that gives them space for a confident and independent lifestyle. For example, 6LoWPAN is employed for maintaining AAL [47]	Glucose level monitoring	A device composed of a glucose collector, computer or mobile phone, and a processor capable of regulating a diabetic patient's glucose level regularly [48]
Mobile-healthcare (m-IoT)	Mobile-based medical computing technologies for healthcare services [49]	Oxygen saturation maintenance	Wearable pulse oximeter developed by Nonin for maintaining a U-healthcare milieu [50]
Medication toxicity and reactions	Adverse drug reactions (ADRs) or medication toxicity is due to excessive dosage of drugs in the body. The iMedPack, a part of the iMedBox, is used to detect these reactions [51]	Blood pressure monitoring	WBAN-based wearable blood pressure body sensor for BP maintenance [52]
Community network healthcare (CNH)	A combination of local clinics, hospitals, and residential areas for forming a cooperative, compatible, virtual networking healthcare unit executed by WBAN [53, 54]	ECG regulation	An IoLT-based ECG regulation setup including an easy-to-handle portable wireless transmitter with a receiving processor enabling automation in detecting cardiac functioning in real time [55]
Body sensor device (BSD)	Harmless accurate body sensors that provide users with blood pressure, heart rate, glucose level, etc. anytime and anywhere [56]. For example, the FitBit wristwatch	Medication management	E-healthcare devices such as I2Pack and iMedBox are efficient IoLT-based applications for management of accurate dosage of medicines prescribed by consultants to patients [57]

Continued

Table 5.2 Services and applications of IoLT in healthcare *Continued*

IoLT service	Description	IoLT application	Description
Sharing medical data	Myriad medical-based semantics and ontologies are shared using IoLT devices, which make it useful for practitioners to collect and share data during emergency situations [50, 58]	Menstruation and fertility tracker	A mobile-based application that keeps track of menstruation cycles and fertility in females [59]
Interconnected networking gateway (ING)	Treats patients, consultants, healthcare providers as network nodes and conjoins them at a single platform marking U-healthcare as a virtual real-time association [60]	Ocular/skin disorder detection	A cloud-based mobile application that is capable of detecting eye or skin diseases using pattern matching and smart camera picture-capturing techniques [61]

MinION, one of the best rapid sequencing devices, could soon be operated by an individual on their mobile phone. IoLT for ONT-based sequencers could unleash a whole new domain of self-actualization and quantification where users won't be dependent on anyone or anything, just their mobile phones. This new technology would be able to quantify blood-based prognostic indicators and track alterations in one's daily routine [62]. IoLT for ONT-based sequencers would promote an active healthy lifestyle in people with a sedentary lifestyle.

It is not new that data is expanding and growing with each passing day. With the advent of novel techniques and IT, researchers are able to extract more and more data on complex phenomenon. Similarly, DNA sequencing also generates humongous data that needs to be stored and utilized in such a manner that it can easily be shared if necessary. ONT devices, especially the MinION, are convenient and smart in this regard as they employ a cloud computing platform called Metrichor [63] for translating the locally produced low-level data into DNA sequence reads (base-calling), thus giving an edge to this veraciously expanding biological data.

This base-caller is exhaustively computer intensive, engaging Recurrent Neural Networks (RNNs) that are optimal for cloud computation [64].

Convergence of ONT-based sequencers with IoLT, cloud computation, and big data marks the beginning of a new path that can make individuals independent and reliable on their handsets for analysis and sequencing of their own biological composition. It will actually make people understand how to maintain a healthy and active lifestyle, protecting them from catastrophic proliferative diseases like cancer. Fig. 5.5 showcases the convergence of the three basic components of interpreting biological composition of an individual and its efficient management therein.

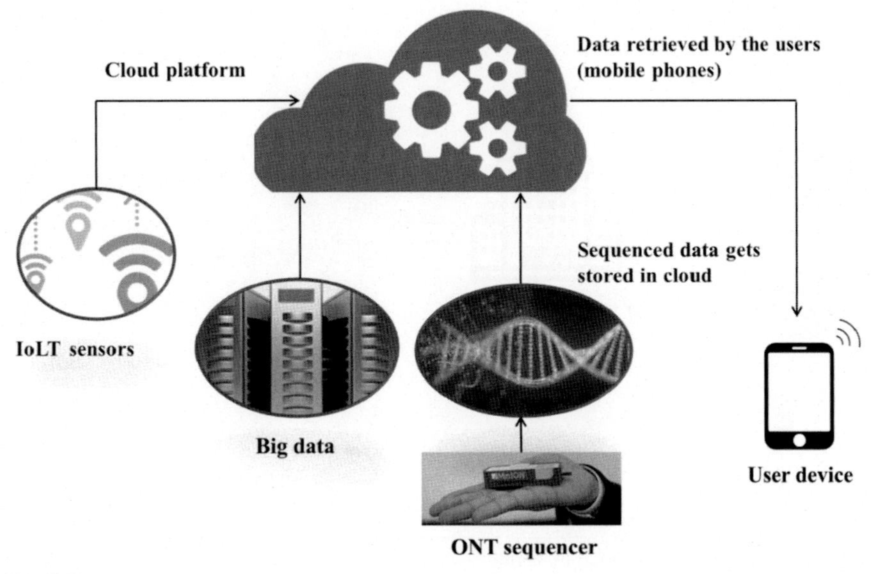

FIG. 5.5

Convergence of ONT with IoLT, big data, and cloud computation.

This magnanimous combination of trending technologies, such as big data analytics, cloud computing, and IoLT sensors for DNA sequencing, is constructive in developing a U-healthcare milieu. People are now motivated towards person-centric healthcare where every individual is treated equally and is prescribed the right medicine at the right dosage at the right time. U-healthcare is a streamlined system of these trending technologies. Ongoing studies suggest that IT-based U-healthcare, which is delivered by employing IoLT and cloud computing, is way better than the traditional methods of healthcare providers. In the U-healthcare industry, cloud computation plays a key role in providing three kinds of web-based services: (1) software as a service (SaaS), namely, ready-to-use software, (2) platform as a service (PaaS), a platform composed of programming languages and tools to develop the software, and (3) infrastructure as a service (IaaS), which is simply the storage of IT resources [65]. IT-based U-healthcare is simple, specific, and provides varied e-infrastructure and facilities for growing healthcare demands [66] along with plasticity of usage for users [67] at affordable prices [68]. Combining IT trends in healthcare is at a boom today and is appreciated for its successful utilization worldwide. Fig. 5.6 summarizes the utility of the IT trends of convergence for U-healthcare.

5.6 Promises, opportunities, and challenges

As repeatedly stated, the convergence of nanopore sequencing, the IoT, sensors, cloud computing, computational intelligence algorithms, and so on are moving us towards self-quantification of the genome sequence. As a result, detection of the

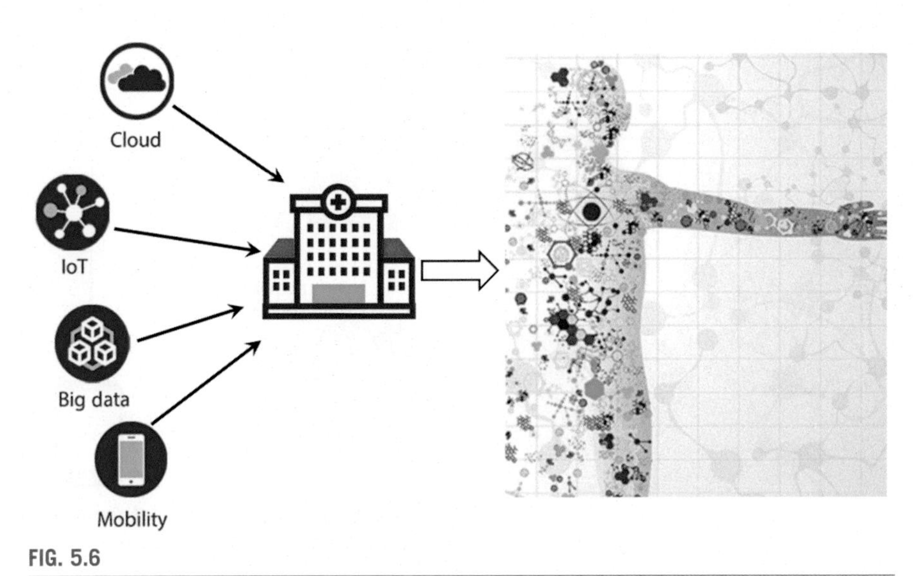

FIG. 5.6

The utility of the convergence in U-healthcare.

transmission paths and mutations of deadly viruses, lethal pathogens, engineered bacteria, and early-stage biomarkers of cancer tumors or viral agents may become possible to accomplish in-house by individuals in a nonlaboratory environment. Further, millions of individuals streaming these data to the cloud will build a powerful system for predictive, preventive, personalized, and participatory (P4) medicine. In a nutshell, genetic identity of any living things will make a new life on the Internet and the genetic world will enter into the age of IoLT. Despite these promises, the application of portable nanopore sequencers and the IoLT framework in U-healthcare has several opportunities and challenges, as described in the following sections.

5.6.1 ONT is expected to revolutionize healthcare

ONT's work is a result of decades of discoveries and convergences of interdisciplinary science that brought the power of genomics to a small smartphone-driven chip. The small size, low cost, and easy sample preparation protocol of the MinION and SmidgION devices are revolutionizing the healthcare sector, allowing nanopore technology applications in small clinics as well as for point-of-care gene sequencing to the onsite field. Some examples of success of ONT's devices are as follows:

(i) In 2016, MinION was deployed to study Zika virus outbreaks.
(ii) In 2017, MinION was used for a same-day diagnostic test of a tuberculosis patient and predicted drug-susceptibility and provided disease surveillance.
(iii) The portability, speed, and accuracy of ONT devices also transformed personalized medicine. For instance, tumor samples may be sequenced

routinely to discover key mutants and then targeted with treatment so that only cancerous cells may be killed, without harming healthy cells.

(iv) In 2017 MinION was used to detect epigenetic modifications, allowing both genomics and epigenomics analysis from a single sequence run.

5.6.2 Need for huge computing infrastructure

The application of portable ONT sequencers, biosensors, and computational intelligence algorithms requires huge computing infrastructure for real-time monitoring of human body fluids, genetic changes, and surveillance of viruses and pathogens. The idea is to connect genetics with the IoLT. Real-time data and environmental changes need to be streamed online to cloud servers for processing and analysis by healthcare providers so that they may make and healthcare recommendations to their patients. This shared data can also be made available to doctors and researchers for sequencing purposes.

5.6.3 IoLT and its convergences with ONT for U-healthcare

The IoLT can be easily incorporated into the healthcare industry. It has provided many motivational strategies to the medical fraternity in unleashing close-fisted and powerful healthcare schemes to the public. It allows the patient's medical information to be stored in compressed files irrespective of distance, aiding in its accession from any geographical region. Furthermore it also provides a common platform for consultants, patients, hospitals, local clinics, diagnostic labs, research laboratories, and many others to interact with one another in real time.

IoLT-based healthcare is focused on the patient's medical information exchange instead of drifting it from one end to a common centralized server. This new trend in healthcare is rolling eyeballs as it enables radiofrequency identification techniques (RFID) to patients, medical consultants, medical staff and nursing, apparatuses, and so on. It also expands to searching for healthcare practitioners, querying and retrieving information from existing healthcare systems. In an IoLT strategy, data is generated by the "thing," which is usually a nonliving thing, on living things in real time. This analogy has so far helped in extracting novel information for maintaining a patient's health status and in preventing that which are predicted by vigorous analyses [41, 69–72].

The IoLT approach combines living things (humans) with nonliving things (medical equipments, sensors, wearable body regulatory devices, etc.) in actual time by means of the virtual world. It also hints at its similarity with virtual reality-based tenets, and since it promises to provide "personalized healthcare," (person-centric healthcare), also assures IoLT-based healthcare is economical in initializing and regulating U-healthcare [41]. IoLT is admired as the best healthcare delivery, although it does have many limitations [72].

The IoLT has some barriers that limit its full potential in the medical fraternity. IoLT is more of a personalized and preventive way of healthcare rather than a traditional reactive healthcare. Its rise has revealed some serious Ethical, Legal, Social Issues (ELSI). The major concern of people today is their privacy and security, so why would anyone freely provide their private information to the internet without any assurance of their information's security? Furthermore standardization issues also exist in IoLT, which refers to the different standards of intercommunication and operability that may vary from location to location. An internet connection and sophisticated devices would be required to maintain integrity all around the globe, cutting short the irregularities in its potential. Another major problem is expandability, referring to expansion in both urban and rural areas. This is currently a big challenge because of lack of basic internet amenities in some developing countries' rural regions [41, 72].

5.7 Discussion

Recent breakthroughs in nanopore sequencing technology, such as the MinION device by ONT, are expected to revolutionize the healthcare sector. The portable, low-cost device is a game changer in the sequencing market; it is expected that it will be used by many people and industries in the future. Other sequencers on the market include PromethION and GridION. These sequencers not only sequence DNA and RNA, but they are also being used to build an IoLT. Sequencing devices can be miniaturized and sensors placed on the body to monitor human health and vital signs. Advancement in sensor technologies have given rise to DNA-reading sensors for almost everything, from food production equipment to water bodies to farms. Under the IoT framework, sequencing devices can be connected to other technical systems such as mobile phones, hospitals, airports, insurance companies, and so on. As a result, scientists, doctors, patients, employees–everyone–will be able to monitor their own DNA on shared cloud computing labs. In addition, equipped DNA-reading sensors can identify the nature, transmission paths, and mutations of deadly viruses, lethal pathogens, and engineered bacteria, and alert to outbreaks at a very early stage. Further, they can be used by individuals to detect the most vital biomarkers of early-stage cancers or viral agents.

U-healthcare is a combination of electronic and mobile healthcare, with person-centric therapy and treatment. All thanks go to the Internet, which revolutionized healthcare by making it mobile, creating a new field of study called *tele-based healthcare systems* [73]. Nanopore sequencing techniques have been appreciated and accepted worldwide. Some of the applications of this technology in U-healthcare include real-time monitoring of body fluids, virus control and surveillance, molecular level understanding of disease mechanisms, wastewater management, and inferring evolutionary relationships, among others.

The convergence of ONT's sequencers with the IoLT, cloud computation, and big data marks the beginning of a new path for individuals to analyze and sequence their own biological composition on their mobile devices. This magnanimous combination of the trending technologies is constructive in developing a U-healthcare milieu. In the U-healthcare industry, cloud computation plays a key role in providing three kinds of web-based services, as previously discussed (SaaS, PaaS, and IaaS) There have been several success stories of ONT applications in healthcare and these are expected to grow further with the hope of several new discoveries and wider applications at the individual level. Despite several promises, ONT and IoLT applications in U-healthcare do have some challenges. Further research is needed to address these challenges.

5.8 Conclusion

Nanopore sequencing technology is expected to revolutionize the healthcare sectors and may prove to be a game changer in the sequencing market. The technology can be used in modern sequencers as well as for creating miniaturized versions that work within the IoT. Under the IoT model, sequencers can be easily connected to other technical systems, which may help scientists, doctors, and patients monitor their DNA on shared cloud computing labs. Sequencers can also be used to predict outbreaks of viral and deadly viruses, pathogens, and engineered bacteria. The convergence of sequencers with the IoT, cloud computation, and big data brings a new avenue in the field of sequencing and big data analytics. This combination of trending technologies is constructive in developing a U-healthcare milieu.

References

[1] M. Eisenstein, Oxford Nanopore announcement sets sequencing sector abuzz, Nat. Biotechnol. 30 (4) (2012) 295–296.

[2] M. Jain, H.E. Olsen, et al., The Oxford Nanopore MinION: delivery of nanopore sequencing to the genomics community, Genome Biol. 17 (1) (2016) 239.

[3] H. Lu, F. Giordano, Z. Ning, Oxford Nanopore MinION sequencing and genome assembly, Genom. Proteom. Bioinform. 14 (5) (2016) 265–279.

[4] Y.I.N. Yuehong, Y. Zeng, X. Chen, Y. Fan, The internet of things in healthcare: an overview, J. Ind. Inf. Integr. 1 (2016) 3–13.

[5] D.V. Dimitrov, Medical Internet of things and big data in healthcare, Healthc. Inform. Res. 22 (3) (2016) 156–163.

[6] K. Raza, S. Ahmad, Recent Advancement in Next Generation Sequencing Techniques and Its Computational Analysis, 2016, arXiv preprint arXiv: 1606.05254.

[7] M. Jain, S. Koren, K.H. Miga, J. Quick, A.C. Rand, T.A. Sasani, et al., Nanopore sequencing and assembly of a human genome with ultra-long reads, Nat. Biotechnol. 36 (4) (2018) 338.

[8] Y. Feng, Y. Zhang, C. Ying, D. Wang, C. Du, Nanopore-based fourth-generation DNA sequencing technology, Genom. Proteom. Bioinform. 13 (1) (2015) 4–16.

[9] F. Touati, R. Tabish, U-healthcare system: state-of-the-art review and challenges, J. Med. Syst. 37 (3) (2013) 9949.

[10] Nanopore Sequencing Offers Advantages in All Areas of Research, https://nanoporetech.com/applications. (Accessed 28 December 2018).

[11] S.L. Butt, T.L. Taylor, et al., Rapid and sensitive virulence prediction and identification of Newcastle disease virus genotypes using third-generation sequencing, bioRxiv (2018) https://doi.org/10.1101/349159.

[12] A.A. Votintseva, P. Bradley, et al., Same-day diagnostic and surveillance data for tuberculosis via whole-genome sequencing of direct respiratory samples, J. Clin. Microbiol. 55 (5) (2017) 1285–1298.

[13] R. Marszalek, Real-Time Genomic Surveillance With Nanopore-Seq. On Biology, Blog Network, BMC, June 1, 2015, http://blogs.biomedcentral.com/on-biology/2015/06/01/. (Accessed 28 December 2018).

[14] L. Gege, Nanopore Sensors Make Breakthrough in Monitoring Health and Disease, Chemistry World, Royal Society of Chemistry, 2018. October 17, https://www.chemistryworld.com/news/. (Accessed 28 December 2018).

[15] S. Juul, F. Izquierdo, et al., What's in My Pot? Real-Time Species Identification on the MinION, 2015, https://doi.org/10.1101/030742.

[16] C. Chiu, Nanopore Sequencing for Metagenomic Diagnosis of Infectious Diseases. Resource Centre, Oxford Nanopore Technologies, May 26, https://nanoporetech.com/resource-centre/, 2016. (Accessed 28 December 2018).

[17] E. Ghedin, Getting the Flu: Exploring Influenza Virus Evolutionary Dynamics by Single Molecule Sequencing. Resource Centre, Oxford Nanopore Technologies, December 3, https://nanoporetech.com/resource-centre/, 2015. (Accessed 28 December 2018).

[18] O.O.Y. Hu, N. Ndegwa, et al., Stationary and portable sequencing-based approaches for tracing wastewater contamination in urban stormwater systems, Sci. Rep. 8 (2018) 11907.

[19] M.H. Andersen, S.J. McIlroy, et al., Genomic insights into *Candidatus Amarolinea aalborgensis* gen. nov., sp. nov., associated with settleability problems in wastewater treatment plants, Syst. Appl. Microbiol. (2018). pii: S0723-2020(18)30235-2.

[20] W.G. Feero, A.E. Guttmacher, F.S. Collins, Genomic medicine—an updated primer, N. Engl. J. Med. 362 (21) (2010) 2001–2011.

[21] D. Branton, D.W. Deamer, et al., The potential and challenges of nanopore sequencing, Nat. Biotechnol. 26 (10) (2008) 1146–1153.

[22] A. Pollack, Company unveils DNA sequencing device meant to be portable, disposable and cheap, The New York Times (2012). February 17.

[23] J.F. Thompson, P.M. Milos, The properties and applications of single-molecule DNA sequencing, Genome Biol. 12 (2) (2012) 217.

[24] D.W. Deamer, D. Branton, Characterization of nucleic acids by nanopore analysis, Acc. Chem. Res. 35 (10) (2002) 817–825.

[25] C. Dekker, Solid-state nanopores, Nat. Nanotechnol. 2 (4) (2007) 209–215.

[26] O.K. Dudko, J. Mathe′, A. Meller, Nanopore force spectroscopy tools for analyzing single biomolecular complexes, Methods Enzymol. 475 (2010) 565–589.

[27] M. Akeson, D. Branton, et al., Microsecond time-scale discrimination among polycytidylic acid, polyadenylic acid, and polyuridylic acid as homopolymers or as segments within single RNA molecules, Biophys. J. 77 (6) (1999) 3227–3233.

[28] G.M. Skinner, M. van den Hout, et al., Distinguishing single- and double-stranded nucleic acid molecules using solid-state nanopores, NanoLetters 9 (8) (2009) 2953–2960.

[29] B. Hornblower, A. Coombs, et al., Single-molecule analysis of DNA-protein complexes using nanopores, Nat. Methods 4 (4) (2007) 315–317.

[30] R.M. Smeets, S.W. Kowalczyk, et al., Translocation of RecA-coated double-stranded DNA through solid-state nanopores, Nano Lett. 9 (9) (2009) 3089–3096.

[31] J. Rosenstein, The Promise of Nanopore Technology, July/August 2014 Issue, 2014.
[32] R.D. Maitra, J. Kim, W.B. Dunbar, Recent advances in nanopore sequencing, Electrophoresis (23) (2012) 3418–3428.
[33] NASA, Next SpaceX Commercial Cargo Launch Now No Earlier Than July 18, US Media Accreditation Remains Open, Media Advisory M16-073, NASA, June 22, 2016.
[34] Sequencing DNA in Space, Press Release: NASA/SpaceRe, http://spaceref.com/nasa-hack-space/sequencing-dna-in-space.html. (Accessed 2 January 2019).
[35] First DNA Sequencing in Space a Game Changer, Space Station, NASA, https://www.nasa.gov/mission_pages/station/research/news/dna_sequencing. (Accessed 3 January 2019).
[36] J. Quick, N.J. Loman, et al., Real-time, portable genome sequencing for Ebola surveillance, Nature 530 (7589) (2016) 228–232.
[37] J.R. Tyson, N.J. O'Neil, et al., Whole genome sequencing and assembly of a *Caenorhabditis elegans* genome with complex genomic rearrangements using the MinION sequencing device, bioRxiv (2017), https://doi.org/10.1101/099143.
[38] S. Gigantea, Picopore: a tool for reducing the storage size of Oxford Nanopore Technologies datasets without loss of functionality, F1000Res 6 (2017) 277.
[39] G. Elhayatmy, N. Dey, A.S. Ashour, Internet of things based wireless body area network in healthcare, in: Internet of Things and Big Data Analytics Toward Next-Generation Intelligence, Springer, Cham, 2018, , pp. 3–20.
[40] N. Dey, A.S. Ashour, C. Bhatt, Internet of things driven connected healthcare, in: Internet of Things and Big Data Technologies for Next Generation Healthcare, Springer, Cham, 2017, , pp. 3–12.
[41] S.M.R. Islam, D. Kwak, et al., The Internet of things for health care: a comprehensive survey, IEEE Access 3 (2015) 678–708.
[42] J. Höller, V. Tsiatsis, et al., From Machine-to-Machine to the Internet of Things: Introduction to a New Age of Intelligence, Elsevier, Amsterdam, The Netherlands, 2014.
[43] G. Kortuem, D. Kawsar, et al., Smart objects as building blocks for the Internet of Things, IEEE Internet Comput. 14 (1) (2010) 44–51.
[44] K. Romer, B. Ostermaier, et al., Real-time search for real-world entities: a survey, Proc. IEEE 98 (11) (2010) 1887–1902.
[45] D. Guinard, V. Trifa, E. Wilde, A resource oriented architecture for the Web of Things, Proc. Internet Things (IOT) (2010) 1–8.
[46] L. Tan, N. Wang, Future Internet: the Internet of things, in: Proc. 3rd Int. Conf. Adv. Comput. Theory Eng. (ICACTE), vol. 5, 2010. pp. V5-376–V5-380.
[47] M.S. Shahamabadi, B.B.M. Ali, et al., A network mobility solution based on 6LoWPAN hospital wireless sensor network, in: Proc. 7th Int. Conf. Innov. Mobile Internet Services Ubiquitous Comput. (IMIS), 2013, , pp. 433–438.
[48] Guan ZJ. (2013). Somatic data blood glucose collection transmission device for Internet of Things. Chinese Patent 202 838 653 U.
[49] R.H.S. Istepanian, S. Hu, et al., The potential of Internet of m-health Things 'm-IoT' for non-invasive glucose level sensing, in: Proc. IEEE Annu. Int. Conf. Eng. Med. Biol. Soc. (EMBC), 2011, , pp. 5264–5266.
[50] A.J. Jara, M.A. Zamora-Izquierdo, A.F. Skarmeta, Interconnection framework for mHealth and remote monitoring based on the Internet of Things, IEEE J. Sel. Areas Commun. 31 (9) (2013) 47–65.
[51] G. Yang, et al., A health-IoT platform based on the integration of intelligent packaging, unobtrusive bio-sensor, and intelligent medicine box, IEEE Trans. Ind. Informat. 10 (4) (2014) 2180–2191.

[52] Xin TJ, Min B & Jie J. (2013). Carry-on blood pressure/pulse rate/blood oxygen monitoring location intelligent terminal based on Internet of Things. Chinese Patent 202 875 315 U.
[53] L. You, C. Liu, S. Tong, Community medical network (CMN): architecture and implementation, in: Proc. Global Mobile Congr. (GMC), 2011, , pp. 1–6.
[54] W. Wang, J. Li, et al., The Internet of Things for resident health information service platform research, in: Proc. IET Int. Conf. Commun. Technol. Appl. (ICCTA), 2011, , pp. 631–635.
[55] Liu ML, Tao L & Yan Z. (2012). Internet of Things-based electrocardiogram monitoring system. Chinese Patent 102 764 118 A.
[56] W.Y. Chung, Y.D. Lee, S.J. Jung, A wireless sensor network compatible wearable U-healthcare monitoring system using integrated ECG, accelerometer and SpO2, in: Proc. 30th Annu. Int. Conf. IEEE Eng. Med. Biol. Soc. (EMBS), 2008, , pp. 1529–1532.
[57] Z. Pang, J. Tian, Q. Chen, Intelligent packaging and intelligent medicine box for medication management towards the Internet-of-Things, in: Proc. 16th Int. Conf. Adv. Commun. Technol. (ICACT), 2014, , pp. 352–360.
[58] R. Tabish, et al., A 3G/WiFi-enabled 6LoWPAN-based U-healthcare system for ubiquitous real-time monitoring and data logging, in: Proc. Middle East Conf. Biomed. Eng. (MECBME), 2014, , pp. 277–280.
[59] https://tampax.com/en-us/period-tracker. (Accessed 9 January 2019).
[60] X.M. Zhang, N. Zhang, An open, secure and flexible platform based on Internet of Things and cloud computing for ambient aiding living and telemedicine, in: Proc. Int. Conf. Comput. Manage. (CAMAN), 2011, , pp. 1–4.
[61] https://www.skinvision.com/. (Accessed 9 January 2019).
[62] https://www.wired.co.uk/article/clive-brown-oxford-nanopore-technologies-wired-health-2015. (Accessed 12 January 2019).
[63] https://metrichor.com/. (Accessed 12 January 2019).
[64] S.Y. Ko, L. Sassoubre, J. Zola, Applications and Challenges of Real-Time Mobile DNA Analysis, (2017). arXiv: 1711.07370v1.
[65] R.E. Herzlinger, Why innovation in health care is so hard, Harv. Bus. Rev. 84 (5) (2006) 58–66, 156.
[66] M. Quwaider, Y. Jararweh, Multi-tier cloud infrastructure support for reliable global health awareness system, Simul. Model. Pract. Theory 67 (2016) 44–58.
[67] M. Paul, A. Das, Provisioning of Healthcare Service in Cloud. Information and Communication Technology, Springer, Singapore, 2018, pp. 259–268.
[68] A.M. Kuo, Opportunities and challenges of cloud computing to improve health care services, J. Med. Internet Res. 13 (3) (2011) e67.
[69] J.A. Fisher, T. Monahan, Tracking the social dimensions of RFID systems in hospitals, Int. J. Med. Inform. 77 (2008) 176–183.
[70] https://ec.europa.eu/digital-single-market/. (Accessed 15 January 2019).
[71] Gartner, Market Share and Forecast: Radio Frequency Identification, Worldwide, 2004–2010 (Executive Summary), 2005.
[72] C.E. Turcu, C.O. Turcu, Internet of things as key enabler for sustainable healthcare delivery, in: The 2nd International Conference on Integrated Information. Procedia—Social and Behavioral Sciences, vol. 73, 2013, , pp. 251–256.
[73] S. Qazi, K. Tanveer, K. ElBahnasy, K. Raza, From Telediagnosis to Teletreatment: The Role of Computational Biology and Bioinformatics in Tele-based Healthcare. Telemedicine Technologies, Elsevier, 2019.

CHAPTER

Internet of things-enabled virtual environment for U-health monitoring

6

Vinay Chowdary*, Vivek Kaundal*, Amit Kumar Mondal*, Vindhya Devella†, Abhishek Sharma‡

Dept. of Electrical & Electronics, UPES, Dehradun, India Dept. of Aerospace, UPES, Dehradun, India† Research Scientist, Department of R&D, UPES, Dehradun, India‡*

Technological advancements in the Internet of things (IoT) have allowed for the ubiquitous monitoring of health [1, 2]. Remote monitoring of an individual's health is possible in today's world with a combination of wearable sensors [3] and a cloud platform where the data measured from the sensors are accessible around the clock [4]. The advancements in the IoT field have filled the gap between sensor data and data networks. There are many open-source cloud takeaways that can be used as IoT platforms for remote health monitoring. Message Queuing Telemetry Transport (MQTT), ThingSpeak, and Thinger.io are but a few IoT platforms that can be used as data networks for uploading sensor data. Time stamping of data to be uploaded allows real-time tracking of the health of any individual from anywhere in the world. This chapter is dedicated to the introduction of such IoT platforms, their technical insight, and real-time implementation. We focus on MQTT platforms using Raspberry Pi (RPi) microcontrollers and provide all the necessary insights required to bridge the gap between sensor data and data networks using MQTT.

6.1 Understanding IoT protocols

An IoT protocol facilitates the movement of data between four different types of connections:

(i) device to device
(ii) device to gateway
(iii) gateway to cloud
(iv) between clouds

An intelligent device is a sensing system that measures the monitoring data. In order to be considered "intelligent" the device needs to be connected to microcontrollers.

A gateway can be software or a physical device that acts as a bridge between the cloud, the controller, and the sensor. It is a pre-processing device used to minimize the data that will be forwarded to the cloud.

Sensors for Health Monitoring. https://doi.org/10.1016/B978-0-12-819361-7.00006-3

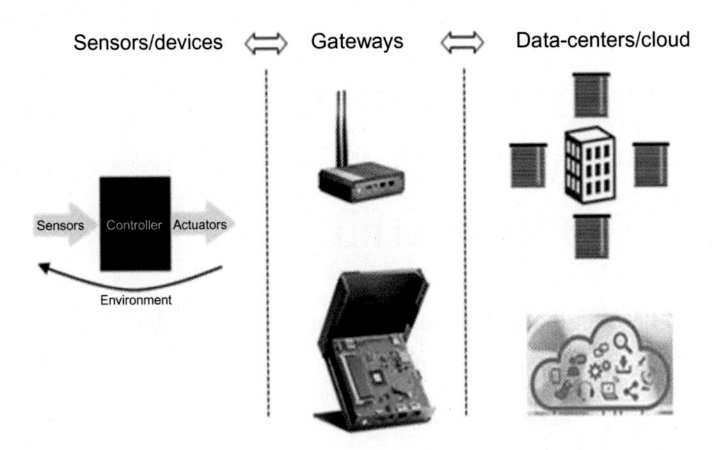

FIG. 6.1

IoT architecture.

A cloud is the platform for storing, processing, and forwarding (optional) the data received from sensors through the gateway. At present, there are many open-source (free) and licensed IoT clouds available.

This chapter is dedicated to a discussion of the Message Queuing Telemetry Transport (MQTT) protocol. Fig. 6.1 shows the complete architecture of the IoT.

6.2 List of open-source and licensed IoT cloud platforms

A complete list of open-source and paid IoT clouds is beyond the scope of this chapter; however, we provide a list of the 10 most famous and widely used cloud platforms in Table 6.1.

Table 6.2 gives a comparison of the different IoT platforms with respect to parameters, protocols support, hardware support [5], programming language, license pricing, time taken to establish connection, and features available.

Table 6.1 List of open-source and licensed IoT clouds

S. No	Open source	Licensed
1	Message Queue Telemetry Transport (MQTT)	Amazon Web Services (AWS)
2	ThingSpeak	Microsoft Azure IoT hub
3	Zetta	IBM Watson IoT platform
4	Open Hybrid	Google Cloud platform
5	Node-Red	Oracle
6	Eclipse Kura	Salesforce
7	Wio Link	Bosch
8	Macchina.io	CISCO IoT Cloud Connect
9	IoTivity	General Electric's Predix
10	Kaa Project	SAP

Table 6.2 Comparison of IoT platforms

IoT platform	Protocols	Hardware support	Operating language	Pricing	Connection time	Features
Thethings.io	Websockets, CoAP, MQTT, HTTP	Hardware agnostic	Python, Node.js MQTT, Node.js HTTP, Node.js CoAP, Jailed Node	According to the number of devices and platform maintenance	Fast	User friendly, customizable, provides deep analytics
IBM Watson IoT	HTTP, MQTT	ARM, Texas Instruments, Raspberry Pi, Arduino Uno	C#, C Python, Java, NodeJS	According to the number of devices, data traffic, and data storage	Moderate	Secure connectivity, information management, risk management
Microsoft Azure	HTTP, AMQP, MQTT	Intel, Raspberry Pi2, Freescale, Texas Instruments	.Net and UWP, Java, C, NodeJS	According to the number of devices and messages per day	Fast	High free storage, user friendly
AWS	HTTP, MQTT, WebSockets	Broadcom, Marvell, Renesas, Texas Instruments, Microchip, Intel	Java, C, NodeJS	Paying for message traffic	Fast	API-based gateway, long-term storage
Google cloud platform	HTTP, MQTT	Freescale, Microchip, Intel, Android	Java, Python, NodeJS, Ruby, Go, .Net, PHP	According to the number of functions requested by the user	Moderate	User friendly, automatic scaling
Artik	MQTT, Thread, CoAP	Texas Instruments, Raspberry Pi2	C, C++, NodeJs	Based on data storage	Fast	Supports local area wireless standards

In the following sections, we provide a detailed architecture of MQTT as well as a discussion of its implementation with RPi.

6.3 Introduction to MQTT

MQTT is a lightweight, open-source, client-to-client or client-to-server messaging protocol. It is used in Machine-to-Machine (M2M), IoT-constrained environments, which require small source code and where communication bandwidth is limited. MQTT was created to implement the IoT concept with minimum battery requirements and limited use of available bandwidth. Andy Stanford-Clark (IBM) and Arlen Nipper invented the MQTT protocol in 1999 (https:/www.hivemq.com/blog/mqtt-essentials-part-1-introducing-mqtt/) to connect oil pipelines with satellites.

The architecture of MQTT is based on the concepts of "publish" and "subscribe." Publish refers to the transmitter and subscribe refers to the receiver. The message or data to be transmitted is called the "topic," and clients can publish/subscribe to a particular topic. MQTT is different from the traditional client-server model in which clients generally communicate with end points. In the MQTT protocol the publish/subscribe (pub/sub) model is used to separate the clients.

A clients that sends messages is called a *publisher*.

A client that receives messages is called a *subscriber*.

The subject to which the clients publishes the message or subscribes to is called the *topic*.

The third-party gateway (IoT gateway) that handles the connection between the clients is called the *broker*.

Fig. 6.2 explains the architecture of MQTT and is a modification of Fig. 6.1 where sensors communicate with the IoT gateway through the pub/sub model.

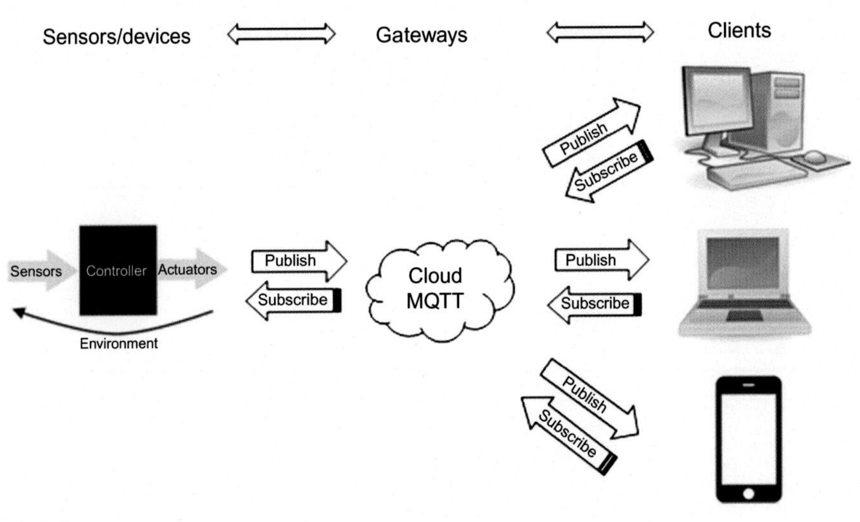

FIG. 6.2

MQTT architecture.

The IoT gateway is the MQTT broker. Clients replace the data centers of Fig. 6.1, where a client can be a mobile phone, a laptop, or a PC. These clients also communicate with the IoT gateway through the pub/sub model. The work of the MQTT broker is to channel the incoming messages, filter them, and convey them to the intended clients. Messages will be delivered only to those clients who have subscribed to the topic, which is same as the topic of the publishers. For example, if the controller connected to the sensor on the left-hand side of Fig. 6.2 publishes data to a topic called "temp," the data/message with this topic name will be sent to the IoT broker, which in this example is the MQTT cloud. The clients on the right-hand side of Fig. 6.2 will receive all the messages from the sensor if and only if they subscribe to the same topic (i.e., "temp"). The reverse is also true; if the client publishes data to a topic called "var," the controller can receive all messages from the clients if and only if it subscribes to the same topic (i.e., "var"). A client can publish and subscribe to more than one topic at the same time. For example, if the controller has temperature, humidity, and CO_2 sensors connected to it and it publishes data from all three sensors (e.g., "temp," "hum," "co"), the client can subscribe their mobile phone to receive all "temp" messages, their laptop to receive all "hum" messages, and their PC to receive all "co" messages.

6.4 MQTT properties

6.4.1 Separation

The clients and sensors are arranged around the MQTT broker in a *star topology*. This means that every sensor and every client can have direct communication between them through the MQTT broker.

The foremost critical viewpoint of the pub/sub model is the separation of messages. The publisher need not be aware of the presence and place of the subscriber, as the broker handles the connection between them. The separation is done at three levels:

(i) *Logical separation:* Publisher and subscriber need not be aware of the presence of each other. Logical separation prevents the exchange of IP addresses and port numbers between the publisher and subscriber.

(ii) *Time separation:* Run-time of publisher and subscriber need not be same.

(iii) *Synchronization separation:* Publisher and subscriber need not be interrupted during their operation.

6.4.2 Scalability

The MQTT pub/sub model scales way better than the conventional client-server model. This is because operations on the MQTT broker are usually parallelized and an event-driven method can be adopted to process the messages. Caching of messages and shrewd routing of messages are frequently unequivocal factors for progressing scalability. In any case, scaling up to millions of devices and managing their connections is a challenge; the same can be achieved with clustering. In a clustered broker, load balancers are used to distribute the data load over individual servers. In a free-to-use MQTT cloud, a total of five brokers can be added.

6.4.3 Message filtering

In the pub-sub process as the broker holds a very essential role, so is the role of broker in filtering the messages when more number of clients are connected. Filtering can be done in three ways:

(i) *Topic-based:* Clients subscribe only to topics of interest and the brokers ensure that clients receive only those messages related to the topics they subscribed to.
(ii) *Language-/content-based:* Brokers perform filtering based on language. Clients subscribe only to topics of interest and reject others. A drawback of this method is that the content of messages must be known in advance, which means messages cannot be scrambled or changed.
(iii) *Event-based:* With object-based programming languages, filtering based on class or type (i.e., event) can be used. Here a client listens to all such messages that are of exception type.

6.4.4 Security

As IoT is the hub for all the devices that can connect to the Internet, security threats inevitably arise. The MQTT protocol always provides a username and password to its clients. Any client that wants to connect to a MQTT broker needs to use this username and password for establishing a secure connection. Although the username and password are sent across the cloud in plain text, a Transport Layer Security (TLS) can be used to authenticate clients. Authentication verifies whether a client is really the same as claimed or if an unauthorized client is sending the messages.

In the MQTT protocol, security is implemented in layers using a Virtual Private Network (VPN) connection. A VPN is a physically secure network that can be used by clients as well as brokers. This is reasonable for gateway scenarios where it is connected to sensors/devices on one side and to the gateway on the other. TLS is used for the purpose of encryption [6]. TLS/SSL (Secure Sockets Layer) provides the authentication to provide authentication on both the client side and gateway side. At the application layer, username and password provides identity of clients. The username and password provided acts as client identifier at application layer.

6.5 IoT framework for U-health monitoring

IoT plays an important part in healthcare applications [7]. It can be used to monitor chronic diseases as well as to predict illness. Devices or sensors equipped with IoT features can be used for wellbeing monitoring and emergency alert notification systems. The main aim of IoT in U-health monitoring [8] is to supply a means to access and control gadgets, resources, and facilities. This section discusses the design and implementation of such a framework. Here the IoT framework is designed using the MQTT cloud along and sensors. The sensors can be interfaced either to an RPi microcontroller or a NodeMCU microcontroller. We present the design steps required in both cases, starting with IoT with RPi for U-health monitoring.

6.5.1 MQTT-based model of U-healthcare system

A ubiquitous healthcare system is a developing innovation that guarantees productivity, precision, and accessibility to therapeutic treatment. An IoT-based U-healthcare framework provides an accessible and convenient healthcare benefit to both patients and caregivers. This helps both in diagnosing the wellbeing of a patient. Individuals can screen their wellbeing without needing to visit a healing center or clinic. Fig. 6.3 represents an example of an IoT-based U-healthcare system. The system is divided into the following:

- Body area network (BAN)
  - wearable sensors for health monitoring
  - placement of sensors
  - controller
- MQTT server
- Clients in a hospital system

6.5.2 Body area network

Technological advancements have led to the miniaturization of portable devices that can be worn by individuals. A Body Sensor Network (BSN) consists of a group of sensors connected to a monitoring device (e.g., cell phone) that monitors physiological parameters [9]; a BSN is also called a Personal Area Network (PAN). A PAN can be converted to a Body Area Network (BAN) so as to include communication. Similarly a Wireless Body Area Network (WBAN) is a special case of Wireless Sensor Networks (WSNs), where sensors are meant only for health monitoring. In a WBAN, integration of different WSNs allows remote monitoring of individual health conditions. A BAN is built from integration of sensors. The sensors monitor things like blood pressure, pulse oximetry, heartbeat, and steps. All these sensors should be of wearable form so as to fit the body of the individual wearing them. Capabilities of a BAN include the following:

- Ubiquitous monitoring of human physiological parameters: The physiological information collected by sensor systems may be put away for a long period of time, and can be utilized for restorative examinations when required. This makes the use of cloud services along with IoT a mandate. The data that will be stored for a long time needs to be time stamped so that a progressive analysis of patient health can be easily carried out. In expansion, the introduced sensors can screen and identify the behavior of elderly individuals and generate an alert in the form of an alarm in case of any emergencies.

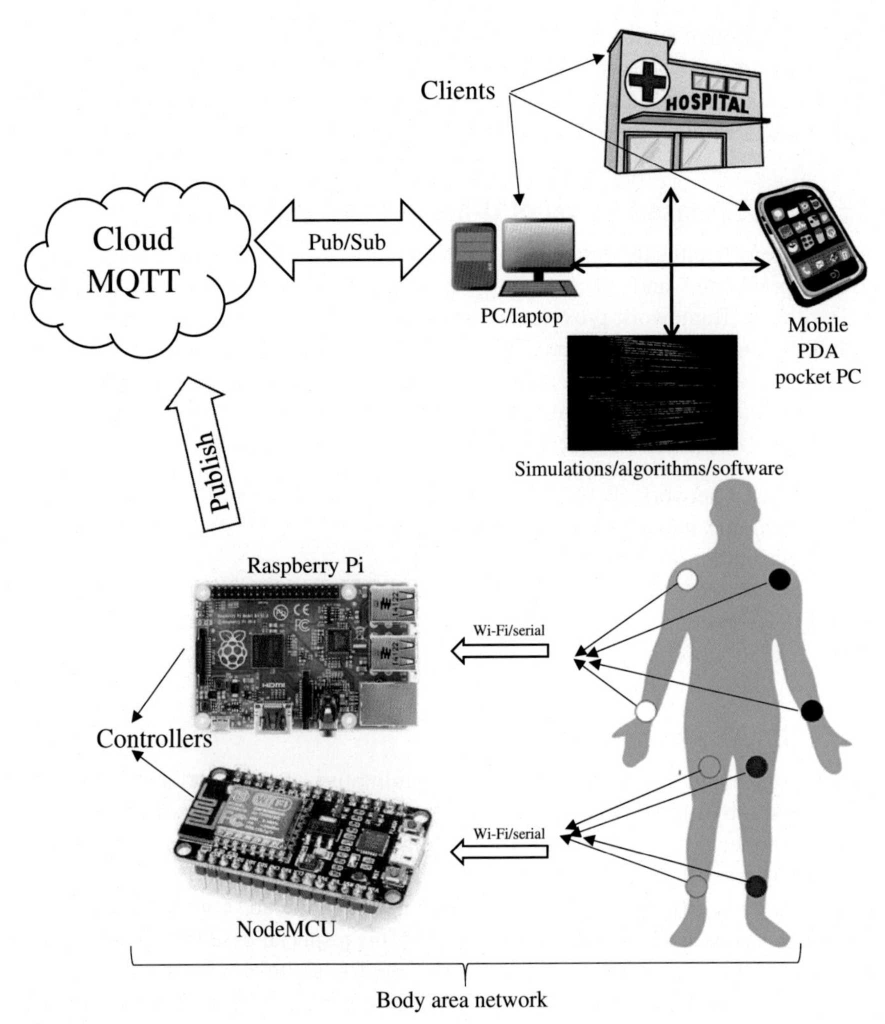

FIG. 6.3

MQTT-based model for U-health monitoring.

All these activities require an intelligent controller with adequate input and output ports. As shown in Fig. 6.3 the controllers used are RPi and NodeMCU.

- Monitoring of patients inside a hospital: Hospital patients can be equipped with a group of sensors that vary based on the sensing parameters. Each sensor in the group will perform a dedicated task and will communicate directly with the controller. For instance, one sensor may be monitoring heartbeat, while the other sensor may be detecting blood level. The location of each patient inside a hospital, at any place other than that hospital, can be found by using RSSI-based localization techniques [10]. The same can be implemented at the doctor's side, which helps the patient to locate doctors and doctors to locate other doctors.

- Drug organization and administration: In situations where a sensor can be attached to pharmaceuticals, the cases of getting and endorsing the off-base medication to patients can be minimized. Hence, patients will have sensor nodes that distinguish their hypersensitivities and required pharmaceuticals. Computerized frameworks [11] have appeared that can offer assistance in minimizing the side effects of drugs.

6.5.3 Wearable sensors for health monitoring

A sensor, in general, converts stimuli at its input to an electrical signal at its output. These electrical signals can be used by healthcare professionals for analyzing the health condition of any individual irrespective of age and gender. Sensors in healthcare play a vital role in increasing the insights of equipment used in the medical field and can empower remote observation of crucial signs and other wellbeing variables [12]. This section describes the types of sensors that can be used in healthcare applications. Table 6.3 gives a few examples.

The sensing of pressure sensors is generally based on piezoelectric and piezoresistive effects; they are used to measure blood pressure. They are also used in endoscopic tools in order to measure the value of pressure in the esophagus. Modern pressure sensors are also being used to detect respiratory disorders such as sleep apnea, asthma, and pulmonary diseases.

Temperature sensors are implemented in digital thermometers to measure a patient's body temperature. They are widely used in ventilators, medical incubators, organ transplant systems, and so on. The sensing mechanism of these sensors is based on pyroelectric and thermoelectric effects [7].

Table 6.3 Sensors used in healthcare

Sensor	Type of measurement	Ease of availability	Sensing based on
Pressure	Korotkoff sounds	Yes	Piezoresistive, piezoelectric
Temperature	Body thermometer, ear thermometer	Yes	Pyroelectric, thermosensitive, thermoelectric
Flow	SBF sensor, Doppler sonography	No	Calorimetric, piezoelectric
Image	Parabolic antenna, IR camera	Yes	Pyroelectric, EM waves
Accelerometer	Mechanical motion, gravity	Yes	Capacitive, piezoelectric
Biosensor	Bioreceptors	No	Antibody, antigens

A flow sensor is generally used to determine the rate of flow of blood. It is used for respiratory monitoring, ventilators, and electrosurgery. Calorimetric and piezoelectric effects are the sensing mechanisms for flow sensors.

Image sensors are used for measuring the surface body temperature, artificial retinas, dental imaging, and cardiology. Usually parabolic antenna and IR cameras are used as the measurement devices. The sensing of image sensors is based on frequency of EM waves and pyroelectric effects.

Accelerometer sensors can be used for blood pressure monitoring, pacemakers, and so on. An accelerometer sensor usually converts mechanical motion to electrical signals that are further processed to fetch the important information related to the patient. The sensing mechanism for accelerometer sensors are piezoelectric and capacitive effects.

Biosensors [13] are used to observe blood glucose levels with the help of biomarkers and bioreceptors. They have also been employed for detecting pregnancy, infectious disease, and drug abuse. In biosensors, antibody and antigen receptors are the basis of the sensing mechanism [14].

6.5.4 Placement of wearable sensors

Biosensors can sense signals such as electrocardiograms and electromyographs, and they can measure body temperature and blood pressure [15]. For example, accelerometers can be used to sense pulse rate, body movement, and movement of muscular activity [16].

The best place for wearable sensors on the human body generally depends on the parameter the sensor is monitoring. An ECG sensor, for example, is used to monitor electrical and muscular activities [17] related to the heart. This permits medical practitioners to interpret and understand the level of physiological excitement or arousal of the patient. It can also be utilized to understand a patient's mental state. Therefore this sensor is best placed at a location close to heart. For example, there are t-shirts that come with inbuilt ECG sensors [18] that are fitted on the upper part of the chest. In the same way, a sensor worn on the wrist is best suited to monitor blood pressure and heart rate. With respect to elderly people, wrist sensors [19] can also be used to detect falls. They can also monitor sleep activity [20]. Sensors placed on an ankle are used to monitor gait patterns. A gait [21] is a movement pattern of the lower body of humans or animals. A person with an abnormal gait pattern may be affected with aches and pains in their ankles or feet. Similarly, sensors placed below the waist are used to monitor the body's fat levels, as fat generally accumulates in the lower body. Fig. 6.4 shows the best possible placement of the aforementioned sensors.

6.5.5 Controllers for IoT-based health monitoring

Table 6.4 presents a comparison of Arduino, NodeMCU, and RPi controllers. Although Arduino is the cheapest and most widely available controller, we do not recommend it.

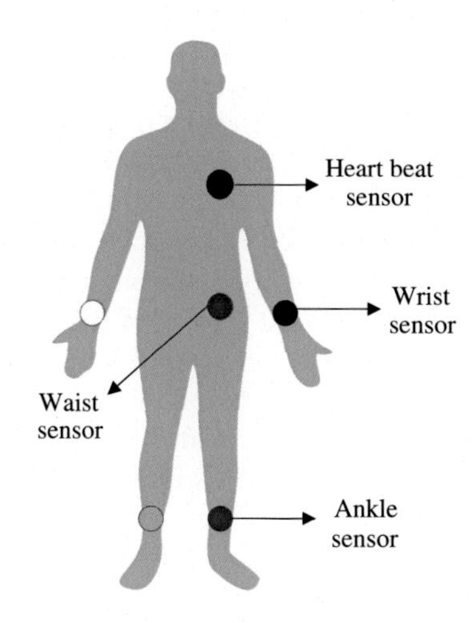

FIG. 6.4

Placement of sensors.

Table 6.4 Comparison of controllers for IoT applications in healthcare

Parameters	Raspberry Pi	Arduino	NodeMCU
Processor/controller	ARM-Cortex SoC	ATMega series	ESP8266 SoC
Memory	1GB	System memory—2 kB, Flash memory—32 kB, EEPROM—1 kB	System memory—45 kB, Flash memory—128 MB
USB ports	Four	One micro USB	One micro USB
Recommended programing language	Python	C/C++	C/C++
Need for Integrated Development Environment (IDE)	Not required	Need Arduino IDE	Need Arduino IDE
Operating system	Need Raspbian OS	Runs on any Windows/Linux OS	Runs on any Windows/Linux OS
Wi-Fi connectivity	On-board	Via external shield	On-board
Ethernet connectivity	On-board	Via external shield	Via external shield
Analog pins for analog sensors	None	More than one	Only one
Requirements for MQTT connection	Mosquito broker installation	Pub-Sub library installation	Pub-Sub library installation
Normal code size for MQTT connection	Minimum	Maximum	Average
Security	SSH security	N/A	N/A

6.5.5.1 Raspberry Pi

RPi is a system on chip (SoC) minicomputer the size of a credit card. The controller is based on Advanced RISC Machine Architecture (ARM), which also houses a Graphical Processor Unit (GPU). It has electrically erasable programmable read-only memory (EEPROM) and RAM on its board, making it a full-fledged, pocket-sized computer. It works on the Raspbian OS, which is a lightweight Linux operating system. Initially when it was released in 2006, the architecture was based on an 8-bit ATmega644 microcontroller. The latest version of RPi has a 32-bit ARM v6 architecture. Fig. 6.5 presents the complete architecture of a sample RPi. A detailed explanation of each block of the controller is beyond the scope of this chapter, however, one can refer to the manual/data sheet provided by the manufacturer.

The stepwise procedure to connect RPi with an MQTT server is as follows.

6.5.5.2 Setting up Raspberry Pi

- **(i)** Download the Raspbian operating system from www.raspberrypi.org/downloads/raspbian/.
- **(ii)** Install an image-burning software; we recommend Etcher, which is open source.
- **(iii)** Insert an SD card into your laptop or PC on which Etcher and Raspbian OS are downloaded, then insert the SD card and burn the image of Raspbian onto it.
- **(iv)** Open the SD card from the PC/laptop and create a new Secure Shell (SSH) text file without any contents; *do not use an extension.* The SSH file is required for RPi and guarantees its security.

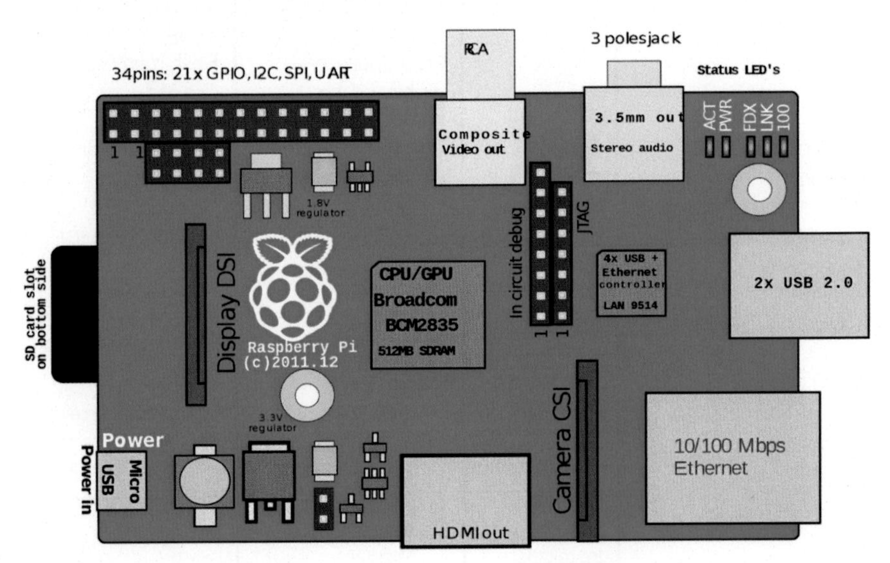

FIG. 6.5

Architecture of Raspberry Pi.

(v) Insert the SD card into the slot provided on the RPi board and power on the RPi. Make sure that all required devices (e.g., mouse) are connected. Also make sure the green light flashes for at least 10 s when booting up. If it does not flash, the RPi is not ready for programming.

(vi) Once the RPi is on, check that is has a secure Wi-Fi connection (use the dropdown menu at the top right of the screen).

(vii) Select the intended connection, provide the username and password for the Wi-Fi (if any), and connect the RPi to the Internet.

(viii) Note the IP address allotted to the RPi; this can be found by running a "sudo ifconfig" command. Alternatively, if the RPi is connected to a mobile hotspot then the IP address can be directly found in the hotspot section by selecting the connected devices.

(ix) Install PuTTY software, which is a serial console, terminal emulator application. It can also be used for file transfer. For RPi it will be used as serial console terminal. Once successfully installed, open PuTTY and provide the IP address of the RPi in the host name. Before establishing connection through PuTTY, make sure that the laptop/PC where PuTTY is installed is also connected to the same Wi-Fi as the RPi. This ensures that laptop/PC acts as local host for RPi and avoids IP looping issues (see Fig. 6.6).

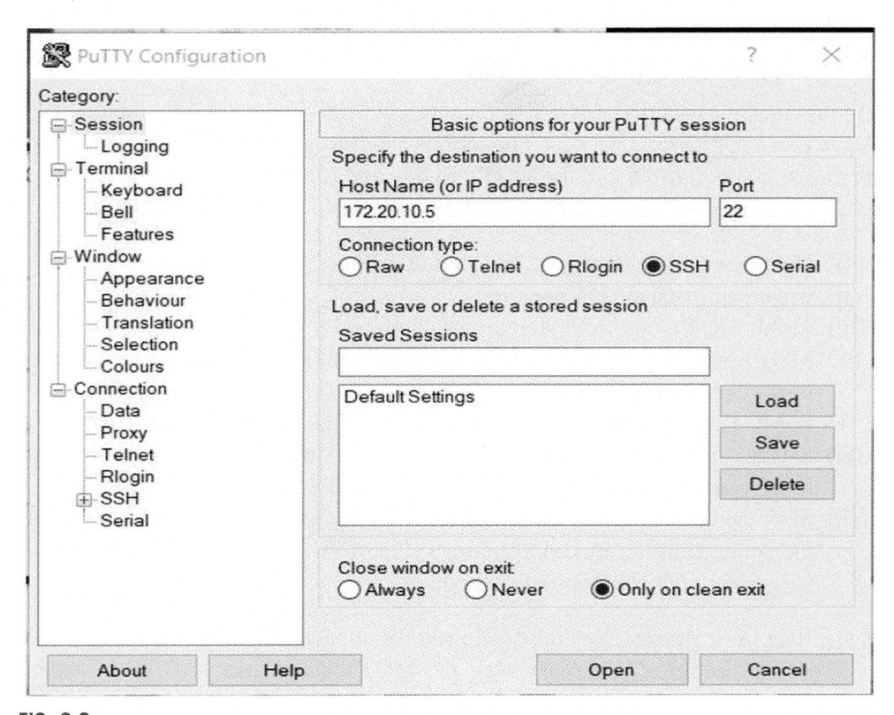

FIG. 6.6

Screenshot of RPi connection with PuTTY.

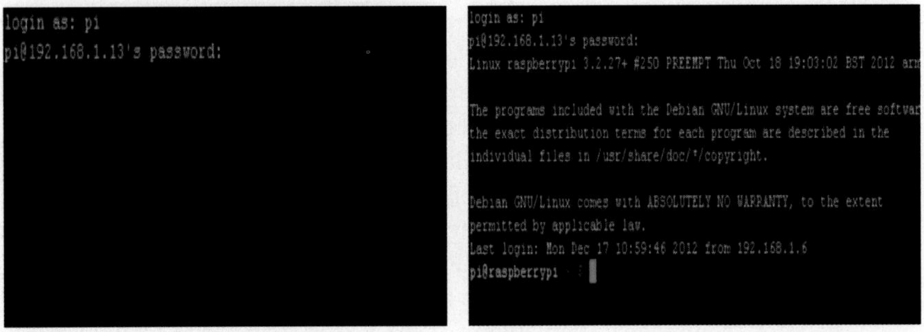

(A) (B)

FIG. 6.7

(A) Login to RPi from PuTTY. (B) Login successful.

(x) In place of the IP address shown in Fig. 6.6 (172.20.10.5), one can also type pi@172.20.10.5, where pi is the RPi username.

(xi) As soon as you click on open, if a connection is established, you will be asked for a password as shown in Fig. 6.7A. The password is raspberry and will connect RPi to PuTTY (Fig. 6.7B). RPi is now ready for programming.

6.5.6 Connecting RPi with MQTT

As previously explained, MQTT is a cloud platform. To use it on a laptop or PC, MQTTBox software is needed. For mobile platforms, the MQTT Dashboard app is available for Android and the MQTT Buddy app is available for Apple. The step-wise procedure for connecting MQTTBox with the cloud MQTT is as follows.

(i) Open www.cloudmqtt.com and sign up and create a profile.

(ii) Create an instance and open it.

(iii) Install MQTT box in PC/Laptop or download the app in the mobile.

(iv) MQTTBox can be downloaded from http:/workswithweb.com/html/mqttbox/installing_apps.html#install_on_windows.

(v) In MQTTBox click on create MQTT client and enter the client name.

(vi) Select mqtt/tcp as the protocol. The host name should be server name followed by a colon and port number (see Fig. 6.8).

(vii) Save the current configuration of MQTTBox. As soon as you click on save, the connection tab should turn green (Fig. 6.9), which is an indication of successful connection establishment.

Note: The same process has to be followed for connection in a mobile app.

To check communication between the MQTT cloud and MQTTBox one can publish/subscribe to topics using the following steps:

(i) In MQTTBox, give a topic name using the “topic to publish” option, for example, “test.”

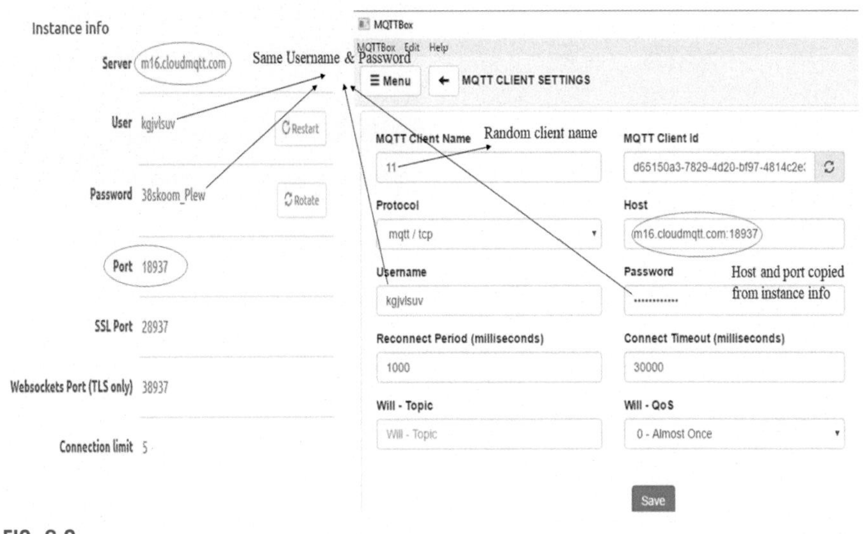

FIG. 6.8

Screenshot for configuration of MQTTBox and cloudmqtt.com.

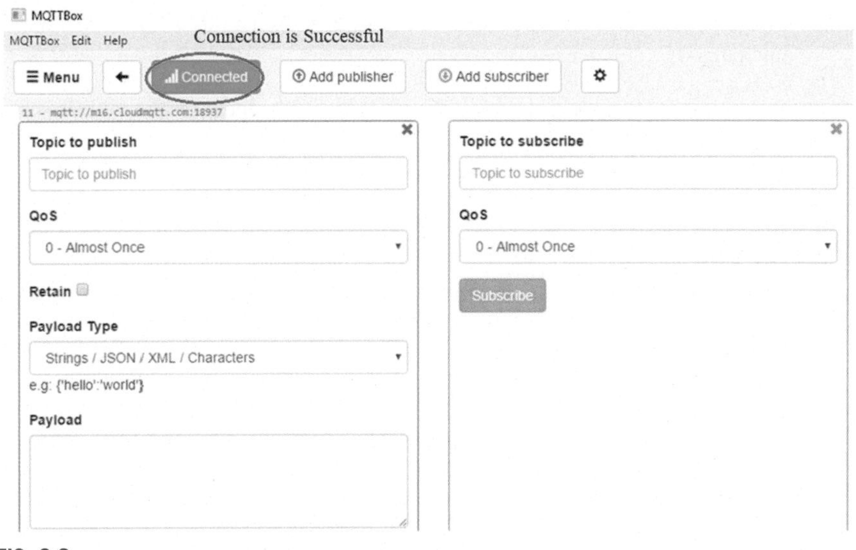

FIG. 6.9

Successful connection between MQTTBox and cloudmqtt.com.

(i) If using a mobile app (we used MQTT Dashboard for Android), follow steps (iv)–(vi) as given above.

(ii) In MQTT Dashboard open the session created, click on subscribe, enter the same topic name (i.e., "test"), and click create.

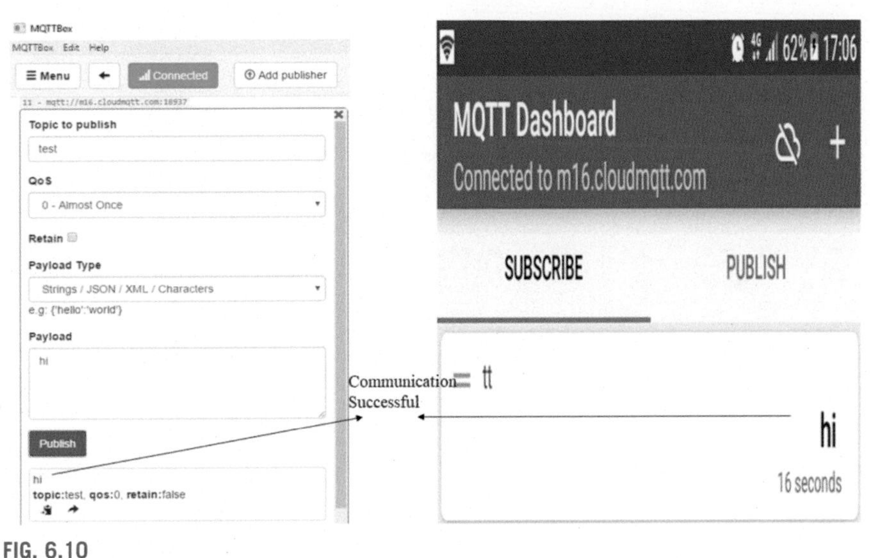

FIG. 6.10

Successful communication between MQTTBox and MQTT Dashboard.

(iii) In the payload option, write the message to be transmitted. Rest options unchanged.

Once the communication is successful, the message will appear on the mobile screen as shown in Fig. 6.10.

References

[1] G. Elhayatmy, N. Dey, et al., Internet of things based wireless body area network in healthcare, Springer, Internet of Things and Big Data Analytics Toward Next-Generation Intelligence, 2018, pp. 3–20.

[2] M. Haghi, K. Thurow, et al., Wearable devices in medical internet of things: scientific research and commercially available devices, Healthc. Inform. Res. 23 (1) (2017) 4–15.

[3] U. Anliker, J.A. Ward, et al., AMON: a wearable multiparameter medical monitoring and alert system, IEEE Trans. Inf. Technol. Biomed. 8 (4) (2004) 415–427.

[4] N. Dey, A.S. Ashour, et al., Developing residential wireless sensor networks for ECG healthcare monitoring, IEEE Trans. Consum. Electron. 63 (4) (2017) 442–449.

[5] M. Mukherjee, I. Adhikary, et al., A vision of IoT: applications, challenges, and opportunities with Dehradun perspective, in: Proceeding of International Conference on Intelligent Communication, Control and Devices, Springer, 2017.

[6] B. Aziz, On the security of the MQTT protocol, in: Engineering Secure Internet of Things Systems, Institution of Engineering and Technology, 2016.

[7] V. Verma, V. Chowdary, et al., IoT and robotics in healthcare, in: Medical Big Data and Internet of Medical Things, CRC Press, 2018, pp. 245–269.

[8] C. Bhatt, N. Dey, et al., Internet of Things and Big Data Technologies for Next Generation Healthcare, Springer, 2017.
[9] S. Asimakopoulos, G. Asimakopoulos, et al., Motivation and User Engagement in Fitness Tracking: Heuristics for Mobile Healthcare Wearables, Informatics, Multidisciplinary Digital Publishing Institute, 2017.
[10] V. Kaundal, P. Sharma, et al., Design, development and deployment of a RSSI based wireless network for post disaster management, Int. J. Eng. Technol. 7 (2.6) (2018) 6.
[11] N. Noury, T. Hervé, et al., Monitoring behavior in home using a smart fall sensor and position sensors, in: 1st Annual International, Conference on Microtechnologies in Medicine and Biology, 2000, IEEE, 2000.
[12] K. Kaewkannate, S. Kim, A comparison of wearable fitness devices, BMC Public Health 16 (1) (2016) 433.
[13] M.H. Sedaaghi, M. Khosravi, Morphological ECG signal preprocessing with more efficient baseline drift removal, in: ASC, Proceedings of the 7th. IASTED International Conference, 2003.
[14] M. Khosravy, M.R. Asharif, et al., Morphological adult and fetal ECG preprocessing: employing mediated morphology (医用画像). 電子情報通信学会技術研究報告, MI, 医用画像 107 (461) (2008) 363–369.
[15] V. Chowdary, V. Kaundal, et al., Implantable Electronics: Integration of Bio-Interfaces, Devices and Sensors. Medical Big Data and Internet of Medical Things, CRC Press, 2018, pp. 55–79.
[16] K.R. Evenson, M.M. Goto, et al., Systematic review of the validity and reliability of consumer-wearable activity trackers, Int. J. Behav. Nutr. Phys. Act. 12 (1) (2015) 159.
[17] K. Altun, B. Barshan, et al., Comparative study on classifying human activities with miniature inertial and magnetic sensors, Pattern Recogn. 43 (10) (2010) 3605–3620.
[18] T. Martin, E. Jovanov, et al., Issues in wearable computing for medical monitoring applications: a case study of a wearable ECG monitoring device, in: The Fourth International Symposium on Wearable Computers, IEEE, 2000.
[19] F. de Arriba-Pérez, M. Caeiro-Rodríguez, et al., Collection and processing of data from wrist wearable devices in heterogeneous and multiple-user scenarios, Sensors 16 (9) (2016) 1538.
[20] A. Henriksen, M.H. Mikalsen, et al., Using fitness trackers and smartwatches to measure physical activity in research: analysis of consumer wrist-worn wearables, J. Med. Internet Res. 20 (3) (2018).
[21] R. Takeda, S. Tadano, et al., Gait analysis using gravitational acceleration measured by wearable sensors, J. Biomech. 42 (3) (2009) 223–233.

CHAPTER 7

Health status from your body to the cloud: The behavioral relationship between IoT and classification techniques in abnormal situations

Md. Shahriar Hassan, Atiqur Rahman, Ahmed Wasif Reza
Department of Computer Science and Engineering, East West University, Dhaka, Bangladesh

7.1 Introduction

The Internet of things (IoT) refers to the interconnectivity of everyday devices and physical objects. For example, using your cell phone to control your air conditioner, riding in a smart car that suggests the shortest route to your destination, or wearing a smart watch to track your steps. We also know that IoT is a giant network of interconnected devices that gather and share data. The IoT uses many kinds of sensors in many types of devices, including mobile phones, electrical appliances, vehicles, medical instruments, and almost everything that an individual uses in their day-to-day life. The IoT is a scenario wherein some devices connect to a network via the Internet and extend computing capabilities to objects, sensors, and other items. This is especially important for healthcare and medical treatment. Nowadays many kinds of treatment are provided via computerized techniques. In recent years, much research has been done using IoT in the biomedical sectors. Clinical decision support systems (CDSS) are improving health and health care by raising knowledge and patient-specific information at appropriate times. Essentially there are two groups of CDSS, namely, knowledge-based CDSS and nonknowledge-based CDSS. Disease like heart disease, diabetes, cancer, retinopathy, and so on can be predicted by CDSS in biomedical science. Ubiquitous healthcare (U-healthcare) services are innovations that guarantee improved, precise, and accessible medical treatment. The U-healthcare insurance framework confers advantages to patients and caregivers, making it simpler to analyze and monitor a patient's wellbeing [1].

Usually medical professionals take some data or information from patients during physical examinations, surgical procedures, and medical tests. However, many times the important data necessary for predicting disease are not found in these contexts. Most of the time, medical decisions are based on doctors' perceptions and experiences. Hidden information may not be recognized thus preventing doctors from

Sensors for Health Monitoring. https://doi.org/10.1016/B978-0-12-819361-7.00007-5

diagnosing diseases at the early stage. Medical diagnosis is a very complex process. It can be made easier if a CDSS is used with machine learning algorithms such as J48 and K-nearest neighbors (k-NN). CDSS are used to improve health and health care by raising knowledge and patient-specific information at appropriate times.

Many kinds of research have been done using artificial intelligence, machine learning, and data-mining algorithms in many sectors including health care. Some researchers proposed a design for a healthcare system in the IoT using a many-network layer system and computer software. Their design is basically an IoT healthcare network (IoThNet) [2]. In their research [3], the authors introduced wearable sensors to monitor a person's physiological activity. They also highlighted that their hardware system, designed by Raspberry Pi, can track and send various health data like temperature, heartbeat, and fall detection to medical caregivers for measuring and monitoring.

Many types of wearable devices for IoT are available for use in improving health care [4]. One author [5] proposed a system that can calculate temperature, blood pressure, and heartbeat and send notifications through a smartphone using a Global System for Mobile communications (GSM). The system uses a wrist band (hardware device) that is connected to the Internet [6]. Other researchers [7] implemented seamless service trials that can be used both online and offline with IoT technology for near-field communication (NFC) tags as well as connect with Bluetooth. The IoT helps create a remote control-based and smart technology in health care [8]. In the future the whole world will connect with the web or the Internet. The IoT is an integrated part of the next generation, but it does have some security issues. Some techniques [9] that are used to combat these issues include access control, hashing control, steganography, cryptography, and hybrid cryptography [10, 11]. Analyses of IoT security and privacy features, including security requirements, treat models and attack taxonomies from the healthcare perspective [12]. Authors have also designed a wearable healthcare context to determine how people get facilities by economic and societies in terms of sustainability. For handling missing values in the C4.5 algorithm in their paper [13], Raja et al. use the statistical methods of hot deck imputation, cold deck imputation, nearest neighbor imputation, substitution, and mean substitution. Saar-Tsechansky and Provost [14] worked with predictive value imputation, the distribution-based imputation used by the C4.5 algorithm, and reduced modeling for a classification tree. The proposed approach only numerical values to impute the missing values that also can extend to handle categorical attributes. Finally, they compared with other factors like time, space, cost, and so on [15]. In their paper, Naik and Samant [16] presented the comparison between k-NN algorithms, decision trees, and naïve Bayes algorithms using Waikato Environment for Knowledge Analysis (Weka), Rapid miner, Tanagra, Orange, and Knime tools on the Indian Liver Patient Dataset. For missing data, Patidar and Tiwari used two categories [17] and introduced their proposed system for dealing with missing values in test datasets and training datasets. In the case of predicting diabetic disease, Iyer et al. [18] showed a comparison with the help of Weka tools.

Dziak et al. proposed an IoT-based information system designed for indoor and outdoor use [19]. They introduced a Design Methodology (DM) that can classify problems and notify the healthcare center or staff immediately. Saravananathan and Velmurugan [20] analyzed the diabetic dataset using a support vector machine (SVM), J48 algorithms, CART, and k-NN for finding classifications. Using k-NN, the authors' accuracy rate for correct classification of instances was 53.39% and their accuracy rate for incorrect classification instances was 46.605%. Huang et al. made a diagnostic classification of diabetic nephropathy using a test dataset of type 2 diabetic patients. Their results show serum triglyceride using a naïve Bayes algorithm had a 59.57% accuracy [21]. Iyer et al. [18] found the solution to diagnosing disease by analyzing the pattern found in the data using decision trees and naïve Bayes algorithms. Using a CDSS, some authors proposed how to predict the risk for heart disease [22]. In their paper [23], Kumar et al. used the C4.5 algorithm to obtain an accuracy rate of 71.4%. They also compared the performance of different types of algorithms.

The main motivation of our work is to develop an IoT-based CDSS to predict disease with better accuracy. Another incentive is to analyze disease datasets according to proposed algorithms. We also wished to design a system where missing and abnormal values can be found and manipulated easily. In this research, we used an algorithm that can manipulate the missing values of datasets.

7.2 Materials and methods

For our proposed system we used a dataset of diabetic patients as the input. Data in this set came from hospital databases, smartphone users, and any other users connected to the system. Users may have also sent data via their wristband sensors or smart watches. All data is stored in a central database. When data comes from a hospital or user it may be unstructured or raunchy and there may be some missing values. Therefore we have to replace any abnormal or missing values with expected values. Our algorithm dynamically identifies these types of situations and replaces abnormal values with expected values. We also save those unstructured values in the central database because we can use that data to find patterns of how users provide data. When we set everything up the way we want, we will save all the data to the central database. This is now the ideal dataset for classifying with any existing algorithm (e.g., C4.5, k-NN). After classifying the data, we can predict whether someone has diabetes or not and, in general, if someone has any complexity or not.

We designed an expert system to predict many kinds of disease. There are four steps in this framework as discussed in the following section (see also Fig. 7.1).

1. *Input section*: Through a wristband, smart watch, or smartphone users can send health data such as plasma glucose, blood pressure, heartbeat, and so on to a database via the internet or mobile app (for the smartphone).

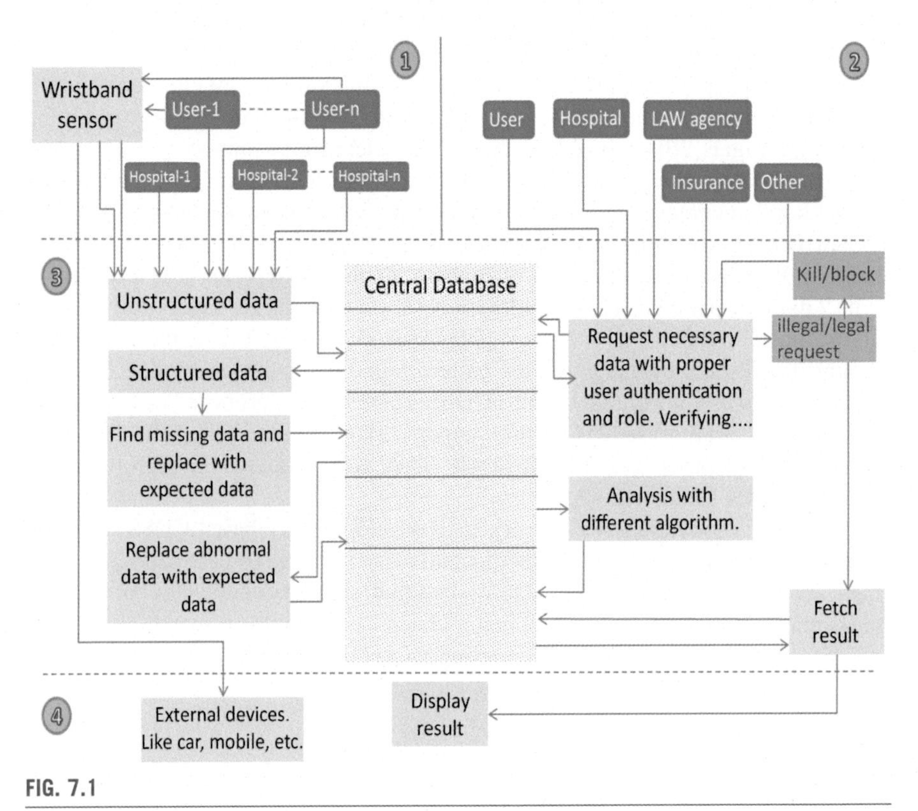

FIG. 7.1

Main framework.

2. *Input section*: In this part, the user/hospital/insurance company can see the predicted result of a patient by sending a request to the server or management system.
3. *Processing section*: This section is the most important section of our proposed system. We use several machine learning algorithms in this section. We collect all data from users and hospitals and place it into what we call a "central database." Users and hospitals can send unstructured data and our system can convert that data into "structured data." If our intelligent system finds any missing or abnormal values, we can use statistical methods to impute them. In this way, the system replaces the abnormal and missing values and stores them in the database. When there are no abnormal or missing values in the system, it will begin analyzing all data according to C4.5, k-NN, or any other specified algorithm and predict the result for patients. The predicted result will then be saved to the database. When the input section requests the database to show the predicted result, it will check for proper authentication and send the final result to the users or hospital.

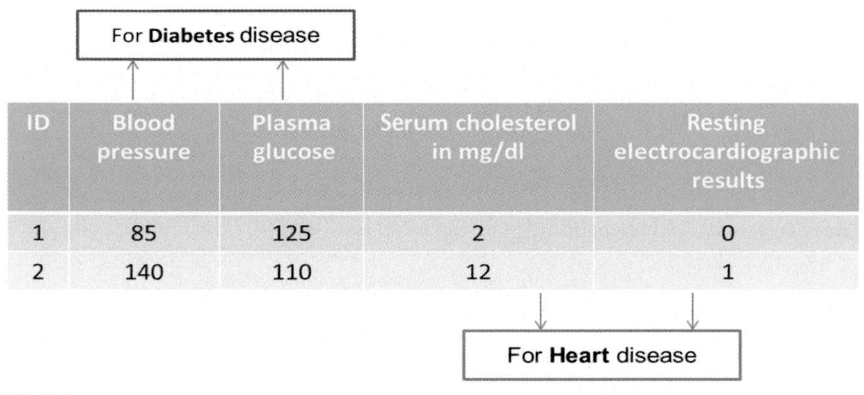

FIG. 7.2

A portion of central database.

4. *Output section*: This is the output section where the results appear on user devices or any other external devices, after which the device takes action according to the output. For example, if a user driving a car suddenly feels chest pain the output section will provide a signal to the car and the car will stop. At the same time a notification will be sent to the nearest hospital, the user's family, or to whomever the user wants to receive the notification.

Fig. 7.2 presents a portion of the central database. This is where all the data is stored. There are many attributes for all the diseases, however, we focus on diabetes and therefore use only those attributes related to this disease. If we wanted to focus on heart disease, we would pick only those attributes related to heart disease.

7.2.1 Dataset collection

We worked with a diabetes dataset, which we collected from the UCI Machine Learning Repository: Pima Indians Diabetes Data Set [24]. There are 768 instances and eight attributes in the dataset. The parameters of the dataset are number of times pregnant, plasma glucose level, diastolic blood pressure, triceps skin fold thickness (mm), serum insulin, body mass of patient, age, and diabetic pedigree. Total numbers of infected diabetic patients is 268; the other 500 are not affected by this disease. This dataset is referred to as being "raw," meaning there are many missing and abnormal values; we find those values by using our techniques. We call this new dataset the "new" dataset.

7.2.2 Classifier models

We used the Weka toolkit to develop the decision models. We put our data in a comma-separated values (CSV) format and then imported this dataset in this tool for preferred access. We also used two classifiers: (1) J48 and (2) k-NN.

7.2.2.1 J48 classifier

J48 is an extension of ID3. J48 is used in the Weka tool, which is nothing but the implementation of a C4.5 algorithm in java programming language. If we want to represent the relative entropy as a mathematical formula it will look like Eq. (7.1).

$$E(X) = -p(P)\log 2p(P) - p(N)\log 2p(N) \tag{7.1}$$

Here, $p(P)$ is related probabilities of positive classes and $p(N)$ is related probabilities of negative class.

7.2.2.2 K-nearest neighbor

k-NN is a very well-known algorithm for classification and regression. It is the simplest classification algorithm that stores all training instances and uses a Euclidean distance function to classify new instances (shown in Eq. (7.2)):

$$\sqrt{\sum_{i=1}^{k}(x_i - y_i)^2} \tag{7.2}$$

7.2.3 Performance measures

The possible net result of this proposed classifier can be represented as follows:

- *TP (true positive)*: Number of items correctly figured out
- *TN (True Negative)*: Number of items correctly discarded
- *FP (false positive)*: Number of items incorrectly figured out
- *FN (false negative)*: Number of items incorrectly rejected

$Recall_M$ or sensitivity is the fraction of relevant instances that have been reclaimed over the total number of instances. Its formula is:

$$Recall_M = \frac{\sum_{i=1}^{C}\frac{TP_i}{TP_i + FN_i}}{C} \tag{7.3}$$

$Precision_M$ is the fraction of relevant instances among the reclaimed instances. Its formula is given as:

$$Precision_M = \frac{\sum_{i=1}^{C}\frac{TP_i}{TP_i + FP_i}}{C} \tag{7.4}$$

$F\text{-}Measure_M$ is a weighted combination of $Recall_M$ and $Precision_M$. Its formula is given as:

$$F\text{-}Measure_M = \frac{(\beta^2 + 1)\, \mathrm{Recall}_M \times \mathrm{Precision}_M}{\beta^2\, \mathrm{Precision}_M + \mathrm{Recall}_M} \tag{7.5}$$

Accuracy is the portion of correctly predicted test results of the total dataset:

$$Accuracy = \frac{(TP+TN)}{(TP+FP+TN+FN)} \times 100\% \quad (7.6)$$

Average accuracy is the portion of correctly predicted test results of the total classes:

$$Accuracy_{Avg} = \frac{\sum_{i=1}^{C} \frac{TP_i + TN_i}{TP_i + FN_i + TN_i + FP_i}}{C} \quad (7.7)$$

Area under the ROC curve (AUC) is used to calculate performance measure for given classification for various thresholds. Its formula is given as:

$$AUC = \frac{1}{2}\left(\frac{TP}{TP+FN} + \frac{TN}{TN+FP}\right) \quad (7.8)$$

7.2.4 Algorithm

```
Step 1: data = dataset.
Step 2: pick a missing or abnormal attribute among all the
        attributes;
Step 3: while (missing or abnormal)
  Step 3.1: if it is nominal then
    Step 3.1.1: data=Different algorithm kNN (data), C4.5
                (data), any other classification algorithm.
  Step 3.2: else
    Step 3.2.1: A=Find the mean of the specific attribute.
    Step 3.2.2: B=Find the mean of those who have diabetics.
    Step 3.2.3: C=Find the mean of those who have not diabetic.
    Step 3.2.4: result=round ((A+B+C)/3).
    Step 3.2.5: data=result.
  end if
 end while
Step 4: Output=Different algorithm kNN (data), C4.5 (data), any
        other classification algorithm
Step 5: Central database = Output
Step 6: send to user
```

7.2.5 Experimental setup

We conducted multiple experiments on training and testing datasets to verify our work. To assess the accuracy of the proposed system we used two different classifier models (J48, k-NN). We also used five different test environments to generate the results.

7.3 Results and discussion

In this analysis, two algorithms (J48 and k-NN where $K=5$) were used for finding accuracy and other values. In each case we show how our algorithm gives results with any existing algorithm that we use. In the experiment we also used several test environments to calculate accuracy in different stages and situations. As a test environment, we used each training dataset for testing, one raw dataset for testing, and a new dataset for training, and vice versa. We used cross-validation with the value of fold of 10. To calculate accuracy we used Eq. (7.6).

We present one detailed accuracy table, one summary table, and one confusion matrix table for each test environment.

Test Environment-1

a. Training data: new dataset
b. Test data: new dataset
c. Algorithm: J48

In this part, we used the new dataset for both training and testing purposes. J48 is the algorithm. Table 7.1 shows the accuracy, Table 7.2 presents the summary of this test, and Table 7.3 presents the confusion matrix for Test Environment-1.

Table 7.1 Detailed accuracy of the new dataset for Test Environment-1

Class	Tested_Negative	Tested_Positive	Weighted avg.
TP Rate	0.939	0.672	0.844
FP Rate	0.328	0.064	0.226
Precision	0.842	0.849	0.844
Recall	0.939	0.672	0.844
F-Measure	0.886	0.750	0.839
MCC	0.648	0.648	0.648
ROC Area	0.894	0.894	0.894
PRC Area	0.920	0.822	0.886

Table 7.2 Summary of the new dataset for Test Environment-1

	New data
Correctly classified instances	648 (84.375 %)
Incorrectly classified instances	120 (15.625 %)
Kappa statistic	0.6386
Mean absolute error	0.2288
Root mean squared error	0.3382
Relative absolute error	50.3421%
Root relative squared error	70.9614%
Total number of instances	768

Table 7.3 Confusion matrix for the new dataset in Test Environment-1

	New dataset	
	Predicted: No	**Predicted: Yes**
Actual: No Tested_Negative	468	32
Actual: Yes Tested_Positive	88	180

Test Environment-2

a. Training data: new dataset
b. Cross-validation with 10 folds
c. Algorithm: J48

In this part, we used a new dataset for training purposes. J48 is the algorithm. Table 7.4 shows the accuracy, Table 7.5 presents the summary of this test, and Table 7.6 presents the confusion matrix for Test Environment-2.

Table 7.4 Detailed accuracy of the new dataset for Test Environment-2

Class	Tested_Negative	Tested_Positive	Weighted avg.
TP Rate	0.808	0.597	0.734
FP Rate	0.403	0.192	0.329
Precision	0.789	0.625	0.732
Recall	0.808	0.597	0.734
F-Measure	0.798	0.611	0.733
MCC	0.410	0.410	0.410
ROC Area	0.751	0.751	0.751
PRC Area	0.809	0.577	0.728

Table 7.5 Summary of the new dataset for Test Environment-2

	New data
Correctly classified instances	564 (**73.4375%)**
Incorrectly classified instances	204 (26.5625%)
Kappa statistic	0.4093
Mean absolute error	0.3161
Root mean squared error	0.4477
Relative absolute error	69.541%
Root relative squared error	93.9366%
Total number of instances	768

Table 7.6 Confusion matrix for the new dataset in Test Environment-2

	New dataset	
	Predicted: No	**Predicted: Yes**
Actual: No Tested_Negative	404	96
Actual: Yes Tested_Positive	108	160

Test Environment-3

a. Training data: new dataset
b. Test data: raw dataset
c. Algorithm: J48

For this part, we used a new dataset for training and a raw dataset for testing. J48 is the algorithm. Table 7.7 presents the accuracy, Table 7.8 shows the summary of this test, and Table 7.9 presents the confusion matrix for Test Environment-3.

Table 7.7 Detailed accuracy of the new dataset for Test Environment-3

Class	Tested_Negative	Tested_Positive	Weighted avg.
TP Rate	0.936	0.668	0.842
FP Rate	0.332	0.064	0.239
Precision	0.840	0.848	0.843
Recall	0.936	0.668	0.842
F-Measure	0.886	0.747	0.837
MCC	0.645	0.645	0.645
ROC Area	0.893	0.893	0.893
PRC Area	0.991	0.818	0.884

Table 7.8 Summary of the new dataset for Test Environment-3

	New data
Correctly classified instances	647 (**84.375%)**
Incorrectly classified instances	121 (15.625%)
Kappa statistic	0.6353
Mean absolute error	0.2309
Root mean squared error	0.34
Relative absolute error	50.7966%
Root relative squared error	71.3321%
Total number of instances	768

Table 7.9 Confusion matrix for the new dataset in Test Environment-3

	New dataset	
	Predicted: No	**Predicted: Yes**
Actual: No Tested_Negative	468	32
Actual: Yes Tested_Positive	89	179

Test Environment-4

a. Training data: new dataset
b. Test data: raw dataset
c. Algorithm: k-NN where $K=5$

In this part, we used a new dataset for training and a raw dataset was used for testing. We used k-NN as the testing algorithm where $K=5$. Table 7.10 shows the accuracy, Table 7.11 presents the summary of this test, and Table 7.12 presents the confusion matrix for Test Environment-4.

Table 7.10 Detailed accuracy of the new dataset for Test Environment-4

Class	Tested_Negative	Tested_Positive	Weighted avg.
TP Rate	0.908	0.623	0.809
FP Rate	0.377	0.092	0.277
Precision	0.818	0.784	0.806
Recall	0.908	0.623	0.809
F-Measure	0.861	0.694	0.803
MCC	0.565	0.565	0.565
ROC Area	0.891	0.891	0.891
PRC Area	0.924	0.772	0.871

Table 7.11 Summary of the new dataset for Test Environment-4

	New data
Correctly classified instances	621 (**80.8594%)**
Incorrectly classified instances	147 (19.1406%)
Kappa statistic	0.5577
Mean absolute error	0.2546
Root mean squared error	0.3554
Relative absolute error	56.0087%
Root relative squared error	74.5617%
Total number of instances	768

Table 7.12 Confusion matrix for the new dataset in Test Environment-4

	New dataset	
	Predicted: No	**Predicted: Yes**
Actual: No Tested_Negative	454	46
Actual: Yes Tested_Positive	101	167

Test Environment-5

a. Training data: new dataset
b. Cross-validation with 10 folds
c. Algorithm: k-NN where $K = 5$

In this part, we used a new dataset for training purposes and a cross-validation with 10 folds. We used k-NN as the testing algorithm where $K=5$. Table 7.13 shows the accuracy, Table 7.14 presents the summary of this test, and Table 7.15 presents the confusion matrix for Test Environment-5.

Table 7.13 Detailed accuracy of the new dataset for Test Environment-5

Class	Tested_Negative	Tested_Positive	Weighted avg.
TP Rate	0.848	0.526	0.736
FP Rate	0.474	0.152	0.326
Precision	0.770	0.650	0.728
Recall	0.848	0.526	0.736
F-Measure	0.807	0.581	0.728
MCC	0.396	0.396	0.396
ROC Area	0.781	0.781	0.781
PRC Area	0.844	0.627	0.768

Table 7.14 Summary of the new dataset for Test Environment-5

	New data
Correctly classified instances	565 (**73.5677%)**
Incorrectly classified instances	203 (26.4323%)
Kappa statistic	0.3914
Mean absolute error	0.3092
Root mean squared error	0.4248
Relative absolute error	68.0353%
Root relative squared error	89.1192%
Total number of instances	768

Table 7.15 Confusion matrix for the new dataset in Test Environment-5

	New dataset	
	Predicted: No	**Predicted: Yes**
Actual: No Tested_Negative	424	76
Actual: Yes Tested_Positive	127	141

7.4 Comparison

A brief or hurried look at Table 7.16 shows that there is a huge number of missing and abnormal values in the "Pima Indiana Diabetic" dataset. In this analysis, we apply our technique to finding those values to make this dataset well organized. We take 40 mm Hg as the lower limit for the diastolic blood pressure [5]. Below 40 mm Hg we get 39 instances that need to be changed and therefore we have changed them.

In the research, we used several test environments to test our framework. We present one summary table, one confusion matrix table, and one figure for each test environment to give a better view of accuracy.

Table 7.16 Comparison of a single attribute (diastolic blood pressure)

Raw data	New data
Diastolic blood pressure (mm Hg) **Missing: 39 (5%)** **Distinct: 43** Unique: 6 (1%) Minimum: 40 Maximum: 122 **Mean: 72.635**	Diastolic blood pressure (mm Hg) **Missing: 0 (0%)** **Distinct: 46** Unique: 6 (1%) Minimum: 40 Maximum: 122 **Mean: 72.557**

Test Environment-1

a. Training data: New dataset and raw dataset
b. Test data: New dataset and raw dataset
c. Algorithm: J48

In Test Environment-1, we used a new dataset for both training and testing purposes. We also used a raw dataset for both purposes. We used J48 as the testing algorithm.

From Table 7.17 we can see that we achieve higher accuracy using our technique. Our accuracy is **84.375%** and the raw dataset accuracy is **82.4219%**. If we look at the relative absolute error then we can see that our error rate is lower than the raw dataset

(Table 7.17). Table 7.18 represents the confusion matrix for both datasets in Test Environment-1.

Fig. 7.3 presents a graph of the accuracy. We can see that the accuracy of the new dataset is higher than the accuracy of the raw dataset.

Table 7.17 Summary of results for both datasets in Test Environment-1

	Raw data	New data
Correctly classified instances	633 (**82.4219%)**	648 (**84.375%)**
Incorrectly classified instances	135 (17.5781%)	120 (15.625%)
Kappa statistic	0.5863	0.6386
Mean absolute error	0.2504	0.2288
Root mean squared error	0.3534	0.3382
Relative absolute error	55.0837%	50.3421%
Root relative squared error	74.1354%	70.9614%
Total number of instances	768	768

Table 7.18 Confusion matrix for both datasets in Test Environment-1

	Raw dataset		New dataset	
	Predicted: No	Predicted: Yes	Predicted: No	Predicted: Yes
Actual: No Tested_Negative	470	30	468	32
Actual: Yes Tested_Positive	105	163	88	180

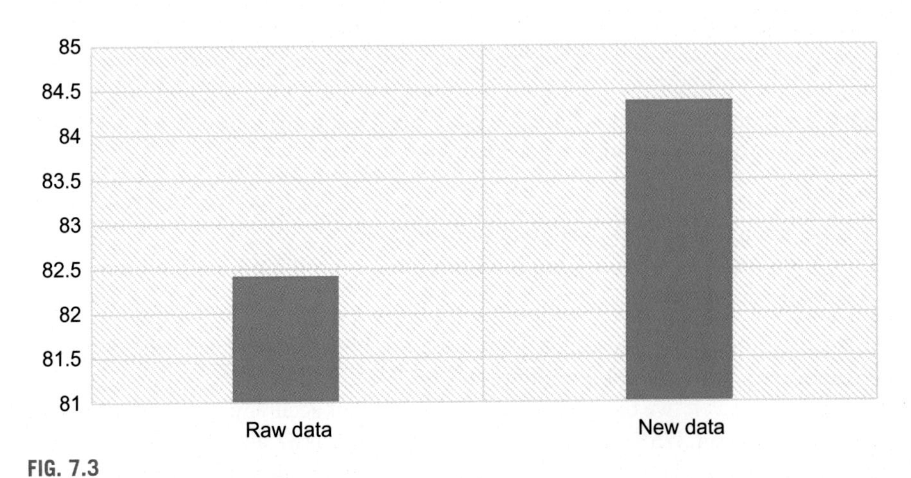

FIG. 7.3

Accuracy graph for Test Environment-1.

Test Environment-2

a. Training data: new dataset and raw dataset
b. Cross-validation with 10 folds
c. Algorithm: J48

In Test Environment-2, we used a new dataset for training and a cross-validation with 10 folds. We also used a raw dataset for training and a cross-validation with 10 folds. We used J48 as the testing algorithm.

From Table 7.19 we can see that we get higher accuracy using our technique. Our accuracy is **73.4375%** and the raw dataset accuracy is **73.1771%**. If we take a look at the relative absolute error then we can judge that our error rate is lower than the raw dataset (Table 7.19). Table 7.20 represents the confusion matrix for both datasets in Test Environment-2.

Fig. 7.4 presents a graph of the accuracy. We can see that the accuracy of the new dataset is higher than the accuracy of the raw dataset.

Table 7.19 Summary of results for both datasets in Test Environment-2

	Raw data	**New data**
Correctly classified instances	562 (**73.1771%)**	564 (**73.4375%)**
Incorrectly classified instances	206 (**26.8229%**)	204 (**26.5625%**)
Kappa statistic	0.4056	0.4093
Mean absolute error	0.3182	0.3161
Root mean squared error	0.4479	0.4477
Relative absolute error	**70.0146%**	**69.541%**
Root relative squared error	93.9623%	93.9366%
Total number of instances	768	768

Table 7.20 Confusion matrix for both datasets in Test Environment-2

	Raw dataset		**New dataset**	
	Predicted: No	**Predicted: Yes**	**Predicted: No**	**Predicted: Yes**
Actual: No Tested_Negative	401	99	404	96
Actual: Yes Tested_Positive	107	161	108	160

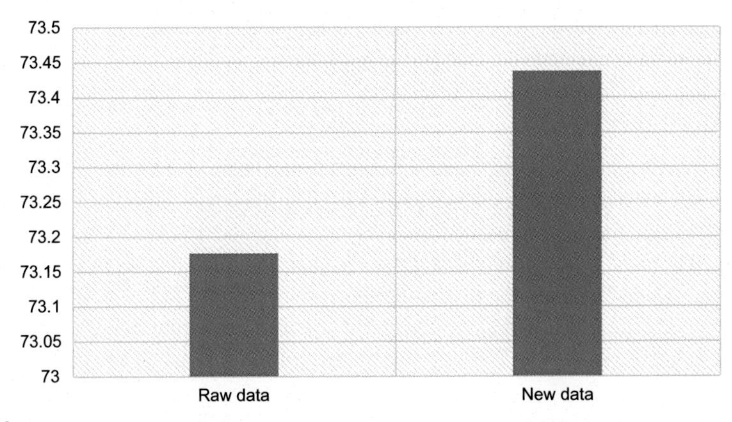

FIG. 7.4

Accuracy graph for Test Environment-2.

Test Environment-3

a. Training data: New dataset and raw dataset
b. Test data: Raw dataset and new dataset
c. Algorithm: J48

In Test Environment-3, we used a new dataset for training and a raw dataset for testing. We also used a raw dataset for training and a new dataset for testing. We used J48 as the testing algorithm.

From Table 7.21 we can see that we get higher accuracy using our technique. Our accuracy is **84.2448%** and the raw dataset accuracy is **82.4219%**. If we look at the relative absolute error then we can see that our error rate is lower than the raw dataset (Table 7.21). Table 7.22 represents the confusion matrix for both datasets in Test Environment-3.

Fig. 7.5 presents a graph of the accuracy. We can see that the accuracy of the new dataset is higher than the accuracy of the raw dataset.

Table 7.21 Summary of results for both datasets in Test Environment-3

	Raw data	New data
Correctly classified instances	633 (**82.4219%)**	647 (**84.2448%)**
Incorrectly classified instances	135 (17.5781%)	121 (15.7552%)
Kappa statistic	0.5863	0.6353
Mean absolute error	0.2506	0.2309
Root mean squared error	0.3539	0.34
Relative absolute error	55.1443%	50.7966%
Root relative squared error	74.2407%	71.3321%
Total number of instances	768	768

Table 7.22 Confusion matrix for both datasets in Test Environment-3

	Raw dataset		**New dataset**	
	Predicted: No	**Predicted: Yes**	**Predicted: No**	**Predicted: Yes**
Actual: No Tested_Negative	470	30	468	32
Actual: Yes Tested_Positive	105	163	89	179

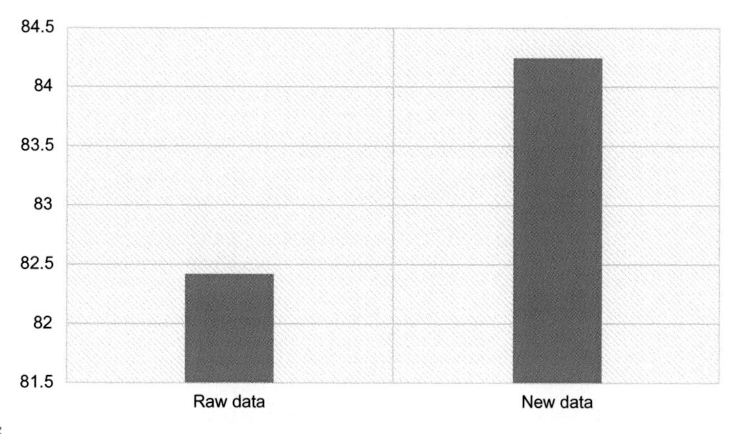

FIG 7.5

Accuracy graph for Test Environment-3.

Test Environment-4

a. Training data: New dataset and raw dataset
b. Test data: New dataset and raw dataset
c. Algorithm: k-NN where $K=5$

In Test Environment-4, we used a new dataset for both training and testing purposes. We also used a raw dataset for both purposes. We used k-NN as the testing algorithm where $K=5$.

From Table 7.23 we can see that we get higher accuracy using our technique. Our accuracy is **81.6406%** and the raw dataset accuracy is **80.9896%**. If we look at the relative absolute error then we can see that our error rate is lower than the raw dataset (Table 7.23). Table 7.24 represents the confusion matrix for both datasets in Test Environment-4.

Fig. 7.6 presents a graph of the accuracy. We can see that the accuracy of the new dataset is higher than accuracy of the raw dataset.

Table 7.23 Summary of results for both datasets in Test Environment-4

	Raw data	New data
Correctly classified instances	622 (**80.9896%)**	627 (**81.6406%)**
Incorrectly classified instances	146 (19.0104%)	141 (18.3594%)
Kappa statistic	0.5603	0.5796
Mean absolute error	0.2553	0.2486
Root mean squared error	0.3568	0.3488
Relative absolute error	56.1806%	54.6916%
Root relative squared error	74.853%	73.1812%
Total number of instances	768	768

Table 7.24 Confusion matrix for both datasets in Test Environment-4

	Raw dataset		New dataset	
	Predicted: No	**Predicted: Yes**	**Predicted: No**	**Predicted: Yes**
Actual: No Tested_Negative	455	45	452	48
Actual: Yes Tested_Positive	101	167	93	175

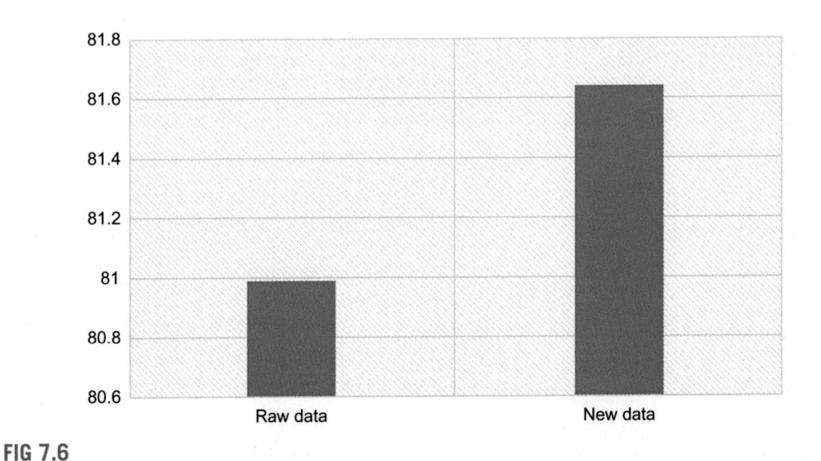

FIG 7.6

Accuracy graph for Test Environment-4.

Test Environment-5

a. Training data: New dataset and raw dataset
b. Cross-validation with 10 folds
c. Algorithm: k-NN where $K = 5$

In Test Environment-5 we used a new dataset for training and a cross-validation with 10 folds. We also used a raw dataset for training and a cross-validation with 10 folds. We used k-NN as the testing algorithm where $K = 5$.

From Table 7.25 we can see that we get higher accuracy using our technique. Our accuracy is **73.5677%** and the raw dataset accuracy is **72.7865%**. If we look at the relative absolute error then we can see that our error rate is lower than the raw dataset (Table 7.25). Table 7.26 represents the confusion matrix for both datasets in Test Environment-5. Fig. 7.7 presents a graph of the accuracy. We can see that the accuracy of the new dataset is higher than accuracy of the raw dataset.

Fig. 7.8 presents a graph of the overall differences in all the test environments. From this graph we can see that we get achieved greater accuracy in every test we performed.

Table 7.25 Summary of results for both datasets in Test Environment-5

	Raw data	New data
Correctly classified instances	559 **(72.7865%)**	565 **(73.5677%)**
Incorrectly classified instances	209 (27.2135%)	203 (26.4323%)
Kappa statistic	0.3642	0.3914
Mean absolute error	0.3165	0.3092
Root mean squared error	0.431	0.4248
Relative absolute error	69.6387%	68.0353%
Root relative squared error	90.4211%	89.1192%
Total number of instances	768	768

Table 7.26 Confusion matrix of both datasets in Test Environment-5

	Raw dataset		New dataset	
	Predicted: No	Predicted: Yes	Predicted: No	Predicted: Yes
Actual: No Tested_Negative	429	71	424	76
Actual: Yes Tested_Positive	138	130	127	141

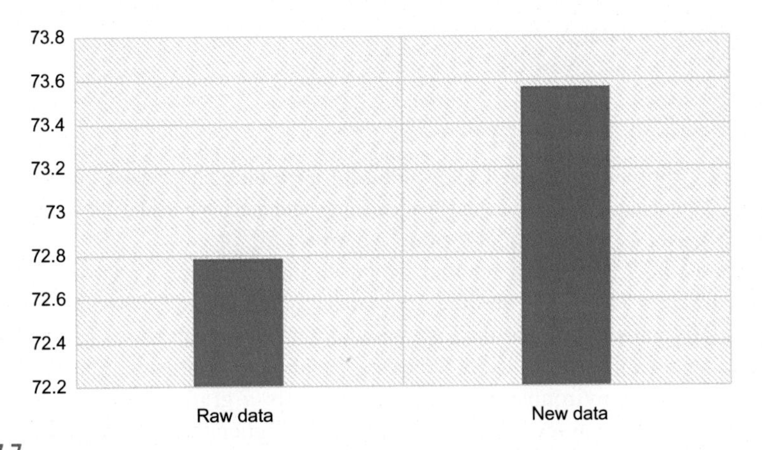

FIG. 7.7

Accuracy graph for Test Environment-5.

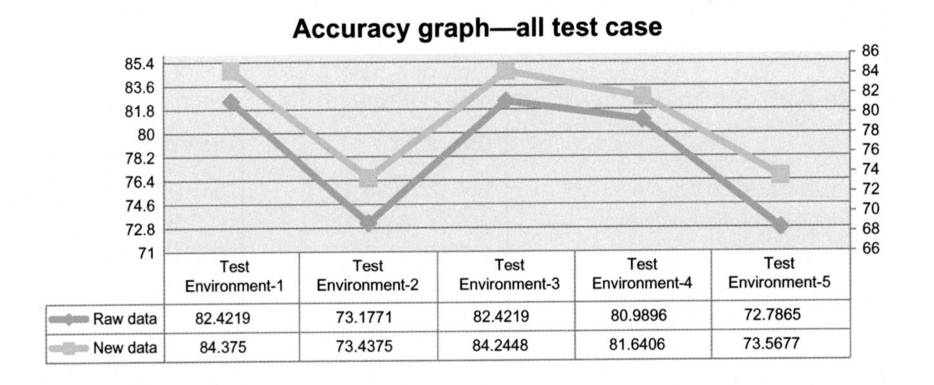

	Test Environment-1	Test Environment-2	Test Environment-3	Test Environment-4	Test Environment-5
Raw data	82.4219	73.1771	82.4219	80.9896	72.7865
New data	84.375	73.4375	84.2448	81.6406	73.5677

FIG 7.8

Accuracy graph for all test environments.

7.5 Conclusion

In this research, we developed an intelligent system for predicting diseases like diabetes, heart disease, cancer, and more. We also developed a system that can be integrated with any existing machine learning algorithms for classification tasks. Finally, we integrated the IoT with this system for better connectivity. Here everything is executed and implemented in a real-life scenario. In this experiment, we designed several test environments and we achieved the better result in every case. For example, in Test Environment-1 we achieved 84.375% accuracy, whereas the raw dataset achieved 82.4219% accuracy. In addition, we were also able to reduce the relative absolute error from 55.0831% to 50.3421%. This represents a 4.7416% error reduction from Test Environment-1. We similarly succeeded in every test environment. This model will be helpful for doctors, clinicians, medical students, and people who work with the IoT.

Every research has led to further exploration and more research and our research is no exception. In this research, we used our own technique so it may be possible that using another technique will lead to better accuracy. However, our technique gives better accuracy among all the studied papers. In the future we will apply more techniques to potentially increase the accuracy of our results. Our study is a prototype, a mini-implemented work based on the research. Perhaps it can be implemented on a larger scale allowing us to also study the effect of electronic devices on the human body. We didn't implement bio data security in this research, so this could be a good study option for the future.

References

[1] S. Begum, H. Parveen, U-healthcare and IoT, Int. J. Comput. Sci. Mobile Comput. 5 (8) (2016) 138–142.

[2] S.F.U. Begum, I. Begum, Smart health care solutions using IOT, Int. J. Mag. Eng. Technol. Manag. Res. 4 (3) (2017).

[3] R. Aarthi, An IoT based wireless body area network for healthcare applications using Raspberry Pi, Int. J. Res. Appl. Sci. Eng. Technol. 6 (4) (2018) 3076–3083.

[4] M. Haghi, K. Thurow, R. Stoll, Wearable devices in medical Internet of things: scientific research and commercially available devices, Healthc. Inform. Res. 23 (1) (2017) 4.

[5] K. Harsh, IOT based calorie calculator for healthcare system, Int. J. Eng. Computer Sci. (2017).

[6] Hicon smart wrist band with social network icons (Online), Indiegogo (2017). Available from: https://www.indiegogo.com/projects/hicon-smartwristband-with-social-network-icons#/. (Accessed 2 August 2018).

[7] A. Park, H. Chang, K. Lee, How to sustain smart connected hospital services: an experience from a pilot project on IoT-based healthcare services, Healthc. Inform. Res. 24 (4) (2018) 387.

[8] K.U. Sreekanth, K.P. Nitha, A study on health care in Internet of things, Int. J. Recent Innov. Trends Comput. Commun. 4 (2) (2016).

[9] L. Yehia, A. Khedr, A. Darwish, Hybrid security techniques for Internet of things healthcare applications, Adv. Internet Things 05 (03) (2015) 21–25.

[10] R.K. Gupta, P. Singh, A new way to design and implementation of hybrid crypto system for security of the information in public network, Int. J. Emerg. Technol. Adv. Eng. 3 (2013) 2250–2459.

[11] L.L. Yu, Z.J. Wang, W.F. Wang, The application of hybrid encryption algorithm in software security, in: 2012 4th International Conference on Computational Intelligence and Communication Networks (CICN), 2012, pp. 762–765.

[12] S. Riazul Islam, D. Kwak, M. Humaun Kabir, M. Hossain, K.-S. Kwak, The Internet of things for health care: a comprehensive survey, IEEE Access 3 (2015) 678–708.

[13] K. Raja, G. Tholkappia Arasu, C.S. Nair, Imputation framework for missing values, Int. J. Comput. Trends Technol. 3 (2) (2012) 215–219.

[14] M. Saar-Tsechansky, F. Provost, Handling missing values when applying classification models, J. Machine Learn. Res. 8 (2007) 1625–1657.
[15] B. Mehala, P. Ranjit Jeba Thangaiah, K. Vivekanandan, Selecting scalable algorithms to deal with missing values, Int. J. Recent Trends Eng. 1 (2) (2009).
[16] A. Naik, L. Samant, Correlation review of classification algorithm using data mining tool: WEKA, Rapidminer, Tanagra, Orange and Knime, Proc. Comput. Sci. 85 (2016) 662–668.
[17] P. Patidar, A. Tiwari, Handling missing value in decision tree algorithm, Int. J. Comput. Appl. 70 (13) (2013) 31–36.
[18] A. Iyer, S. Jeyalatha, R. Sumbaly, Diagnosis of diabetes using classification mining techniques, Int. J. Data Mining Knowl. Manag. Process 5 (1) (2015) 01–14.
[19] D. Dziak, B. Jachimczyk, W. Kulesza, IoT-based information system for healthcare application: design methodology approach, Appl. Sci. 7 (6) (2017) 596.
[20] K. Saravananathan, T. Velmurugan, Analyzing diabetic data using classification algorithms in data mining, Indian J. Sci. Technol. 9 (43) (2016).
[21] G. Huang, K. Huang, T. Lee, J. Weng, An interpretable rule-based diagnostic classification of diabetic nephropathy among type 2 diabetes patients, BMC Bioinform. 16 (1) (2015) S5.
[22] P. Anooj, Clinical decision support system: risk level prediction of heart disease using weighted fuzzy rules and decision tree rules, Open Comput. Sci. 1 (4) (2011).
[23] D. Senthil Kumar, G. Sathyadevi, S. Sivanesh, Decision support system for medical diagnosis using data mining, Int. J. Comput. Sci. Issues 8 (3) (2011).
[24] UCI Machine Learning Repository: Pima Indians Diabetes Data Set, Archive.ics.uci.edu (Online) 2017. Available from: https://archive.ics.uci.edu/ml/datasets/pima+indians+diabetes, (Accessed 2 August 2018).

Further reading

[25] STAY WELL WORLD | Dignity of the Human Person (Online), Available from: https://www.staywellworld.org/single-post/2017/06/03/LOW-BLOOD-PRESSURE-HYPO TENSION, 2017. (Accessed 2 August 2018).
[26] B. Shepard, News—low diastolic blood pressure linked to higher risk of heart failure (Online), Uab.edu (2017). Available from: https://www.uab.edu/news/latest/item/1681-low-diastolic-blood-pressure-linked-to-higher-risk-of-heart-failure. (Accessed 2 July 2018).
[27] Blood Pressure: Q. What is low blood pressure? (Online), Bloodpressureuk.org (2017) Available from: http://www.bloodpressureuk.org/microsites/u40/Home/facts/Whatislow. (Accessed 2 August 2018).
[28] R.P. Bhandari, P. Sapna Yadav, A. Shyam Mote, P. Devika Rankhambe, A survey on predictive system for medical diagnosis with expertise analysis, Int. J. Innov. Res. Comput. Commun. Eng. 4 (2016).
[29] U. DBD, Confusion Matrix, Www2.cs.uregina.ca (Online), Available from: http://www2.cs.uregina.ca/~dbd/cs831/notes/confusion_matrix/confusion_matrix.html, 2017. (Accessed 2 August 2018).

CHAPTER

Intelligent energy-efficient healthcare models integrated with IoT and LoRa network

8

Ratula Ray*, Debasish Kumar Mallick†, Satya Ranjan Dash†

School of Biotechnology, KIIT Deemed to be University, Bhubaneswar, India School of Computer Applications, KIIT Deemed to be University, Bhubaneswar, India†*

8.1 Introduction: The era of smart healthcare

The field of healthcare is rapidly transitioning from traditional healthcare system toward adapting smarter solutions with the development of sensor technology merged with data analysis. The current scenario in healthcare offers perfect opportunities for researchers and practitioners alike to invest their time and efforts toward surfacing out better options, which is cost effective and can cater to the necessities of the patients.

Remote monitoring and communication in healthcare have been made possible with the concept of Internet of Things (IoT), which interconnects many devices together to form an interlink between them. New patented concepts, such as Long Range (LoRa) technology, have been effective in such scenarios, which consumes lesser power and has good receiver sensitivity.

Real-time monitoring of patients' symptoms has proven to be invaluable in emergency cases where the life of the patients depends upon it. Taking proactive measures in those cases offers a better chance at survival and saves time. Overall cost of the treatment is also greatly reduced with smarter healthcare options and the care quality of the patients is highly improvised. Constant monitoring also results in gathering of more data, which creates a strong medical profile that the doctors can analyze, which helps them predict the underlying problems.

Wearable sensors are widely used in healthcare systems, which monitor certain physiological parameters in the body such as motion, vital signs, temperature, and pressure. This helps a wireless body area network (WBAN) to create an environment, which is used for efficient communication between various smart devices that make predictions [1]. Identification of biomarkers for certain diseases is also another important criterion that is essential to be taken into account for assessing the health condition of the patient.

The advent of Machine Learning has put forward a plethora of opportunities for data scientists to explore and work closely in association with medical professionals

Sensors for Health Monitoring. https://doi.org/10.1016/B978-0-12-819361-7.00008-7

in order to build up strong mathematical models, which help in prediction of diseases based on the observed parameters. Sensors allow data to be collected and create a network with smart devices, which can then be modeled with powerful Machine-Learning algorithms (like Artificial Neural Network) for predictive analysis. Combining the power of Machine Learning with sensor technology facilitates the designing of smarter healthcare devices, which aid in constant monitoring of the patients and are particularly helpful in cases of elderly people for creating an assisted ambient living (AAL) environment for them [2].

In this book chapter, we have proposed three different case studies aimed toward solving three very different health problems and which are of extreme importance in the current scenario. Our aim has been focused toward creating an interlink between the sensors used for collecting the data, using smart wearable healthcare devices.

8.2 Embedded intelligence and its applications

Embedded intelligence suggests the embedding of intelligent algorithm and device like sensors into a system, which can help track important parameters associated with a particular case study. This enables the system to create a sensitive, adaptive, and responsive system, which can understand human necessities and respond accordingly. It also allows for creating of an intelligent computing environment, which can interact with the users with higher efficiency and produce much better results.

Applications of embedded intelligence can cater toward planning of smart homes, self-driving cars, and smart healthcare systems. Smart options in healthcare have led to the concept of personalized medicines, which is aimed toward creation of a tailored medication regime for the patients. Exploiting the information available from human gene sequencing and utilizing relevant specifications from data analysis have led to the creation of Electronic Health Record (EHR), which is specific for particular patient. This can offer much more efficient healthcare options and the chances for better recovery.

Ability of the machine to be able to learn from predefined sets of data related with a particular case of interest is acquired through Supervised Machine-Learning techniques. On the other hand, prediction made from unlabeled dataset is done by recognizing the pattern of occurrences of the instances in the dataset.

The application in relation with embedded intelligence is catered toward areas such as continuous physiological and behavioral monitoring associated with healthcare, detection of emergency situations and technology associated with assisted living for the elderly [3].

8.3 Case studies

8.3.1 Smart asthma-monitoring device

Asthma is a chronic condition, which affects almost around 300 million people worldwide. It is associated with the constriction of the respiratory pathways by inflammation of the tracts, which can cause difficulty in breathing. Sometimes, building up of mucus is also responsible for blocking of the airway.

Given the frequency of the occurrence of the asthmatic attacks and the severity of the cases, asthma can be divided into four different classes [4]:

- intermittent stage
- mild persistence of wheezing and coughing
- moderate persistence of asthmatic symptoms
- severe persistence of cough and wheeze over several days and weeks

Various factors can trigger the cause of asthma such as indoor and outdoor allergens, sensitivity toward certain chemicals, smoking, extreme physical activities, to name a few. Sensitivity toward certain medications and certain pollutants in the atmosphere as a result of rapid urbanization can also effectively contribute toward this condition to arise.

Learning the triggers that can lead to this condition and keeping in mind what all to avoid are the two main keys to avoid asthmatic attacks.

8.3.1.1 Problem statement

The main concern associated with designing a smart wearable device for keeping track of asthma is the detection of the onset of the symptoms and to be able to differentiate between the different stages of severity associated with it. Warning the people against the triggers that can lead to cause asthma is the key to build a better model with a higher efficacy. Apart from keeping a check on the environmental factors responsible, being able to differentiate between the asthma and certain other diseases due to closely related symptoms is another factor to keep in mind.

Other factors that are to be noted while designing the device are an interactive cloud-based platform for general public to access; maintaining a digital diary associated with the frequency of the occurrence of the attacks, which can help in designing a personal medication regime for a particular patient; and an alert system, which will keep close contacts on the list of the patient and the doctor referred updated about the health of the patient.

8.3.1.2 Proposed design

Our design mainly consists of two parts: a wearable digital bracelet on the wrist, which can detect the biomarker associated with asthma (that can give a better prediction for the condition) and track the vital signs and a patch with an acoustic sensor attached on the chest for detection of the wheezing sound. A web-based application connected with these two devices will help keep the track of occurrences and severity of the symptoms, like the rate of coughing events, and respiratory rate.

The smart digital bracelet is designed with an embedded graphene-based nanosensor that can detect the presence of nitrite, a biomarker associated with asthma [5]. Literature suggests that when the lining of the respiratory walls swells up during an asthmatic attack, an increased level of inducible Nitric Oxide synthetase (iNOS) is produced along the lining of the bronchioles. L-arginine gives out nitrite, which is primarily produced from Nitric Oxide (NO) in the presence of NO synthetase.

Exhaled Breath Condensate(EBC) of the patient consists of nitrite, which can be detected by the graphene-based electrode placed on the smart bracelet. Electrochemical detection of the nitrite in the EBC comprises the reaction of nitrite with reactive oxygen species(ROS) to give out more stable oxides such as nitrate(NO_3^-), that can then be detected. Along with it, small circuits of carbon nanotubes can also be integrated in the wearable device, which can detect the toxic gases in the atmosphere and can send immediate signals in the smartphones regarding the intensity of the triggers related to asthma in the environment using LoRa technology and a risk assessment of the patient being exposed to it can be made. Along with detection of the biomarkers, another important function of the smart bracelet is to keep track of the heart beat of the patient, which can help in cases such as pulmonary edema and is associated with building of fluid inside the lungs because a part of heart stops working. Also referred to as cardiac asthma at times, this condition has overlapping symptoms with those of bronchial asthma and its primary signs can often be misdiagnosed. In such cases, tracking of the heart rate becomes very important and the misdiagnosis can be avoided.

Wheezing sound is an indication of labored breathing and absence of sufficient air getting inside the respiratory tract. Our proposed device will allow the sound waves to be transduced into electrical signals (through an acoustic sensor) by passing it through the preamplification circuit and high pass band filter to acquire the proper breathing sounds without interference of the noise [6]. Placed on the anterior surface of the chest cavity, the sensor will acquire the breathing sound and transmit the amplified signal through a microprocessor and provide wireless connection with a smartphone to analyze the breathing pattern. An effective algorithm for breathing sound analysis can be used by studying the characteristic differences between normal breathing (frequency range between 100 and 1000 Hz) and wheezing sound (frequency range between 250 and 800 Hz) using supervised Machine-Learning classifiers such as Support Vector Machine (SVM), and Neural Networking. The features are extracted and studied to reach the conclusion about the severity of the asthma and help us design a medication regime accordingly. Fig. 8.1 describes the layout of the proposed design for real-time asthma monitoring.

Along with the proposed design, an effective asthma-monitoring app can also be developed to maintain the medical history of the patient, and regular updates from the devices used can also be documented. The monitoring app should also be equipped with breathing sound analysis algorithm, which can detect the severity of the condition by analyzing the features associated with asthma and send an alert message to the referred physician in case of emergency. Information acquired from both the smart bracelet and the acoustic sensor can be stored and analyzed for future references using the app. Fig. 8.2 gives an overview regarding the different functions that the asthma-monitoring application can cater to. Such platforms are rapidly growing to provide more precise and personalized care to the patients.

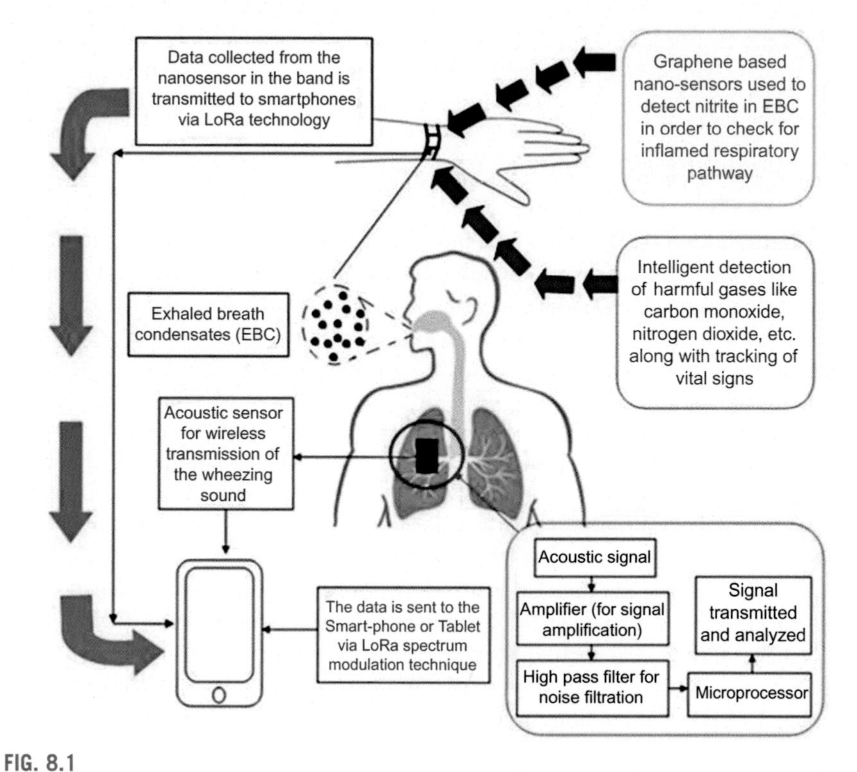

FIG. 8.1

Asthma monitoring layout using smart healthcare options.

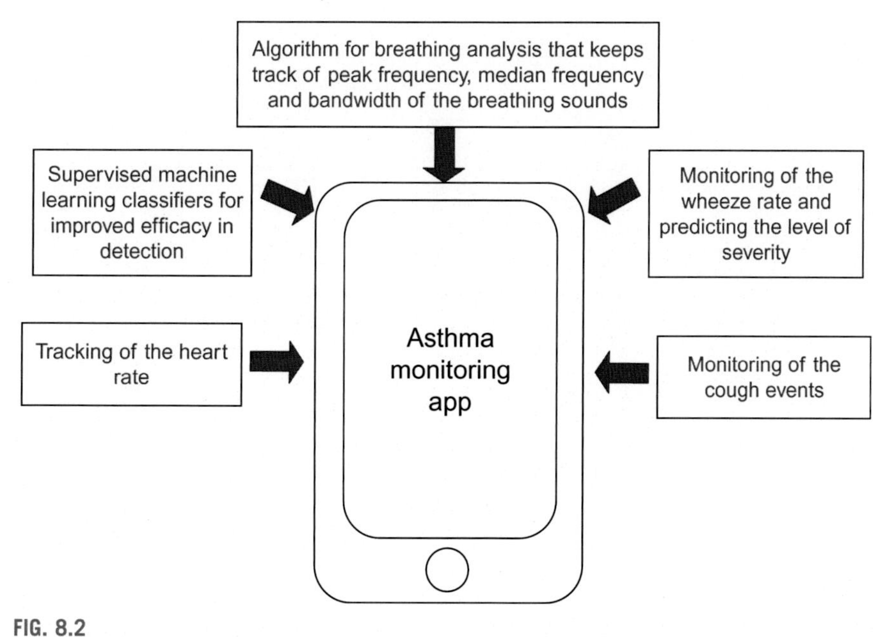

FIG. 8.2

Overview of the functionalities of the asthma-monitoring app connected with the sensors to collect patient's data.

8.3.2 Detecting symptoms of Parkinson's disease using wearable tech

Parkinson's disease (PD) is one of the most common neurodegenerative disorders in the elderly, which impairs their cognitive functioning. Tremor of the hand is one of the most significant signs associated with this disease, which is again related to complex cerebello-thalamo-cortical phenomenon. Symptoms might include significant changes in the gait cycle concerning the upper and lower limbs and change in toe-off dynamics.

Twitching movements involved with involuntary contraction and relaxation of the muscles is termed as Tremor. Early signs of tremors can go unidentified in cases of patients with PD. Tremors can again be categorized into three major types: Resting tremor (RT), Postural tremor (PT), and Essential tremor (ET), and each of them is discussed in detail in Fig. 8.3.

Understanding the abnormalities in the gait cycle associated with the patients of PD is the key to develop newer solutions for assessment of the disease. Figs. 8.4–8.6 depict the biomechanical anomalies associated with a patient of PD in their gait cycle when they are off their medication versus in their motion in healthy state. The gait assessment depends upon three important parameters:

- Walking speed (m/s)
- Cadence (steps/min)
- Stride length (distance between two successive placements of foot)

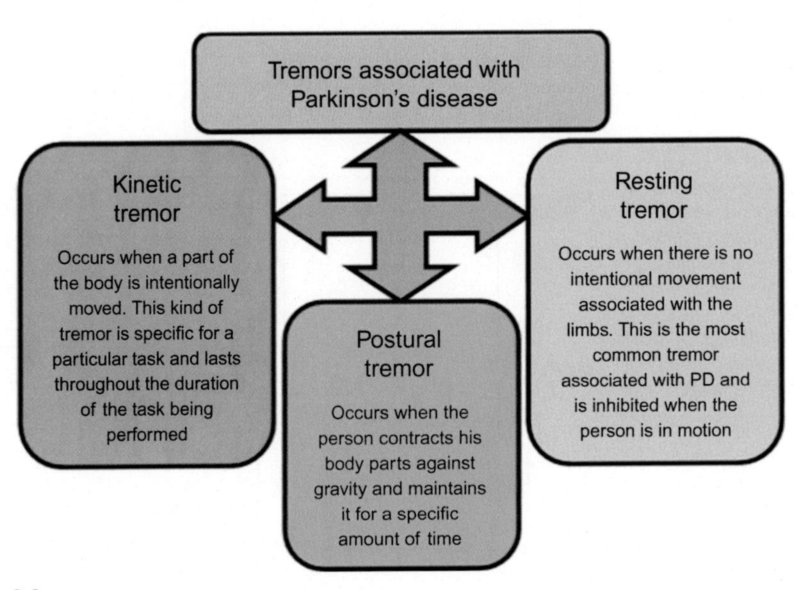

FIG. 8.3

Overview of the types of tremors associated with Parkinson's disease and period of occurrence.

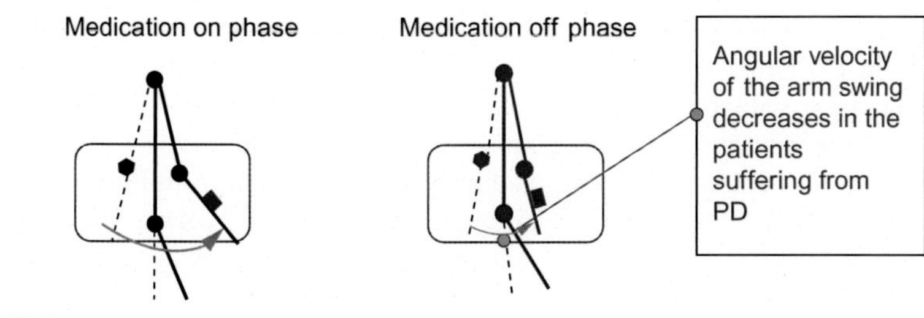

FIG. 8.4

Upper-limb anomaly in the gait cycle in patients suffering from Parkinson's disease.

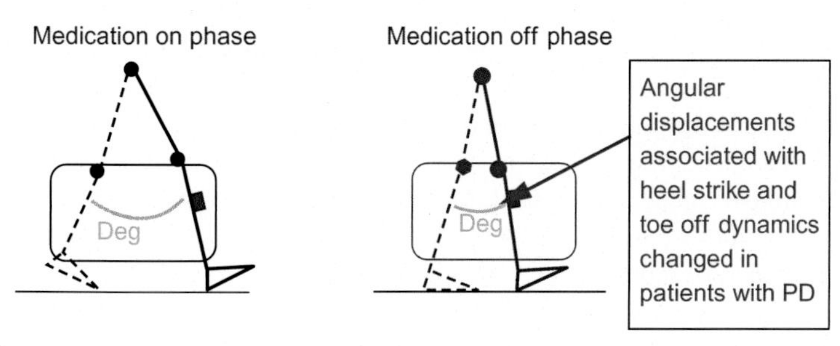

FIG. 8.5

Lower-limb anomaly in the gait cycle in patients suffering from Parkinson's disease.

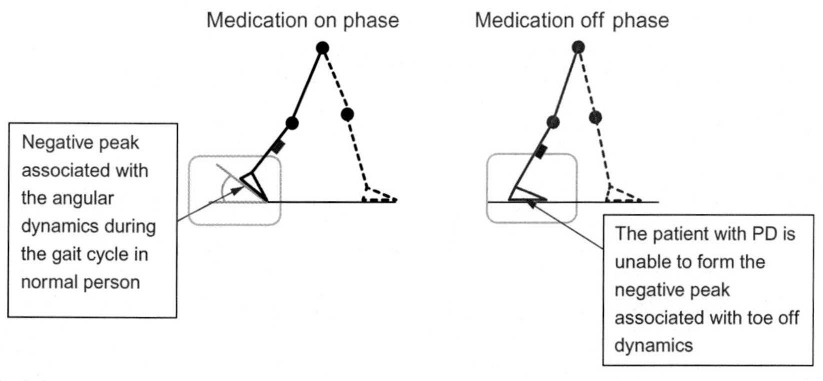

FIG. 8.6

Toe-off dynamics observed in patients suffering from Parkinson's disease.

Gaining an in-depth perception regarding the biomechanical changes observed in patients with PD helps us design the smart wearable technological devices, which can collect these valuable data that help us track the progression of the disease and take right measures at the right time.

Another significant anomaly observed in patients suffering from PD is the event of Freezing of gait (FoG), which is experienced when there is a blockage in the motor system and the person cannot continue with the motion, which can last from few seconds to almost a minute. It might lead to fall in certain cases and loss of independent movements. Evidence suggests that FoG is caused by the disruptions in the frontostriatal-parietal networks [7] and can lead to problems associated with central drive and automaticity.

8.3.2.1 Problem statement

Tremor analysis is one of the most crucial investigative studies to understand the progression of PD in the patient and helps in the correct identification of the treatment options available for him. Neurologists often rely on visual examinations to identify the tremor type in the patients and this process of examination is always not effective. Identification of the tremor helps in designing a therapy for the disease with a probability of higher success rate. Origin of each of the tremor type is different and depending upon this particular aspect, the treatment regime is being decided. There is another type of tremor called Essential tremor (ET), which has characteristics similar to the tremors associated with PD patients and thus often results in misdiagnosis. Research shows that the temporal fluctuations associated with kinetic and resting stages in the body are different in patients of PD and ET [8]. Knowledge of this gives us scope to design a better protocol for developing the sensors used for tremor analysis in case of PD and provide solutions with better efficacy.

Identification of a FoG event is another important symptom to track in the patients with PD. It has been seen that Rhythmic Auditory Signals (RAS) are effective in helping the patients resume their gait following a FoG event.

Our aim is focused toward developing smart solutions for collecting data from these tremors and anomalies in the gait cycle associated with patients of PD like FoG from sensors and analyzing them correctly.

8.3.2.2 Proposed design

Our proposed design is essentially a smart monitoring wearable tech consisting of two main parts: an Electromyographic (EMG) sensor and Inertial Measurement Unit (IMU) device. EMG measures the electric potential generated by the cells of the muscles upon activation. Our proposed IMU consists of three parts: a 3-axis accelerometer (used for measurement of instantaneous acceleration), a 3-axis gyroscope (used for measuring the angular displacement and velocity of the forelimbs and hind limbs), and a 3-axis magnetometer (measurement of the magnetic field around the body). For gaining maximum information about different types of tremors associated with PD, triaxial inertial sensor is more beneficial than uniaxial sensors since it can gather data from all the three axes.

The ratio of the temporal fluctuations (calculated from the angular velocities from IMU) between RT and KT in case of patients suffering from Parkinson's Disease is relatively higher, and this helps in differentiating between the tremors of PD and ET. The tremor recognition can further be made more specific by integration of the IMU device along with an EMG sensor so that data from both the angular dynamics and activation of the muscle cells can be collected to reach the conclusion. Integrated EMG and IMU sensors placed on the muscle of the forearm and on the anterior muscle of the leg can help in keeping the track of the tremors as shown in the figure. Fig. 8.7 shows the placement of the sensors on the body of the patient for collecting the data during motion. The data collected from the Extensor Digitorium muscle of the forearm and Tibialis Anterior muscle of the leg help to track the abnormalities in the gait. Placement of the sensors is very important to collect the relevant information without any interventions. The sensors should be placed between two motor points or between the motor point and the tendon insertion.

Another important aspect to keep track of using the smart wearable monitoring system for the patients with PD is the FoG detection as discussed earlier. The modulation of the internal clock involved with the BG (Basal Ganglia)-SMA (Supplementary Motor Area)-PMC (Premotor Cortex) circuit helps in planning of the cued tasks [9]. External auditory stimulus allows for the recalibration of the internal clock in cases of patients with PD experiencing FoG event.

The FoG detection algorithm consists of a dataset with data (angular acceleration instances) from and normal people and from patients suffering from PD and a supervised machine-learning model trained upon it [10]. Upon getting the prediction, analysis of the statistical and kinematic features associated with the gait cycle in patients with PD results in taking of immediate actions to prevent it. Research and clinical

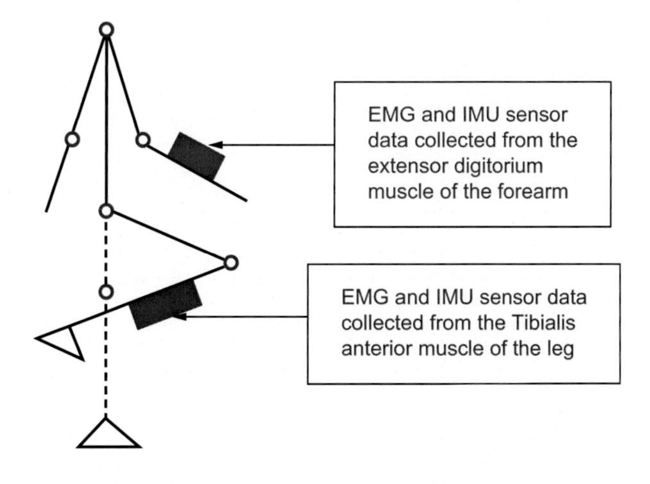

FIG. 8.7

Placement of sensors on the forearm and leg to track the motion of the patient.

trials have shown that rhythmic cueing or RAS helps in resuming the gait after a FoG event has occurred. Our design consists of an auditory signal output facility on detection of the FoG with cueing rates ranging from very slow, slow, fast to very fast, depending upon the necessity.

Our design also consists of a camera for capturing the motion during the gait cycle, which focuses on certain aspects of the body and helps to keep track of the movements associated with PD. Fig. 8.8. depicts the layout for the design for monitoring of symptoms associated with Parkinson's Disease using components of IoT and data gathered from different sensors.

Wi-Fi connectivity and Bluetooth connection are also present, which can connect to nearby devices such as Smartphone, Laptops, or Tablets in order to update the digital library of the patient and assess the performance and response to the medication associated with PD.

Our design is aimed toward providing a comfortable easy-to-wear technological device at the muscle of the forearm and the anterior muscle of the lower limb to collect the necessary data and keep the track of the symptoms associated with the patients with PD and help them to take decisions in real time.

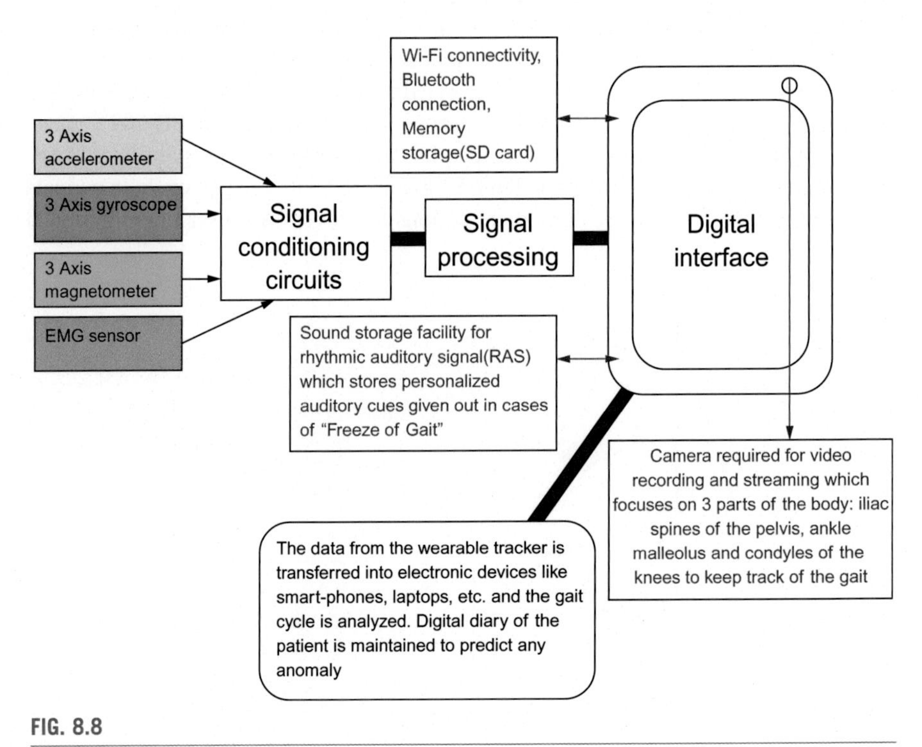

FIG. 8.8

Layout for IoT-based symptom detection of Parkinson's disease.

8.3.3 Smart band-aid to detect drug dosage

Inspired from the work carried out by Harvard Medical School, University of Nebraska, and MIT, our third case study deals with a smart bandage, which is made up of hydrogel that can sense certain parameters related to the wound and can track the drug dosage in real time. Along with the function of predicting drug dose, our proposed model also highlights the imaging of wound dressing in future by the use of holographic sensors.

The parameters that we are focusing on right now for tracking of the wound include: pH, moisture content, oxygen concentration, temperature, odor, and hemoglobin. Sensors, which are able to sense these particular features, are embedded inside the hydrogel-based bandage. A titanium wire, which can make the gel conductive, is passed through the matrix of the hydrogel. To eliminate the problems related to the healing of the chronic wound, which includes improper timing of the drug release, we are focused toward creating a model, which will put the necessities of the patients foremost.

Microparticles, which are generally in the range of 1–1000 μm, serve as carriers and reservoirs for drug to be delivered at the site of the wound. Microparticles, which can again be classified into microspheres and microcapsules, can act as effective means for providing accurate delivery of the drugs at a particular location and with the exact concentration as required [11]. Our aim is to incorporate these microparticles-carrying drugs into the hydrogel patch, which, upon response to certain stimuli, will get triggered and deliver the drug at certain location. The electric stimulator receives signal based on the change in parameters and the drug dosage is being decided accordingly. This allows for tracking the medical dose of the drug required in real time.

Our reason for choosing hydrogel as the matrix rather than a thread-based patch is generally due to the flexibility it offers and how it can mimic the natural movement of a person. Its ability to stretch and bend at points, which would have been difficult otherwise, accounts for the additional advantage that is provided with the model. The connection between different sensors is built using a Wireless Sensor Network (WSN), and this connection can be controlled remotely by the user using a smartphone. Upon tracking of any abnormality associated with a parameter in the chronic wound, an alert message is sent to the phone to check for the dressing and adjustment to the drug dose is done accordingly.

8.3.3.1 Problem statement

The main problem in the current scenario exists in providing the drug is targeted delivery with respect to the necessity at different stages of the healing process of the wound. This problem can be addressed by using the sensor-embedded smart bandage. Also, the process associated with every wound healing is different for different stages and, thus, not every wound will require the same treatment regime. The different parameters associated require adjustments, which are given by separate dosage profiles of the drugs and can be hard to maintain if real-time tracking is not possible. Sustained drug release is another criterion that needs to be focused on, and the issue can be addressed by using of the smart bandage.

Taking care of the wound at the point when an issue arises with it is also very necessary; otherwise, it can give rise to certain infections as well. This remote controlling of the drug dosage is also made possible by using this model, which can send an alert message even to the doctor appointed to the case as and when required so that the affected area can be managed right at that point of time.

8.3.3.2 Proposed design

Our proposed design targets certain parameters associated with the wound-healing process. The association of these biomarkers in relation to the wound-healing process is discussed in detail:

pH: The pH range of a healthy skin lies between 4.0 and 6.0. Acute wound healing is associated with acidic pH due to causes such as hypoxia and production of lactic acid at the site of the infection. The disruption of the Extracellular Membrane (ECM) in case of a nonhealing wound occurs when the skin is exposed to certain infections and inflammation is caused as a result. The disruption of the ECM results in release of degrading protease at the site of infection, which helps in elevating the pH level around 7.4, making the environment alkaline [12]. This change in pH can work as an effective biomarker in the healing process of chronic wounds and can work as an indicator for real-time monitoring in progression of the wound.

Hypoxia: checking for the level of oxygen at the site of wound is another indicator, which can be associated with the healing process. Hypoxia or the lower oxygen concentration at the site of the wound regulates certain cellular pathways, which is monitored by Hypoxia-Inducible Factor (HIF). HIF is directly associated with the functioning of certain genes, which can play active role in the wound-healing process. Hypoxic condition at the site of wound promotes proliferation of fibroblasts in human and for angiogenesis. Hence, oxygen concentration also acts a biomarker in cases associated with wound healing [13].

Moisture content: Depletion of moisture content on the skin surface for a person recovering from a wound can result in delayed wound-healing process by halting the cell migration and lowering the blood oxygenation level. However, excess of water content surrounding the surface of the wound can result in maceration of the wound tissue. Hence, there is a necessity to maintain a balance of the moisture content associated with the wounded area. The hydration sensor designed for this purpose is basically a capacitor and built based on impedance of the skin. More conductivity is produced when the moisture conductivity and dielectric constant of the skin is high [14].

Bacterial infection: The ability to form biofilm in cases of bacteria such as *Pseudomonas aeruginosa* makes them an active agent to be present frequently during infection at the area of the chronic wound. Formation of biofilm prevents the precise delivery of the drug at the particular location, and thus decreases the efficacy of the drug dosage. The redox potential of pyocyanin (a secondary metabolite produced by Pseudomonas aeruginosa) can be exploited to be used as a potential marker in cases of identification of infection in the wound [15].

Odor: Certain bacterial infection at the site of the wound can give rise to a characteristic smell, which can be determined by using odor detection sensor, which works basically by analyzing certain gases given out as a result of the infection. Metal oxide sensors can serve as effective means for detection of odor, which works on the principle of absorption and desorption of gases related with resistance modulation dependent upon the temperature sensing.

Hemoglobin content: Though a state-of-art sensor with the principle of vibrational spectroscopy has not been developed yet, future prospects might include developing such models, which can detect the presence of oxygen in the hemoglobin. Literature suggests that Near-Infrared (NIR) spectroscopy helps to determine oxy-hemoglobin concentration by using NIR fiber optic probe. Based on the reasoning that oxygen level in the hemoglobin is relatively high in cases of nonhealing wound, this model of a sensor can work in putting oxy-hemoglobin content as an effective biomarker in wound healing [16].

Temperature: Temperature modulation in relation with intensity of inflammation at the site of wound can also work as an effective indicator in the wound-healing process. Resistive temperature detector (RTD) can be used as temperature sensors based on a negative nonlinear relationship between temperature and resistance. Work has also been done in developing thermistor sensor based on carbon nanotubes for detecting the temperature at the wounded site.

Fig. 8.9 describes the entire workflow of our proposed design. The integration of different sensors has resulted in formation of a network, which works on WSN. The data are collected in the smartphone and advanced Machine-Learning tools help the patient to predict the symptoms associated with any anomaly in the wound-healing process and take action immediately.

Another aspect of the smart bandage model, which have been proposed as a vision for the near future, includes imaging of the wound in real time using holographic sensors, which can be integrated in the hydrogel matrix. Holographic sensors can diffract light between infrared and ultraviolet region, which is used for detection of certain physical parameters associated with the wound-healing process. This technology can be utilized for constructing highly sensitive optical sensors, which can analyze the image by a visual readout of the wound without disturbing the affected area. Work has already been done in creating glucose-responsive holographic sensor, which is functional in conjunction with 4-vinylphenylboronic acid (4-VPBA) [17]. Fig. 8.10 shows the idea of a holographic sensor in imaging of wounds in case of foot ulcer.

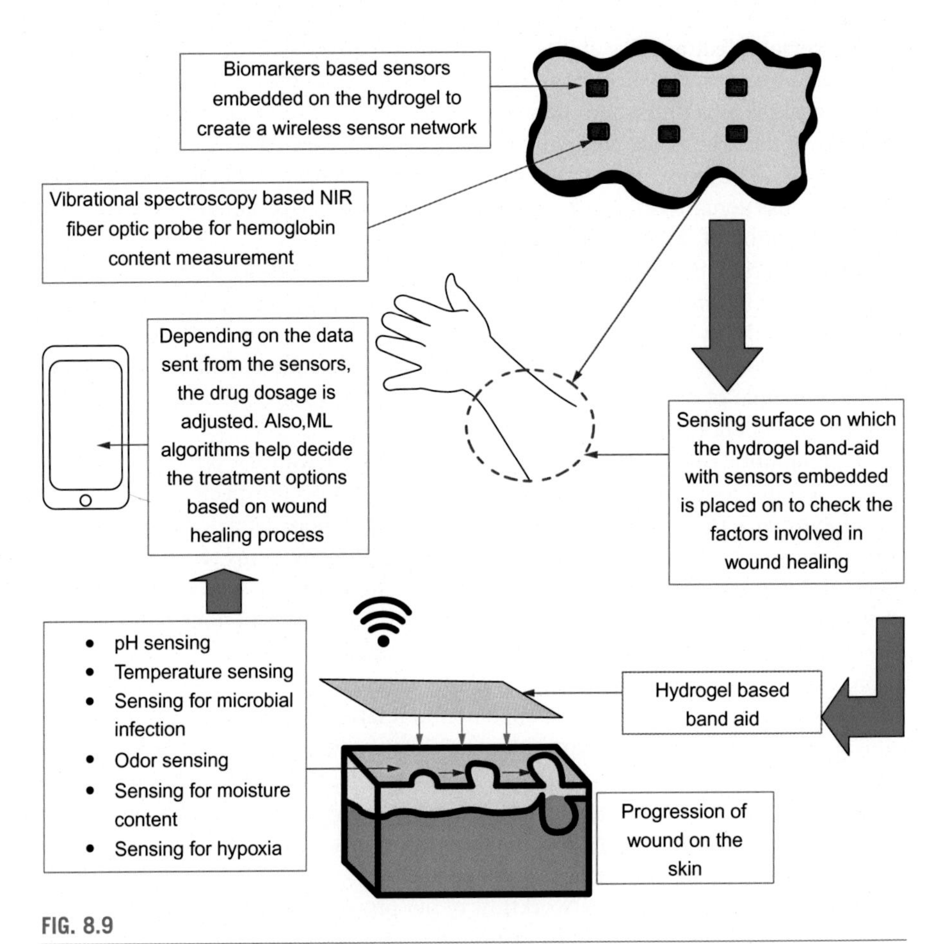

FIG. 8.9

Layout for biomarker detection using smart bandage for monitoring drug delivery.

8.4 Creating energy-efficient models

In this chapter, we discuss about one of the most widely used IoT networking models in recent times, which is energy efficient and consumes lesser power. Termed as LoRa (Long Range) technology, the main purpose of this wireless connections in IoT is to connect the individual devices in the network with the internet using lower radio frequencies for long-range communication. Though Wifi (Wireless Fidelity) provides similar function, it consumes a lot of energy in comparison to the amount of data being transmitted and is not fit for processing data from so many devices at once and processing them.

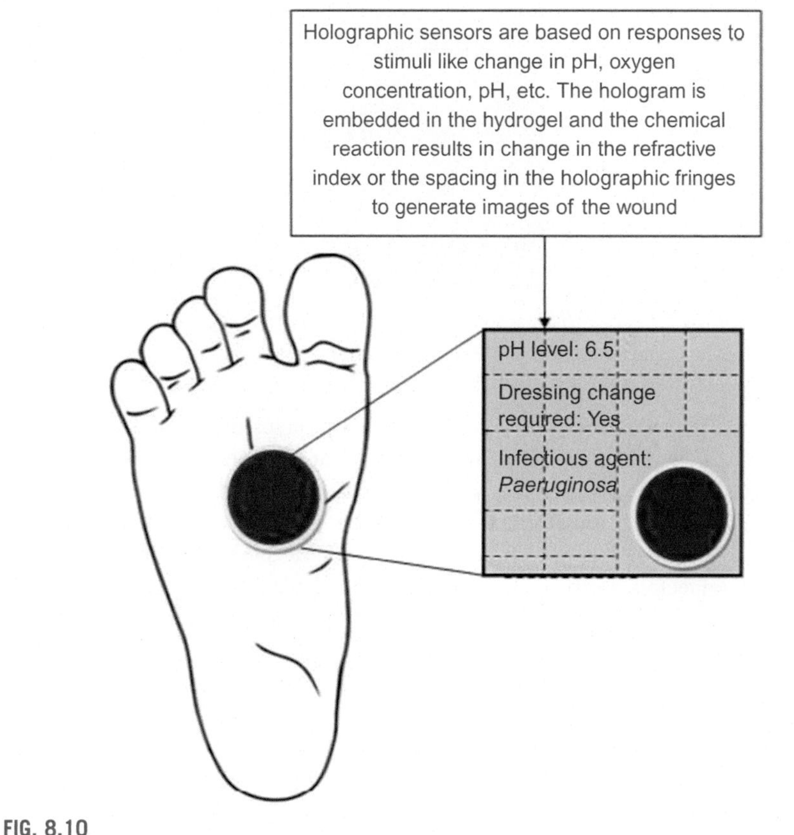

FIG. 8.10

Futuristic wound imaging technique using holographic sensors to get a visual readout of pH, infection feedback, and indicator displaying if the dressing change is required.

Developed by a company called SemTech in 2012, LoRa technology has gained considerable impetus and has established itself as a reliable networking option in connecting different battery-powered devices of IoT. The most impressive aspect of this technology is its ability to handle multiple nodes in the network at the same time and can span over significant distance. The technology also features different layers, which are stacked one upon the other to form a complete protocol. Fig. 8.11 depicts the LoRa Protocol stack.

The Application Layer consists of 3 broad parts: Frame Header (function includes device address identification, frame control, and frame counter), Frame Port, and Frame Payload. The Frame Header also includes other options such as changing the data rate, transmission power, and connection validation. The MAC (Media Access Control) layer consists of MAC Header and MAC Payload, which is used for authentication of the message at the end node by using the MIC (Message Integrity Code). The Physical Layer consists of a preamble, which depicts the modulation

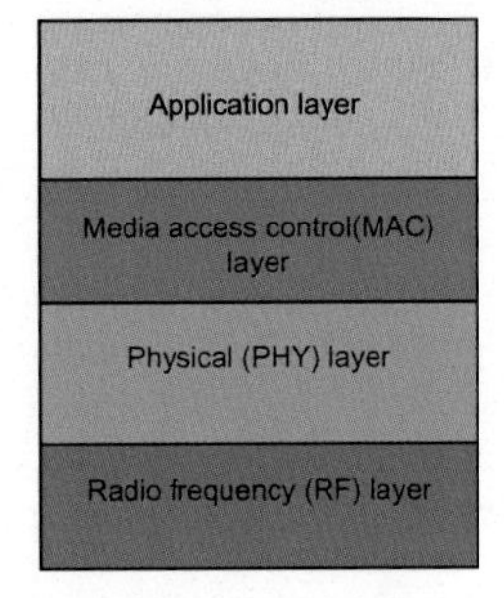

FIG. 8.11

LoRa protocol stack diagram to show the different layers involved in the technology.

scheme of the package. The RF (Radio Frequency) layer uses a spread spectrum modulation, which utilizes pulses of wide band linear frequency. The data transmitted are encoded by the level of increase or decrease in the frequency.

LoRaWAN, which stands for Low Power Wide Area Network, uses a star architecture for communication, which consists of a central node to which all the end devices are connected. LoRa is a chip-based spread spectrum modulating system, which results in transmission of data over long range and utilizes minimum energy.

8.5 Scope for the future

The scope of future in smart healthcare is tremendous and there is a lot of room for improvisation. The emergence of the concept of Electronic Health Record (EHR) has also given rise to interesting ideas such as smart healthcard for patients, which can be implemented and can serve as invaluable information source at the time of need. The Next-Generation Genome Sequencing has brought about a revolutionizing change in the world of genomic research and researchers are attempting to translate it into routine clinical practice through the EHR. The vision for a robust EHR system can be thought of as a platform, which can skillfully manage health profiles of individuals right from childbirth and use it for faster and more effective medical decision-making processes. Along with the data collected from smart healthcare devices, which can track important parameters such as the vital signs of the patient, physiological changes associated with a particular disease, etc., EHR platform can help us analyze the disease from both the molecular standpoint and also from the symptoms that arise. Embedded sensors, which can offer results with better sensitivity, are also coming up, which will help in getting more precise outcomes. The data collected from the healthcare device can be updated regularly to the EHR and deviation from the normal course of health can be easily detected. This helps in preparing a precise plan for managing the abnormalities and allows popularizing the concept of tailor-made medicines, which would be specific to the genetic makeup of the patient and offer a much precise form of diagnosis for the disease. The idea in creating a single integrated platform for healthcare management is mainly focused toward two important points: time and money. Time becomes an extremely crucial factor in case of certain disease diagnosis where the patient's life depends upon it. A fast and effective diagnosis helps you to reach at a decision quicker and allows you in managing the situation in a better way. Economy is another most important factor when it comes to medical health. Performing a number of tests to diagnose a particular disease costs an large amount money, and most of the time, the patient's family is not able to afford it, thus preventing the patient from receiving a good treatment. Creating a good EHR and making it a standard protocol for each and every person in the country can prevent many such complications and make proper medical treatment, one of the basic rights of the people, regardless of their economic background.

Along with this, future of smart healthcare also includes the concepts of telerobots, which are still at trial stages and have not been implemented commercially. Portable and multifunctional robots equipped with skills to perform minor surgical procedures and biopsies can become important in scenarios where accessibility is an issue.

Smart pills equipped with nanobots for monitoring the parts of the patients which is not detected through normal imaging techniques are also coming up rapidly. Advancements in response speed using the concepts of IoT can also help in saving lives of patients during emergency situations. Controlling of prosthetic limbs with human thoughts to provide the patient with a more natural feeling and smart beds in hospitals for patient safety requirements are also some of the few areas where the concepts of smart healthcare are being explored rapidly.

8.6 Conclusion

In this chapter, we have highlighted three major case studies associated with smart healthcare options, which offer interesting insights into the emerging trends of ubiquitous healthcare [18]. Advantage of using Information and Communication Technology (ICT) is to create better diagnostic tools, which can work in providing a better quality of life to the patients. The data collected from the devices can act as an elaborate database, which can help researchers to develop better healthcare options for the future. Along with the smart healthcare devices in use, various mobile apps are also being developed, which contributes in a big way to the transformation process. These forms of technological communications allow the doctors to remain updated about the progress rate of the patient and take care if any problems arise, without wasting much time.

Proactive health-monitoring devices are paving their way into an exciting future where there is an ample scope of growth and demand. These devices have a growing market potential as people are more aware these days and are looking for instant results. Integration of Big data analytics, Machine Learning, IoT, and Cloud Computing can provide invaluable contribution in the field of healthcare. Power-efficient low-cost networking solutions such as LoRa technology can be used effectively in such cases for providing wireless connectivity. Taking advantage of the automated decisions facilitated by these smart technologies, these devices coupled with the power of data analysis can account for successful diagnosis and better patient service in the future.

References

[1] G. Elhayatmy, N. Dey, A.S. Ashour, Internet of things based wireless body area network in healthcare, in: Internet of Things and Big Data Analytics Toward Next-Generation Intelligence, Springer, Cham, 2018, pp. 3–20.

[2] S.B. Baker, W. Xiang, I. Atkinson, Internet of things for smart healthcare: technologies, challenges, and opportunities, IEEE Access 5 (2017) 26521–26544.

[3] G. Acampora, D.J. Cook, P. Rashidi, A.V. Vasilakos, A survey on ambient intelligence in healthcare, Proc. IEEE 101 (12) (2013) 2470–2494.

[4] D. Oletic, Wireless sensor networks in monitoring of asthma, in: 2nd International; ICST Conference, 2009.

[5] A. Gholizadeh, D. Voiry, C. Weisel, A. Gow, R. Laumbach, H. Kipen, … M. Javanmard, Toward point-of-care management of chronic respiratory conditions: Electrochemical sensing of nitrite content in exhaled breath condensate using reduced graphene oxide, Microsyst. Nanoeng. 3 (2017) 17022.

[6] S.H. Li, B.S. Lin, C.H. Tsai, C.T. Yang, B.S. Lin, Design of wearable breathing sound monitoring system for real-time wheeze detection, Sensors 17 (1) (2017) 171.

[7] E. Heremans, A. Nieuwboer, S. Vercruysse, Freezing of gait in Parkinson's disease: where are we now? Curr. Neurol. Neurosci. Rep. 13 (6) (2013) 350.

[8] C. Thanawattano, R. Pongthornseri, C. Anan, S. Dumnin, R. Bhidayasiri, Temporal fluctuations of tremor signals from inertial sensor: a preliminary study in differentiating Parkinson's disease from essential tremor, Biomed. Eng. Online 14 (1) (2015) 101.

[9] A. Ashoori, D.M. Eagleman, J. Jankovic, Effects of auditory rhythm and music on gait disturbances in Parkinson's disease, Front. Neurol. 6 (2015) 234.

[10] S. Mazilu, U. Blanke, M. Hardegger, G. Troster, E. Gazit, M. Dorfman, J.M. Hausdorff, Gait Assist: a wearable assistant for gait training and rehabilitation in Parkinson's disease, in: Pervasive Computing and Communications Workshops (PERCOM Workshops), 2014 IEEE International Conference on, March, IEEE, 2014, pp. 135–137.

[11] K.R. Parida, S.K. Panda, P. Ravanan, H. Roy, M. Manickam, P. Talwar, Microparticles based drug delivery systems: preparation and application in cancer therapeutics, Cellulose 17 (2013) 18.

[12] E.M. Jones, C.A. Cochrane, S.L. Percival, The effect of pH on the extracellular matrix and biofilms, Adv. Wound Care 4 (7) (2015) 431–439.

[13] W.X. Hong, M.S. Hu, M. Esquivel, G.Y. Liang, R.C. Rennert, A. McArdle, … M. T. Longaker, The role of hypoxia-inducible factor in wound healing, Adv. Wound Care 3 (5) (2014) 390–399.

[14] S. Yao, A. Myers, A. Malhotra, F. Lin, A. Bozkurt, J.F. Muth, Y. Zhu, A wearable hydration sensor with conformal nanowire electrodes, Adv. Healthc. Mater. 6 (6) (2017) 1601159.

[15] H.J. Sismaet, A. Banerjee, S. McNish, Y. Choi, M. Torralba, S. Lucas, … E.D. Goluch, Electrochemical detection of Pseudomonas in wound exudate samples from patients with chronic wounds, Wound Repair Regen. 24 (2) (2016) 366–372.

[16] T.R. Dargaville, B.L. Farrugia, J.A. Broadbent, S. Pace, Z. Upton, N.H. Voelcker, Sensors and imaging for wound healing: a review, Biosens. Bioelectron. 41 (2013) 30–42.

[17] A.M. Horgan, A.J. Marshall, S.J. Kew, K.E. Dean, C.D. Creasey, S. Kabilan, Crosslinking of phenylboronic acid receptors as a means of glucose selective holographic detection, Biosens. Bioelectron. 21 (9) (2006) 1838–1845.

[18] N. Dey, A.S. Ashour, C. Bhatt, Internet of things driven connected healthcare, in: Internet of Things and Big Data Technologies for Next Generation Healthcare, Springer, Cham, 2017, pp. 3–12.

CHAPTER

Wearable fitness band-based U-health monitoring

9

Ravi Kumar Patel*, Akash Gupta†, Vinay Chowdary‡, Vivek Kaundal‡ Amit Kumar Mondal‡

**Research Centre, R&D Department, UPES, Dehradun, India †Dept. of Mechanical Engineering, UPES, Dehradun, India ‡Dept. of Electrical & Electronics, UPES, Dehradun, India*

9.1 Introduction

Health care is one of the most important aspects of our lives. It not only directly affects us as people but it also affects the economy. Therefore it is not surprising that the wearable technology market is gaining momentum. Health care via wearable devices brought changes in the field of medicine, as well as in research and development in different areas. Adding to this, today's advanced communication system has ensured real-time sharing of data around the clock. This new trend has opened up a new field called Mobile Health and according to the World Health Organization (WHO), it has enabled health check-ups to be conducted via smartphones and other types of wearable devices [1].

It's true that wearable healthcare devices have made people aware of their physical state by promoting a healthy lifestyle and removing the need to regularly visit healthcare clinics. At the same time, however, they have introduced grave privacy and security issues to our society. Wearable healthcare devices have to be designed in such a way that they ensure ergonomics, ease of use, and simple battery charging as well as protection of user data. Clearly there is a potential for misuse of health information, which must be addressed in the form of policies and regulations.

Wearable devices have fascinated both academic and industry communities. This high level of interest has made these devices very popular in day-to-day life. These health devices can be worn on the body and monitor an individual's activity without any interrupt or limitation on the user's motion [2]. Wearable devices are small and part of the computers that users wear on various parts of their bodies, such as clothes [3], smartwatches, bracelets or wrist bands [4], and glasses [5]. Today there is a wide range of fitness devices available that are helpful in tracking daily activities. These devices mainly consist of pedometer sensor for tracking the number of steps taken by an individual depending upon their motion of hands or hips. The devices that make use of these pedometers are found to be better activity life trackers. This makes the device smarter with high accuracy and calculates more than just walking distance [6]. The area of developing fitness monitoring devices and frameworks is moving

Sensors for Health Monitoring. https://doi.org/10.1016/B978-0-12-819361-7.00009-9

towards miniaturizing devices that measure vital signs with secure communication between the devices and a server. Newer compact devices track and monitor fitness statistics, such as walking and running distance, calories consumed and expended, heart rate, sleeping duration, and so on, and then sync the data with a computer or smartphone. Some smartphones also have their own ability to track health activities [3] (Fig. 9.1).

The ubiquitous monitoring of human health in the coming generation can be done with the help of wearable devices [7]. Fig. 9.2 depicts the wearable sensor nodes that connect to the human body and are helpful in recording one or more potential physiological parameters. These recorded signals are sent wirelessly to the base station/electronic world such as smartphones, personal computers, and so on.

FIG. 9.1

Wearable fitness device tracking and monitoring features.

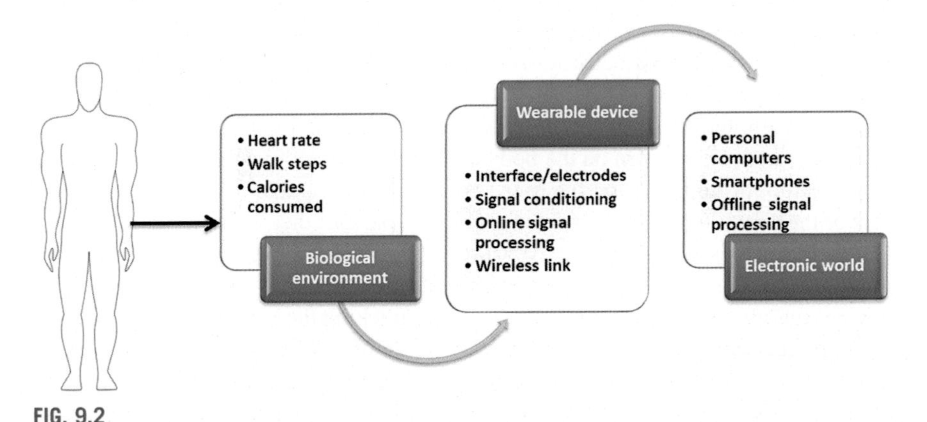

FIG. 9.2

Wearable device connects the biological environment (heart rate, walk steps, calories, etc.) to the electronic world (personal computers, smartphones, offline signal processing) and wearable devices (interface/electrode(s), online processing, signal conditioning, and wireless link).

9.2 Interfacing

The whole of fitness-centric tech products has seen substantial demand over the past few years [8]. This is basically due to the boom in smartphones and mobile gadget systems. Wearable fitness bands available in the market mostly focus on interfacing with smartphones via Bluetooth. These fitness trackers are a cheaper way to monitor your basic health in an optimal way. The most common feature of these fitness trackers is their ability to count steps, calories burned, heart rate, sleep quality, and so on. Accomplishments in the area of sensors and their day-by-day miniaturization have to lead us to a single device for basic health monitoring without having to visit a physician. These devices act as wireless sensor nodes that collect data and transfer it via a suitable wireless sensor module. These sensors can be chemical and biochemical sensors or physical transducers. Chemical sensors are much more complex to work with as compared to physical transducers [9].

Fig. 9.3 shows the interfacing of a wearable fitness device with smartphones and PCs. The device gathers required data and temporarily stores it in the buffer memory and then transfers the data via wireless transmitter when the connection is established [10]. Fig. 9.4 shows the different layers present in the architecture of the device's connection with biomedical signals.

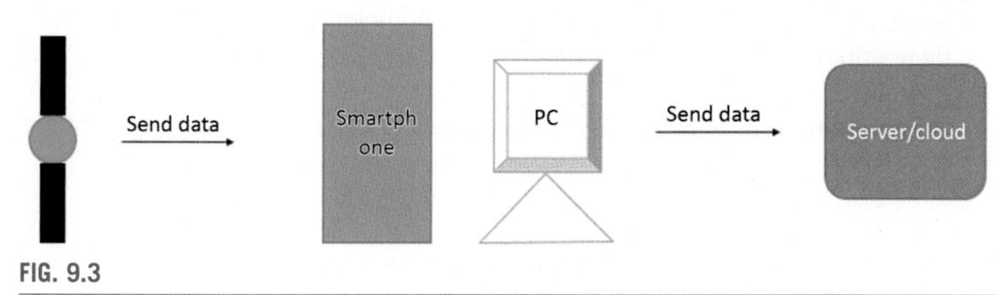

FIG. 9.3

Interfacing of wearable sensor nodes with smartphones and PCs [10].

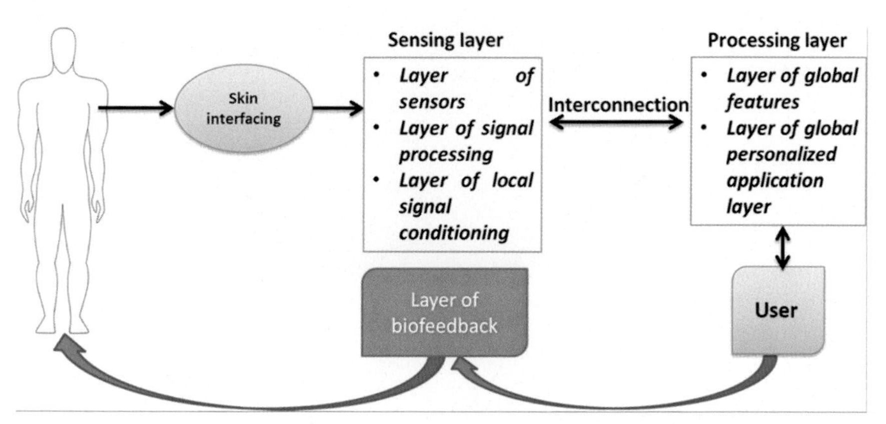

FIG. 9.4

Architecture of device's connection with biomedical signals.

9.3 Features

The most important challenge in the field of wearable sensors is their wearability. Wearing a physical sensing device offers various constraints that have been downplayed by fitness tracking bands that have replaced conventional wristwatches. With an option of remote tracking, these activity trackers have also found their application in medical science. These tracking devices can be helpful for doctors to track the basic health parameters of their patients remotely through smart devices, thus reducing the burden on patients to visit hospitals and clinics on a regular basis [11–13].

These fitness tracker bands use different types of sensors, which in turn provide various features to the device [14]. Fig. 9.5 shows percentage-wise distribution of availability of the sensors in fitness bands. Some types of sensors are as follows:

- *3-Axis accelerometer:* Tracks the movements of the person in every direction. It is used to record the number of steps taken by the individual wearing it as an activity tracking device.
- *Gyroscope:* Mainly used for the measurement of angular velocity because it is capable of measuring rotations and orientations. When combined with a 3-axis accelerometer, it provides motion tracking in six degrees of freedom.
- *Altimeter:* Used for measuring altitude. It helps mountaineers and skydivers track their altitude while climbing or diving. However, only a few devices contain this instrument.
- *Temperature sensor:* Reads the body temperature attained or reached during a workout as well as in general environmental conditions.

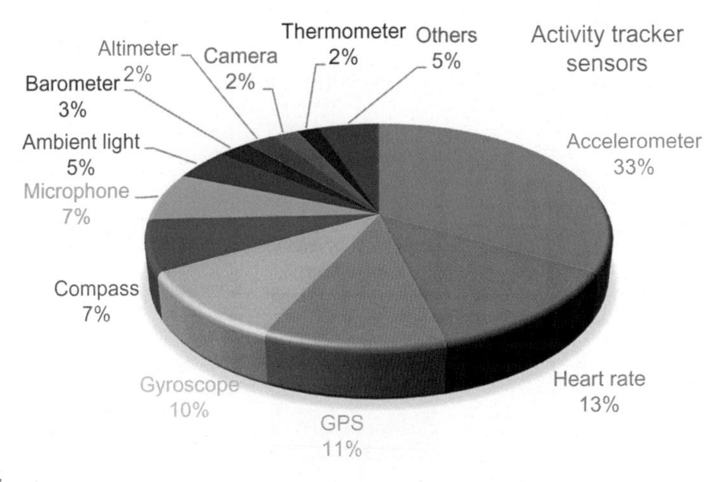

FIG. 9.5

Commonly used sensors in fitness trackers [15].

- *Global Positioning System (GPS):* Available on most of modern smartwatches that also act as activity trackers. The main advantage of these smartwatches is that they have an inbuilt operating system (Android or iOS).
- *Heart rate monitor:* The heart rate monitor uses a simple technique of flashing light to measure your pulse [16]. This technique is called photoplethysmography (PPG) [17].
- *EMG sensors:* These are used to track muscle activity by acquiring the small signals sent to the muscles by the brain. These sensors are placed over the muscles in order to track the signals. Most full-body fitness-tracking clothes use this technology to track muscle function [18].
- *Electrocardiograph (ECG):* An ECG is used to identify cardiac abnormalities. To perform an ECG multiple electrodes are placed on the human skin close to the heart. The electrodes measure all the electrical activities produced by the heart as it contracts. The produced electrical activities are then sent to a receiver where the information is recorded, and at this stage the heart's rhythm can be analyzed and irregularities detected. Researchers [19] studied the effective morphological ECG baseline and drift correction algorithm that is applied for adult and fetal ECG preprocessing. The study showed that the quality was better with less error in baseline drift approximation. The major benefit of ECG wearables is improved accuracy as these wearables don't require any plastering (like a chest strap with ECG technology) on the human body with all those electrodes. These wearables work with smartphones and other wearables (like watches). The recorded signals using ECG wearables can be sent to a receiver through wireless technology [20]. Some available ECG wearables include the Apple Watch Series 4, KardiaBand, QardioCore, and Hexoskin Smart Shirts.

The most important feature of these devices is calibrating the number of calories burned throughout the day. These bands are also helpful while performing exercise. When combined with your smartphone applications, they keep the record of your workouts [21], shown in Fig. 9.6A and B.

Table 9.1 compares the different features available in some of the most popular fitness-centric devices.

9.4 Placement of bands

Activity trackers (smart bands or fitness bands) are normally worn on the wrist or hip and are usually dedicated to physical activity tracking. An activity tracker is usually cheaper than a smartwatch because of its less expensive hardware components and fewer sensors. It generally results in better battery life and a limited user interface for showing tracking results [22].

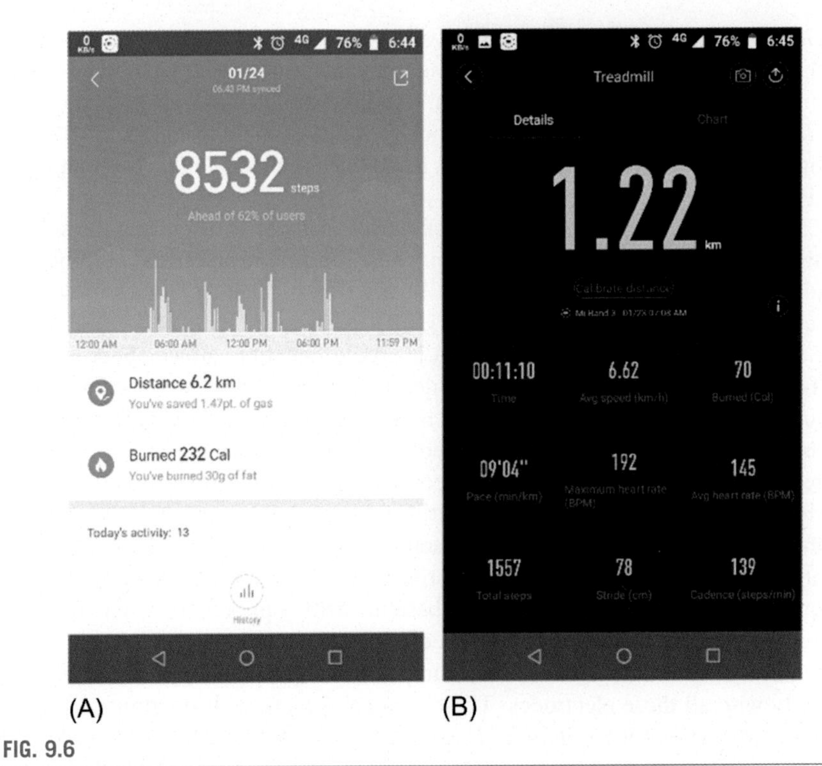

FIG. 9.6

(A) and (B) shows the features of the fitness band smartphone UI applications.

Table 9.1 Comparison of different features in various fitness bands [21]

Features	Fitbit flex	Misfit shine	Mi Band 3	Withings pulse
Design	Good and sleek design, ideal for any sport	Fashionable and attractive design	Fashionable, light and attractive design	Unattractive design
Display	Easy to activate the screen just by tapping	Clock display can also be used as a watch	0.78-inch OLED touchscreen display that shows app notifications, SMS, step count, calories burned etc.	Large display and shows activity tracking

Table 9.1 Comparison of different features in various fitness bands [21]
Continued

Features	Fitbit flex	Misfit shine	Mi Band 3	Withings pulse
Water resistance	Waterproof and can be used in the shower	It is water resistant and specially designed for swimming	Resistant up to 50 m, it features an IP67 rating for water resistance	Not a water-resistant band
Metric function	The main function is food tracking	The main function is goal tracking	Various functions like accurate step counting when compared to its predecessor, the band also features continuous heart rate monitoring	The main function is heart rate measurement
Battery	Battery indicator to check battery status	It is comfortable, no need to charge the battery	Up to 20 days of battery life with 110 mAh battery capacity and can be charged via USB cable	Battery indicator to check battery status
Synchronization	Always loses connection and slow synchronization	Synchronization is fast	Connectivity is provided via Bluetooth 4.2 that provides excellent connectivity	Data transfer via Bluetooth and Wi-Fi, fast synchronization

9.5 Design of fitness bands

Bands and user interface (UI) apps are designed in order to attract a large number of users. As discussed in Table 9.1, the design must be attractive and the band material must be sturdy. The band material must be created from non-allergenic, sturdy, and water/sweat proof materials. The most common materials used are plastic, rubber, leather, and aluminum alloys. The modern bands have attractive displays, many of them even have a touch display feature. The bands are made in such a way to fit most people and provide an adjustable strap that can be sized according to different users. The UI of the band can be black and white or colorful depending upon the price range. Many such smartwatch cum fitness bands like Fossil and Samsung are preloaded with Android apps that directly connect with your smartphone. The bands also support sleep tracking and alarm functions with on-screen notifications and vibrations. However, the disadvantages of these devices include inadequate waterproofing and complex battery chargers.

9.6 Energy requirement for fitness bands

Batteries are selected from three available battery size groups: (1) cylindrical cell (CYC), (2) button cell (BC), and (3) coin cell (CC). These battery sizes have the potential for powering wearable fitness bands. The specifications mainly considered for low-power wireless design are nominal voltage (V), maximum current (mA), minimal current (mA), minimal capacity (mAh), and size (diameter and height (mm × mm)). The device volume is dominated by the physical size of the battery depending upon its energy storage capacity (mAh). The battery available has its own internal resistance due to which all stored energy cannot be fully discharged in all the potential loads. Due to this condition, additional battery parameters are defined in Table 9.2.

Fig. 9.7 shows the various parts of a wearable device that require a battery source for its effective operation. The parts require a defined range of voltage such as those outlined in Table 9.3.

9.6.1 Input voltage

Initially, the battery voltage provided is essential for meeting the task necessities of the electronic hardware utilized in a fitness band. Provided voltage isn't equivalent to the battery minimal voltage due to internal voltage drop of battery because of inner resistance, and this fall will differ depending upon the current draw. The available low-controlled microcontrollers and handsets require voltage somewhere around 1.8–3.6 V. The batteries that provide a voltage outside the required range must be utilized with a voltage converter (DC to DC) or arranged so to build the transferred voltage. Nevertheless, these strategies lessen the lifetime of the battery due to both additional power utilization from extra hardware or because of expanded inner impedance.

9.6.2 Maximum current (continuous)

Furthermore, the maximum current (additionally indicated as the maximum usual current source), I_{avg}(max), confines normal current usage from battery a. In principle, a battery of 200 mAh can give about 1 mA for 200 h, or about 200 mA for 60 min. In exercise, every battery really has extreme maintained current usage and if extra

Table 9.2 The summary of battery utilized for powering the sensor nodes installed in fitness bands

Category	Minimal voltage (V)	Max. current (mA)	Minimal current (mA)	Minimal capacity (mAh)	Size (mm × mm)
Round LiPoL	3.7	–	–	220–1500	30–50
$LiMnO_2$	3.0	4	0.5	285	24.5 × 3.0
$LiSOCl_2$	3.5	50	1	1200	14.5 × 25.2
$Zn(OH)_4$	1.4	16	2	600	11.6 × 5.4
$Zn(OH)_4$	1.4	6	0.9	290	7.9 × 5.4
(CF)n/LI: poly-carbon monofluoride (BR) [BR3032]	3.0	–	–	500	30 × 3.2
MnO_2/LI: Manganese dioxide (CR)					
CR2032	3.0	10	0.19	225	20 × 3.2
CR1632	3.0	6.8	0.19	140	16 × 3.2
CR2330	3.0	10	–	265	23 × 3.0
CR2354	3.0	12	–	560	23 × 5.4
CR2450	3.0	15	0.39	620	24.5 × 5.0
CR3032	3.0	15	–	500	30 × 3.2
CR2025	3.0	10	0.19	170	20 × 2.5
PD3048	3.7	9	–	300	30 × 4.8

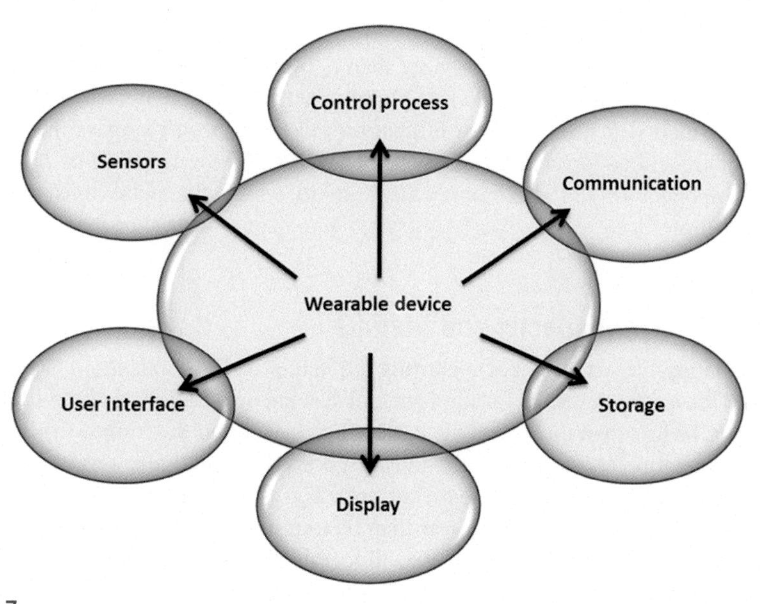

FIG. 9.7

Wearable device consisting of control process, communication, storage, display, user interface, sensors.

Table 9.3 Wearable device parts with required voltage ranges

Parts	Particulars	Voltage range (V)
Control processor (ARM M-Series)	Logic	1.8
	Core	1.2
Communication	GPS, BTLE	1.8
Storage	RAM, FLASH	1.8
Display	LCD, OLED, TR	12
User interface	Touch, Haptics, Buttons	3
Sensors (ECG, OHR, Inertial)	LED	5
	Analog	3.3
	Core	1.2

current is used as per defined then the operative limit won't be exactly 200 mAh. The normal current required by the framework must be lesser than the comparing I_{avg}(max) for the particular battery. Additionally, if more than one battery is utilized then it should be in parallel, perhaps organized with the diodes to keep any nonbattery-powered batteries from coincidentally charging. Once more, the battery must have the capacity to give this current lacking a substantial fall in its voltage source in view of inner impedance.

9.6.3 Pulse current capability

For brief times a battery can give more than I_{avg}(max), up to a greatest pulse value I_{pulse}(max). I_{pulse}(max) should be sufficiently vast to ensure that varieties in source voltage, which will fall if current utilization increases, won't surpass the working scope of the circuit present in the wearable device. The estimation of I_{pulse}(max) is a component of the maximum extent up to which the pulse must be given (the hold time).

9.6.4 Effective capacity and lifetime

Consolidating the above impacts permits the viable battery limit (C_{eff}) to be established. This is a basic factor for long-span observing applications and it will, for the most part, be less than the ostensible battery limit, except if the framework's normal current utilization (I_{sys}) matches the maker's prescribed rate (also referred to as the minimal current or the normal release current). The estimation of C_{eff} is subsequently an element of I_{sys} and it will be lower than its minimal rate when I_{sys} is greater than the minimal current. Depending on this, all together for a specific battery to be a reasonable alternative for a particular framework it must meet the framework prerequisites regarding not only size and supply voltage, but also the normal and most extreme powerful flows that the framework needs to work. When the important situations are met the lifespan of the battery, $LT\ (I_{sys})$, can be resolved as

$$LT\left(I_{sys}\right) = \frac{C_{eff}\left(I_{sys}\right)}{I_{sys}}$$

9.7 Standardization/regulation

In many ways, we can say that wearable fitness bands are just another facet of the Internet of things (IoT), with a specific application in the areas of e-health, lifestyle, fashion, and so on. Very often IoT is said to be disruptive in nature [23] and anything that is said to be so leads to tough standardization. Although IoT was familiar under the broad area of machine to machine (M2M) interfacing, it was less evocative. Standardization of hardware like chips, printed circuit boards, and the software framework to be used has lowered the cost of mobile Internet devices. The same trend can be expected in wearable fitness bands, starting with all the major chip suppliers.

Regarding regulation of data, consider fitness bands that record data pertaining to heart rate and number of steps taken and report this information to the manufacturer who stores it for targeted advertisement to its customers [24]. If the user registers (with name, gender, social networking accounts, etc.) with the manufacturer, the user's account details and health information can be used by the manufacturer to identify him or her. However, if the user just signs up for heartbeat and steps tracking only (and does not officially register), then the manufacturer will have no way to identify him or her and hence the user's data is protected.

9.7.1 Fitness apps

There is a large number of fitness-related apps available for mobile devices. From the outside this looks good, as it adds innovation and quality of service. However, consumers need to understand that their health-related data is with a third party and is being tracked, which can have significant security and privacy implications. In the coming years, there will be an inevitable need for some form of user awareness or standardization from the industry or government side to increase the security and reliability of apps that manage personal information.

9.7.2 Material standardization

Due to the digital revolution and availability of 3D printing, the price of wearable fitness products has decreased. However, the quality of the material has to be taken as top priority because wearable fitness products like fitness bands remain in contact with the wearer's skin for the long term.

9.7.3 Wireless standardization

Most of the available fitness bands support Wi-Fi and Bluetooth for interoperability and sharing of data. But the products should be under the acceptable level of radiofrequency exposures. Also, fitness bands are continuously worn but need charging and hence many of these products come with wireless chargers. In the case of wireless charging a lot of work has been done, but there is still no standardized framework. However, we do have three competing standards for low-power wireless charging: (1) the Power Matter Alliance, (2) the Wireless Power Consortium, and (3) the Alliance for Wireless Power. While it is a clear positive sign for innovation in the short term, it adds complexity for manufacturers and confusion for potential users unless product manufacturers support all of the competing standards in one product.

There are many organization that help in regulating the norms, but they are all country specific. For example, Europe has the ERC 7003 document from the European Research Council (ERC) and the European Telecommunication Standards Institute (ETSI) for electromagnetic field (EF)-based pollution, France has the L'Agence nationale des fréquences (ANFR), the Autorité de Régulation des Communications Électroniques et des Postes (ARCEP), and the French Agency for Food, Environmental and Occupational Health & Safety (ANSES), and globally there is the International Commission on Non-Ionizing Radiation Protection (ICNIRP).

In short, we can say that "regulation" consists of a series of official documents/rules made by an organization that may or may not be attached to a "State" or a "community of States," whose respect is made mandatory through requirements, rules and regulations, laws, by-laws and/or decrees, and other legal texts that regulate a social activity. To this day, the regulatory constraints in relation to wearables, specifically fitness bands, generally follow the following five fundamental rules:

1. controlled use of radiation and radio frequency
2. controlled human exposure to non-ionizing EFs
3. following "Medical Device" regulations
4. respect for private life and notions of privacy
5. proper waste management

9.8 Discussion

Although there are no defined guidelines available, in the view of authors, fitness bands and the endpoints derived from the data they collect have the potential to be suitable for use in the public domain, if they follow a certain basic set of important rules. Groundwork related to the understanding of the potential of wearable-derived endpoints can be assessed from earlier literature.

Commenting on the Indian perspective, the country is only just beginning to use wearable fitness band technology. The Indian fitness band market comprises 90% of the wearable market. The major players in the Indian market are GOQii, Xiaomi, and

Fitbit [25]. Fitness bands are, in general, robust and can work with little maintenance. The most intuitive devices of the incumbent players will be the ones to capture the major market share of Indian consumers [26].

The fitness band market is in a state of tug of war in providing value to their products, be it features such as heart rate monitoring, waterproofing, or sleep tracking. Going beyond, companies like GOQii are providing free expert personalized coaching, diet advice, and other value-adds for a few months to keep hold of its customer base. There are a lot of Chinese vendors also available in this segment, which pressures companies to lower prices. Eventually the next stage of the industry will be commoditization unless players continue to add more features and experience. Fitness bands should be smart and be connected to the Internet to provide a seamless experience. They should have a minimum amount of self-processing power to produce data to users in an ambient way.

The basic objective of fitness bands is to create value on the basis of their daily usage, without interrupting the user in implementing his own tasks. Fitness bands have the additional benefit of tracking the user's mental state, whether relaxed or stressed, and can provide responses accordingly.

There is a lot room for innovation and improvement of existing fitness bands, such as gamification, social incentives, augmenting services, and many others.

9.9 Conclusion

Mobile health care is a booming market and surely it will continue to flourish in the near future. However, it has to be mandated that the user is given full control of their data, even if a third party owns the application or device. Thus issues are likely to arise when one party decides to share information with different companies for profit. Hence, in the context of the healthcare system, it is necessary to apply privacy and safety rules [27].

Technology and aesthetics must go in hand in hand; sleek and modern lightweight design, waterproofing, different options for charging, repeatability and accuracy (even for climbing and descending stairs), and ability to measure vital parameters (heart and pulse rate, body temperature, respiration, etc.) should be there or added. Nevertheless, the current pace of wearable fitness band development is accelerating rapidly with the release of new gadgets and next-generation technology.

References

[1] M. Kay, J. Santos, et al., mHealth: New Horizons for Health Through Mobile Technologies, vol. 64 (7), World Health Organization, 2011, pp. 66–71.

[2] H. LeHong, A. Velosa, Hype Cycle for the Internet of Things, 2014 (Internet), Gartner Inc., Stamford, CT, 2014 (cited 25 January 2017).

[3] R. Scoble, S. Israel, et al., Age of Context: Mobile, Sensors, Data and the Future of Privacy, Patrick Brewster Press, USA, 2014.

[4] T. Martin, E. Jovanov, et al., Issues in wearable computing for medical monitoring applications: a case study of a wearable ECG monitoring device, in: The Fourth International Symposium on Wearable Computers, IEEE, 2000.

[5] E. Jovanov, P. Gelabert, et al., Real time portable heart monitoring using low power dsp, in: International Conference on Signal Processing Applications and Technology ICSPAT, 2000.

[6] W. Devices, Wearable Technology and Wearable Devices: Everything You Need to Know, 2014.

[7] N. Dey, A. Ashour, et al., U-Healthcare Monitoring systems, vol. 1, 2019. Available from: http://public.eblib.com/choice/PublicFullRecord.aspx?p=5518679.

[8] Techradar.com, Best Fitness Tracker in India: The Top Activity Bands in 2018, Available from: https://www.techradar.com/news/best-fitness-tracker-in-india-the-top-10-activity-bands-on-the-planet.

[9] S. Coyle, V.F. Curto, et al., Wearable bio and chemical sensors, in: Wearable Sensors, Elsevier, 2014, , pp. 65–83.

[10] G. Chen, E. Rodriguez-Villegas, A.J. Casson, Wearable algorithms: an overview of a truly multi-disciplinary problem, in: Wearable Sensors, Academic Press, 2014, pp. 353–382.

[11] N. Dey, A.S. Ashour, C. Bhatt, Internet of things driven connected healthcare, in: Internet of Things and Big Data Technologies for Next Generation Healthcare, Springer, Cham, 2017, pp. 3–12.

[12] N. Dey, A.E. Hassanien, C. Bhatt, A. Ashour, S.C. Satapathy (Eds.), Internet of Things and Big Data Analytics Toward Next-Generation Intelligence, Springer, Berlin, 2018.

[13] S. Majumder, T. Mondal, et al., Wearable sensors for remote health monitoring, Sensors 17 (1) (2017) 130.

[14] L. Cashmere, Types of Sensors in Wearable Fitness Trackers, Available from: https://www.news-medical.net/health/Types-of-sensors-in-wearable-fitness-trackers.aspx, 2018 (Retrieved 22 January 2019).

[15] F. de Arriba-Pérez, M. Caeiro-Rodríguez, et al., Collection and processing of data from wrist wearable devices in heterogeneous and multiple-user scenarios, Sensors 16 (9) (2016) 1538.

[16] R. Rettner, How Well Do Fitness Trackers Monitor Heart Rate? Available from: https://www.livescience.com/44170-fitness-tracker-heart-rate-monitors.html, 2014 (Retrieved 22 January 2019).

[17] A. Henriksen, M.H. Mikalsen, et al., Using fitness trackers and smartwatches to measure physical activity in research: analysis of consumer wrist-worn wearables, J. Med. Internet Res. 20 (3) (2018).

[18] K. Vanhemert, Coming Soon: Workout Gear That Monitors Your Muscles, Available from: https://www.wired.com/2013/12/these-smart-gym-clothes-are-the-future-of-wearable-computers/, 2013 (Retrieved 9 February 2019).

[19] M.H. Sedaaghi, M. Khosravi, Morphological ECG signal preprocessing with more efficient baseline drift removal, in: Proceedings of the 7th IASTED International Conference, ASC, 2003.

[20] N. Dey, A.S. Ashour, et al., Developing residential wireless sensor networks for ECG healthcare monitoring, IEEE Trans. Consum. Electron. 63 (4) (2017) 442–449.

[21] K. Kaewkannate, S. Kim, A comparison of wearable fitness devices, BMC Public Health 16 (1) (2016) 433.

[22] V. Chowdary, V. Kaundal, P. Sharma, A.K. Mondal, Implantable electronics: integration of bio-interfaces, devices and sensors, in: Medical Big Data and Internet of Medical Things, CRC Press, 2018, pp. 55–79.
[23] N. Jesse, Internet of things and big data—the disruption of the value chain and the rise of new software ecosystems, IFAC-PapersOnLine 49 (29) (2016) 275–282.
[24] S.P. Cohn, Privacy and Confidentiality in the Nationwide Health Information Network, Available from: http://www.ncvhs.hhs.gov/060622lt.htm, 2006.
[25] A. Ahaskar, Wearables Market Growing in India, Goqii Remains at Top: IDC, Available from: https://www.livemint.com/Technology/9I6p4AJRI29cN6D1n0krjJ/Wearables-market-growing-in-India-Goqii-remains-at-top-IDC.html, 2017 (Retrieved 24 January 2019).
[26] N.P. Shreyas Joshi, The Future of Wearable Tech in India Bengaluru, IIM, Bangalore, 2017.
[27] M. Al Ameen, J. Liu, et al., Security and privacy issues in wireless sensor networks for healthcare applications, J. Med. Syst. 36 (1) (2012) 93–101.

CHAPTER

Role of trust in the ubiquitous healthcare system: Challenges and opportunities

10

Vartika Kulshrestha, Seema Verma

Department of Electronics, Banasthali Vidyapith, Vanasthali, India

Up until recently, health and medicine was basically a hit or miss affair (...) All of that has now changed, and will dramatically change clinical practice by the early 2020s.

Inventor and Futurist at Google, Ray Kurzweil (2013)

10.1 Introduction

Individuals are progressively eager with the shortcomings of health administrations to convey attitudes of national inclusion that meet expressed demands and dynamical needs and with their inability to deliver benefits in a manner that compare to their desires. Some would argue that smart health frameworks must respond more quickly and efficiently to the difficulties of an evolving world, and the ubiquitous healthcare system can do that. The Alma-Alta Declaration of 1978 [1] activated an "Essential Health Care Movement" of experts and establishments, governments and civil organizations, scientists, and grassroots associations that embraced to handle the "politically, socially and financially unsatisfactory" [1] health imbalances around the world. To deliver healthcare for all, the smart healthcare continuum system should include not only doctors' facilities, long-term care offices, and home care, but also smart medical services such as mobile apps and wearable sensors, all of which are part of the IoT.

For instance, a few healing centers have started using "smart beds" to observe patient behavior and provide data over the Internet to medical attendants [2]. The beds can also self-acclimate to make sure that proper pressure and support is applied to the patient without having to be physically balanced by the medical caretakers. Another space where smart healthcare technology is being used is with home medication providers to mechanically transfer information to the cloud once the medication isn't taken [3].

Today, in general, individuals are wealthier and live longer than they did 30 years ago. If people were dying at 1978 rates, 2006 would have seen the passing

Sensors for Health Monitoring. https://doi.org/10.1016/B978-0-12-819361-7.00010-5

of 16.2 million people [4]. However, the actual number was closer to 9.5 million. This is comparable to 18,329 lives being saved each day. This demonstrates that progress is feasible. The worldwide wellbeing economy is rising quicker than GDP, having augmented its share from 8% to 8.6% of the world's value between 2000 and 2005.

10.1.1 Growing expectations for better health performance

A rapidly aging populace has highlighted the significance of economical medical services frameworks [5] and encouraged novel analysis of the convergence of human services, information analytics, wireless communication, installed (embedded) frameworks, and data security. Implantable and Wearable Medical Devices (IWMDs) that encourage noninvasive interference, early identification, and constant treatment of medical disorders are visualized as key elements of contemporary healthcare services [6]. There is no singular definition of smart healthcare; however, *our comprehensive elucidation of ubiquitous healthcare is that besides* ***clinical*** *utilization, it conjointly uses IWMDs to collect, store, and process varied forms of physiological information amid everyday exercise.* The Internet plays a major role in ubiquitous health care in that it provides the full advantages of wireless connectivity to external resources (e.g., computational/capacity assets accessible on nearby gadgets or the cloud). Hence, ubiquitous smart healthcare services offer a proactive way to deal with early detection and even avoidance of any medical conditions. The smart healthcare environment also helps healthcare practitioners to monitor patient health at the patient's home with internet-based healthcare services. This ubiquitous healthcare setup reduces the need for institutionalization and hospitalization and is of great help to the elderly as well as people with disabilities. It also scales back healthcare prices considerably and increases the personal satisfaction of patients.

The smart healthcare system has revolutionized the traditional medical system and provided a chance to substitute in-hospital services with web-linked IWMD-based frameworks, with the emergence of cloud computing, wireless sensor networks, IoT, and fog computing. Healthcare organizations also comprehend the requirement of IT-enabled facilities and digitalization to unlock or cure some crucial medical issues and have considered this as a part of their business models [7]. Due to this, the processing power, energy capability, and interacting competences of IWMDs have improved considerably within the last decade, whereas their sizes have reduced radically. The first IWMD was introduced in 1958 and since then various versions of IWMDs have been industrialized, ranging from a perspiration analyzer device [6] to web-based, multi-sensor wellbeing observing systems [5]. IWMDs are still evolving with the use of tiny devices that can be worn on the body or entrenched or mounted in patients' homes and workstations to monitor heart rate, blood pressure, plasma, and urine biochemical levels.

10.1.2 From the packages of the past to the reforms of the future

With an aging population comes illness, like diabetes, joint inflammation, cardiovascular disease, cancer, and other conditions [8]. With no changes, this circumstance exhibits a stressing future prospect with extreme financial results. To avoid the risks of increasing chronic issues, it is very important to have a healthier way of life that includes a nutritious diet and regular exercise [9].

Due to the technical revolution, individuals are more inclined to use smart phones and tablets instead of laptops and computers [10]. According to a 2016 survey, three out of four individuals across all age groups owns a cell phone. Previously, the cell phone was almost exclusively used to communicate. Today, however, the smart phone is "the supreme ubiquitous customer electronic gadget on the planet" because of its multitude of dynamic features. These mechanical advancements and improvements have made cell phones the most used tool in the ubiquitous healthcare system (health and wellness apps). The reason for using cell phones in ubiquitous health services is because they contain all of an individual's personal data [11]. There are mobile apps to track glucose levels, blood pressure, pregnancy, menstrual cycles, sleep, state of mind, and so on [11]. Around the world, people are keeping track of their daily exercise, fitness, diet, and wellbeing on cell phone apps, which is encouraging trust in ubiquitous healthcare services [12]. In the United States, 60% of adults self-track their weight, diet, and workout performance and around 33% monitor their glucose levels, blood pressure, and sleeping patterns.

Ray Kurzweil, an innovator and futurist at Google, has stated that progressive high-tech healthcare architectures empower the business to twofold its ability yearly for a similar expense [13]. This enhancement also fathoms the absolute biggest operational difficulties: an aged populace, urbanization, lack of personnel, and rising medical expenses. By using the IoT, all the gadgets and items are associated with the system to encourage information gathering and data sharing [14]. Wearable health sensors, for example, the Nike FuelBand or the Jawbone UP2, gather information on various factors, including action levels and pulse [8]. Some of the portable examination entails the procedures of assembling, handling, scrutinizing, and envisioning such large-scale sensor information. New advances can enhance the healthcare system by upgrading the productivity and pellucidity of the wellbeing sector. Therefore the principle viewpoint is for the complaisant and intelligent use of data and communication to improve general human health services [15].

Delivering good healthcare services to all the aged throughout a country is a challenge [16]. For example, according to one survey, almost 650,000 Norwegians are older than 67 years, with this number expected to reach one million by 2030. This forecast highlights a pressing need to either train more medical services specialists or change the plans of action in the business. However, this is a complex circumstance since conventional business models are reliant on expert specialists who are required to complete long educational training. This is more reason to trust the ubiquitous healthcare framework.

The ubiquitous healthcare framework is computerized and shrewd. This technology is increasing the proficiency of medical employees as well as patients who can progressively engage with regards to their individual wellbeing. Patient engagement can likewise increase the odds of a healthy life [16]. So, trusting the digital platform of health not only helps improve medical issues but also increases the economy of the new business models. There is a huge market in the digital healthcare framework to deliver care as the number of individuals requiring help is growing. For instance, it is anticipated that by 2020 there will be almost 117 million Americans relying on medical assistance, but the general number of unpaid caretakers is only expected to reach 45 million. This equates to one voluntary caretaker for every 2.6 people requiring medical help [8]. This marks a huge business prospect for those individuals who are on the web and can utilize ubiquitous healthcare services. However, there are not enough online smart healthcare applications available that can cater to the prerequisites of philanthropic healthcare. According to the report by Project Catalyst and the Health Innovation Technology Laboratory (HITLAB), only 71.5% of caregiver volunteers are interested in using the technology.

This augmented revolution in technology is bourgeoning the perspective in health improvement and health knowledge transformation in an improved cultured and revolutionizing global civilization.

10.2 What is an ubiquitous healthcare

Before discussing the ubiquitous healthcare framework, it's important to understand the concept of the conventional, eHealth, and mHealth systems.

10.2.1 Conventional, eHealth, and mHealth systems

A conventional medical system can be portrayed as the process of a patient going to a general physician's workplace that doesn't involve information and communications technology (ICT) [17]. As advanced knowledge, like the IoT, began to effect the wellbeing business, a rising field in the connection of therapeutic informatics, general wellbeing, and commercial industry began to emerge. This rising area is also known as eHealth (electronic health) [10]. In contrast to the customary human services framework, eHealth utilizes ICT, electronic wellbeing archives or electronic healthcare records (EHRs), and online catalogs to collect the medical information of patients. eHealth is considered as a remarkable projectile in the medical industry because of its cost-efficient electronic hardware, increased productivity, and better quality administrations for patients.

Despite the fact that the idea of eHealth has been in use for a couple of years, this concept is being misconducted with the web and advanced information, therefore separating eHealth from the more extensive field of medical informatics that consolidates complex technologies. eHealth is "a developing field in the convergence of medicinal informatics, general wellbeing and business, alluding to wellbeing administrations and data conveyed or upgraded through the web and related advancements" and it should not be mistaken for vending medications on the web (i.e., a computerized trade) [18].

In recent times, mobile health or mHealth has extended the eHealth framework. Two noteworthy accomplishments have added to the advancement of mHealth: first, the introduction of the 3G network, and second, the launch of Apple's iPhone in 2007, which revolutionized mobile apps [19]. mHealth is characterized as "the utilization of portable registering and communiqué advancements in the medicinal framework and general wellbeing" [20]. mHealth has empowered communicative administrations to be conveyed through versatile specialized gadgets, which has reclassified medical systems by offering access to various administrations in a modified way. For example, a patient can recommence treatment from their cell phone, or somebody in the midst of a furlough can have a video discussion with their specialist at home via cell phone. The contents of portable innovation have extraordinary potential for medical applications. Mobile applications can connect groups of people (e.g., patients and doctors) and fulfill an assortment of needs, like helping people lose weight, quit smoking, become more active, and so on. The greatest favorable circumstances of utilizing cell phones for wellbeing are that the gadgets are smart, associated, customized, and are always with individuals. So, patients are able to use the online gadgets in various circumstances, like to book an appointment, in recapitalization, during hospitalization, or on a regular, everyday basis. It has been also demonstrated that the utilization of cell phone innovation can enhance analysis, patient data, and progress regulatory productivity [21].

10.2.2 The ubiquitous healthcare

Savvy urban communities are pushing the wellbeing business to become more astute by expanding the utilization of highly mechanical hardware [22]. Authors trust that these smart frameworks can bolster the cardinal collection, handling, storage, broadcast, and distribution of resident information. Notwithstanding enhancing the administration and transmission media in the wellbeing business, ICT and intellect plays a significant role in creating defensive, prescient, customized, and participatory medical services frameworks. Lee et al. [20] say that urban communities can modernize the current healthcare services framework (e.g., mHealth, eHealth), creating a new concept called ubiquitous smart health care. Smart health care is a comparatively novel idea and can be characterized as a collaboration between mHealth and smart urban areas.

Ongoing developments in Internet-based technology facilitate envisioning innovative ubiquitous healthcare services that protect social life and furthermore improve it [11, 12] by providing premature recognition for specific disorders and expanding specialist patient proficiency. WSN-based ECG observing systems, RFID observing systems [23], Wearable Medical Sensors (WMSs), and Implantable Medical Devices (IMDs) are revolutionary innovations that will help to decrease the costs of health care.

The ubiquitous healthcare framework as shown in Fig. 10.1 includes various vitality and source-inhibited gadgets, which are naturally designing, self-observing, self-mending, and powerful in nature to a server in erratic conditions with clamor, motion loss, and catastrophes [13–15]. Moreover, the gadgets can be portable, for example, crisis gadgets, or immovable, like temperature sensors and so forth. Ubiquitous sensors should have the following primary characteristics:

(a) insignificant corporeal dimensions
(b) truncated power utilization
(c) constrained processing power
(d) diminutive range communiqué capability
(e) small stockpiling capacity

Using a Wi-Fi sensor network equipment for ubiquitous healthcare service is more arduous than using it for other continuous frameworks since a smart medical system deals with various kinds of waveform information. IoT renovates the medical information for more intelligent patient consideration. Medical services are currently more mechanically progressed and are tied in with correlating possessions. Therefore, IoT is vital in medical services. By utilizing gadgets, such as wearable sensors, users can transfer their data to the cloud where a specialist can observe it in real time.

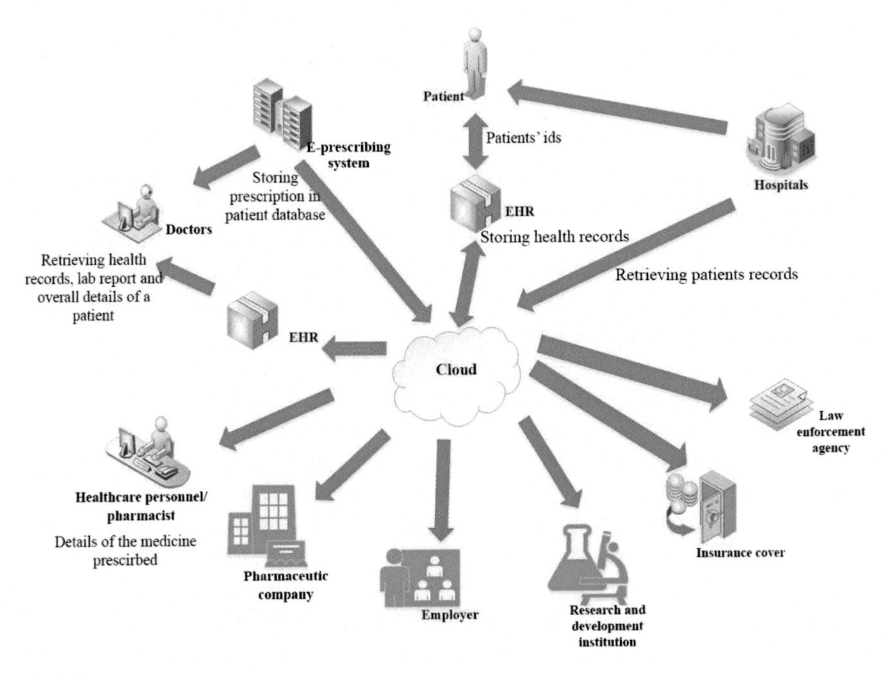

FIG. 10.1

IoT-based medical service framework.

The IoT can provision possibly life-saving applications by gathering information from bedside gadgets, monitoring patient data, and diagnosing in real time [24].

The new patterns and descriptions in the medical service system are still in the early phases of exploration. Clancy [25] proposed that in the near future, ICT would revolutionize the medical services industry. Authors [22] identified that savvy organizations in the wellbeing and welfare industries can be portrayed as firms creating IT-administrations for the wellbeing business.

10.2.3 The ubiquitous smart healthcare framework

The ubiquitous smart healthcare framework can be classified into two categories [26]: (i) regular medical care and (ii) scientific medical care. There are five noteworthy tasks that should be done in ubiquitous medical service as shown in Fig. 10.2.

- *sickness counteractive action:* (1) regular aversion
- *sickness identification:* (2) regular and (3) medical analysis
- *sickness treatment:* (4) medical and (5) regular medications

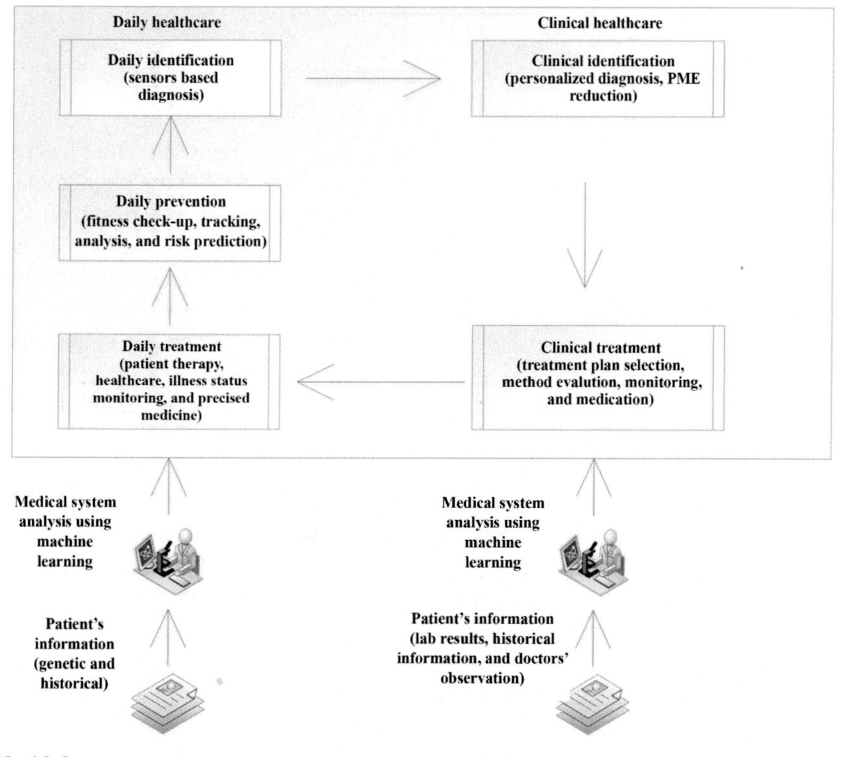

FIG. 10.2

Ubiquitous medical smart framework.

These five *responsibilities* are part of the three most precarious difficulties of ubiquitous healthcare services: (1) preclusion, (2) analysis, and (3) medications of human sickness. Each *undertaking* establishes an effervescent and dynamic research arena that comprises challenges like aptness following, regular disease analysis, general practitioner alteration decrease, medication proposal assortment, and precision drug. These five *undertakings* should be done in a successive way (i.e., alluded to as the ubiquitous smart medical service *circle*.) We call the medical service "savvy" when the framework can take a decision depending on the patient's condition (i.e., the leadership ability). This capacity is empowered by information analytics and this data sorting may help decision making. Information types shift from physiological and ecological readings in the regular setting to doctor perceptions and research center test results in the doctor's facility/center. This information must be effectively captured, prepared, and safely transmitted to the upper dimensions of the human services framework, helped by machine learning motors, for example, Waikato Environment for Knowledge Analysis (WEKA) [9] and TensorFlow [27], to extricate wellbeing inductions.

10.2.4 Gaps in the literature on ubiquitous healthcare

As ubiquitous healthcare services are still developing, existing literature has principally engaged in elucidating essential features and describing the phenomena as a savvy prototype for urban areas. In the scientific literature, there are no technical articles consolidating the two ideas of ubiquitous smart medical models and commercial models, showing that the investigation of smart wellbeing from a commercial point of view remains unexplored. There are only a couple of articles examining the commercial model for eHealth organizations, and these models are only hypothetical. There have been a couple of efforts to examine the intersection of commercial prototypes in eHealth on a progressively diagnostic dimension [28], but they do not ponder the impacts of outside issues or aspects of commercial prototypes from a static view. Furthermore there is no exploration of the inspiration driving ubiquitous wellbeing organizations forward (that we are aware of), so it is essential to explore the commercial point of view in smart wellbeing. The healthcare services system is confronting some difficulties from organizations, laws, and regulations. Consequently, it is urgent that both commercial chiefs and policy designers become conscious of what ubiquitous smart healthcare services resemble. As there is a tremendous assortment in the administrations that smart healthcare services organizations provide, not all organizations create, convey, and catch an incentive in a similar way. Therefore, it can be expected that there will be distinctive commercial models in savvy wellbeing [29]. To inspect this gap, the ubiquitous smart healthcare framework should be examined from a commercial point of view.

10.3 Role of IoT/WSN role in ubiquitous healthcare

Casagras [3] defined the IoT as "a universal network framework, connecting corporeal and simulated entities through the manipulation of information capture and correspondence abilities. This framework incorporates web and system advancements. It also offers exact article identification, sensor and connection capability as the basis for the development of independent cooperative services and applications. These will be characterized by a high degree of autonomous data capture, event transfer, network connectivity and interoperability."

In the healthcare system, IoT offers alternatives to remote checking, early anticipation, and restorative treatment for patients. With IoT individuals or objects can be furnished with sensors, actuators, Radiofrequency Identification (RFID) labels, and so forth. Such gadgets and labels encourage access by patients' caretakers. For instance, RFIDs are discernible, unmistakable, locatable, and controllable by means of IoT applications [5].

IoT has vigorous competencies to deal with the difficulties that people or medical systems face by connecting Device-to-Machine (D2M), Object-to-Object (O2O), Patient-to-Doctor (P2D), Patient-to- Machine (P2M), Doctor-to-Machine (D2M), Sensor-to-Mobile (S2M), Mobile-to-Human (M2H), and Tag-to-Reader (T2R). Intelligently this associates people, savvy gadgets, operating devices, and dynamic frameworks to deliver viable medical services [6, 19].

The World Health Organization (WHO) [2] has stated that the total populace of individuals older than 65 years will exceed two billion by 2050. To avoid overpowering medical services, it is necessary to enable these individuals to live freely and independently at their own homes. This not only enhances their personal satisfaction, but also lessens the cost of living. The amalgamation of shrewd functionalities and Ambient Intelligence (AmI) at the household level has led to the concept of Ambient Assisted Living (AAL) [30], which is the use of AmI techniques, processes, and technologies to help elderly and people with disabilities to gain more independence in their daily lives for a long as possible, without intrusive behaviors [9, 11].

10.3.1 IoT healthcare system design and planning

The architecture of the IoT comprises numerous strata, beginning from the edge innovation layer at one end to the solicitation layer at the other end, as shown in Fig. 10.3 [20, 21]. The two lower strata add to information gathering, while the two higher layers are accountable for application information use. The utilities of the layers (from bottom to top) are as follows:

(a) *Edge innovation layer (perception layer)*: This is an equipment layer that incorporates information gathering via, for example, wireless sensor networks (WSNs), RFID frameworks, cameras, smart terminals, electronic information interfaces (EDI), and global positioning systems (GPS). This equipment delivers documentation and data stowage via RFID labels, data accumulation

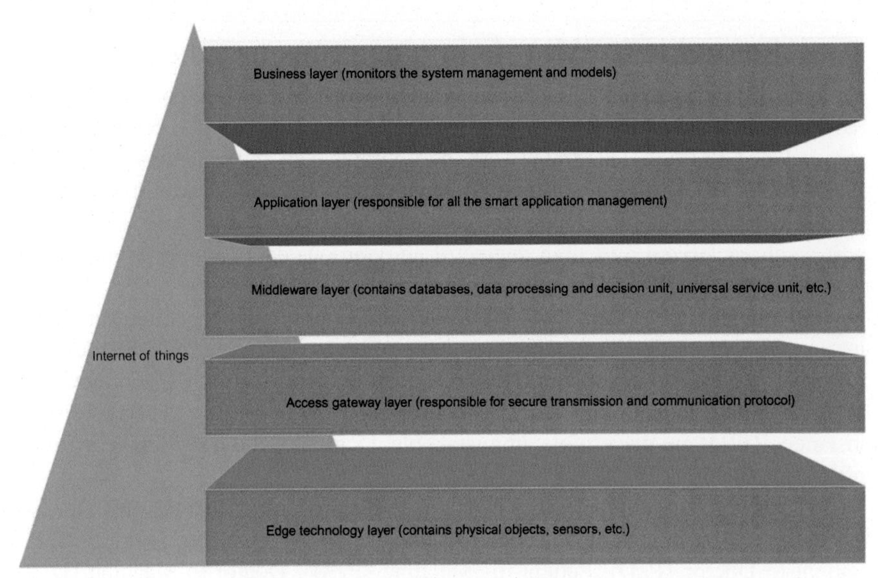

FIG. 10.3

IoT architecture.

through sensor systems, data handling by using the installed edge processors, transportations, and regulators, and actuation through robots [29].

(b) *Access gateway layer (network layer or transport layer)*: This layer is capable of information sharing, including information transmission, message steering, and distributing the memos. It directs to the middleware-level data, which is received from the edge layer utilizing Wi-Fi, Li-Fi, Ethernet, Global System for Mobile Communications (GSM), WSNs, and Worldwide Interoperability for Microwave Access (WiMax) [20, 21].

(c) *Middleware layer*: This layer does not require permission from the user to process the information further and also offers numerous administrations, for example, gadget disclosure and management, information filtering, information conglomeration, semantic information examination, access control, and information revelation through Electronic Product Code (EPC) or Object Naming Service (ONS) [21].

(d) *Applications layer*: This layer is in charge of delivering the different applications to distinctive IoT clients. It is also responsible for delivering the Quality of Service, distributed computing advancements, information processing, M2M administrations, and interfacing to end clients [21].

(e) *Business layer*: This layer deals with external commercial management.

10.3.2 The Internet of things addressing health, social care, and wellbeing challenges

The three innovative web-based techniques of M2M, Machine-Type Communications (MTC), and the IoT are the coinciding connotations and do not have concrete definitions, so these are considered as the adopted technologies.

M2M can be characterized as the arrangement of remote and wired correspondence between automatic or electronic gadgets [11], including communication between remote machines and focal administration applications [12]. The M2M idea has been used to measure, convey, process, and respond to data in an independent design [13]. M2M and MTC are viewed as being equivalent [14]. However, MTC is the working mechanism utilized by 3GPP and frequently viewed as the portion of M2M continued cell systems [15]. These two telecommunication expressions have a solid spotlight on the system and network perspectives. IoT is a more extensive idea contrasted with M2M and MTC. It alludes to a lot of advanced technologies, standards, frameworks, and applications related to Internet-associated objects [16]. It was first used to portray a "universe of consistently associated gadgets that would spare us time and cash," [18] based on the interconnection of the corporeal domain with the simulated domain of the Internet [17].

IoT allows for opportunities to remotely detect and control substances by means of correspondence systems; delivering immediate incorporation of the physical world into PC-based frameworks. The IoT has been demonstrated as gainful for enterprises, each having diverse prerequisites and conditions. In health, social care, and wellbeing (HSCWB), IoT can bolster a change from long-winded medicines of sickness to preventive consideration and prosperity arrangements.

The European Commission has been sure about the Digital Agenda for Europe regarding health and maturing, "ICT can be our most dominant partner for good and reasonable social insurance" [20]. Utilizing ICT to enhance medical services is not new, yet the utilization of present-day communication advancements and the accessibility of advantageous gadgets empowers the formation of better administrations and enhances the way care administrations are overseen. IoT is an advanced technology inside the territory of ICT. To deliver progressively proficient health and social considerations, IoT can be utilized in the context of HSCWB in three categories:

- using nonintrusive gadgets
- social consideration and homecare frameworks
- wellbeing and preventive framework to aware and inspire people for healthier practices.

10.4 IoT healthcare features and opportunities for people with disabilities

The absence of experts and offices for people with disabilities is particularly apparent in certain countries [7]. This lack of resources forces people with disabilities to search for social insurance benefits at expensive doctors' facilities or other high-cost medical establishments. IoT is a chance to expand legitimate human services administrations to people with disabilities at a low cost.

10.4.1 Types of disabilities

Like everyone else, people with disabilities must engage in their regular life assignments, including activities of daily living (ADLs), instrumental activities of daily living (IADLs), and electronic activities of daily living (EADLs), to preserve their individuality and fitness. To take part in or practice these, however, can be difficult because of physical or cognitive disabilities [31]. The classifications of disabilities are as follows:

(a) Physically disabled
(b) Intellectually disabled: perceptual disabilities and cognitive disabilities
(c) Independently living disabled
(d) Homebound disabled
(e) Institutionalized disabled

10.4.2 Responses from the changing world

There is a requirement for universal social insurance to enhance human prosperity. Modern medical services frameworks have become more advanced in terms of information analytics and progressively inescapable dependent on the quick organization of IoT [32]. There are numerous frameworks from the medicinal domain:

(a) IBM Watson
(b) Open mHealth
(c) HDSS
(d) SoDA
(e) A vitality proficient framework

10.4.2.1 IBM Watson

IBM Watson [13, 15] is a cognitive framework that combines deep Natural Language Processing (NLP), assumption group, and active erudition to produce confidence-based reactions. It is skilled at gathering information from amorphous and raucous data, including logical articles, course books, client manuals, rules and regulation notes, regularly requested queries, plans, research facility notes, newscast, and copyrighted information. The removed information base is put away as a Watson quantity. Watson creates a novel corpus for each target space. Because of its huge NLP capacities, Watson has been used in the engineering, medical, law, and financial areas [32].

In the medical domain, Watson has the ability to respond to wellbeing-related inquiries in both regular and clinical situations. It can examine and break down the content from a wide scope of restorative assets, such as logical diaries, licenses, medication and illness related ontologies, clinical preliminaries, EHRs, research facility and imaging information, genomic information, claims information, and web social substance [33]. This has prompted three noteworthy utilizations of Watson in social insurance: (1) oncology, (2) drug discovery, and (3) genomics.

10.4.2.2 Open mHealth

The Open mHealth venture is meant for everyday chronic sickness anticipation and management by using the collected information from medical applications in cell phones and gadgets [33]. These applications empower patients to electronically record and track their fundamental physiological signs every day. For example, WellDoc is a cell phone-based diabetes management app that utilizes messages and stimulates to follow the glucose level of its clients.

10.4.2.3 Health decision support system

The health decision support system (HDSS) [2] empowers disease analysis in all scenarios through the incorporation of WMS information to CDSSs. HDSS has a multilevel structure, beginning with a WMS level, supported by strong machine learning. It consecutively structures the data system for everyday wellbeing monitoring, starting clinical checkup, nitty-gritty clinical examination, and postanalytic choice help. HDSS has two noteworthy parts to help everyday and medical services: (1) pervasive decision support system (PHDS), and (2) PHDS-assisted CDSS (CDSS+).

10.4.2.4 Stress detection and alleviation system

Stress is connected with different medical issues, from cardiovascular illness to sleep illness and cancer. Reducing the risk of these serious medical problems requires stress monitoring. The stress detection and alleviation system called SoDA [5] is designed to deal with this issue. It is a software that gathers physiological signs utilizing WMSs and achieves machine learning implications on them. It delivers consistent and client state oriented feeling of anxiety tracking (green way) and instructing (red way).

10.4.2.5 Energy-efficient health monitoring system

Conventional medical screens (e.g., bedside ECG and oxygen immersion observing frameworks), accumulate, store, and transmit information with no (or negligible) on-gadget handling. Moreover, such frameworks are normally controlled from the electrical outlet, rather than IWMDs that depend on batteries. Recent technology in gesture processing, low-power gadgets, communication procedures, and, specifically, low-power radiofrequency transmission modules, have empowered Wi-Fi availability to energy-constrained medical gadgets, such as Wireless Body Area Networks (WBANs) [27].

Apart from the medical frameworks, there are numerous ubiquitous IoT medicinal applications:

- **(i)** Blood pressure monitoring
- **(ii)** Rehabilitation system
- **(iii)** Oxygen saturation monitoring
- **(iv)** Wheelchair management
- **(v)** Healthcare solutions using smartphones
- **(vi)** General health care apps in smartphones

(vii) Calorie counter
(viii) Hear-rate monitor
(ix) Body temperature
(x) Pedometer
(xi) Water your body
(xii) OnTrack diabetes
(xiii) Skin vision
(xiv) EyeCare plus
(xv) Asthma trackers and log
(xvi) Cardiomobile
(xvii) Pill remainder
(xviii) Fall detector
(xix) Helo wristband
(xx) AliveCor heart monitor
(xxi) Allb baby monitor.

10.5 Challenges of a ubiquitous healthcare system

Society's craving and innovative capacity to utilize organizing advancements dependably surpass their capacity to control the security of that technology. Organized restorative gadgets are not exclusion. They give colossal advantages to the cutting edge social insurance framework, so engineers and adopters close their eyes and do whatever it takes not to see genuine security holes in new items. The circumstance will continue as before or deteriorate if security authorities and gadget creators don't make the required strides now. It could happen that an immense blast of medical zero-day adventures and security gaps without protection will touch the market. Along these lines, there are fundamental and closed zones of human services that the IoT concerns [2].

10.5.1 Accidental catastrophes

Inadvertent disappointments [2] are the clearest issue and diminish the faith of clients. Provided that any extraordinary disappointment occurs, society will deny organized restorative gadgets, deferring the advancement of human services and drugs for a considerable length of time. Individuals can confide in specialists, yet it is in every case hard to confide in the machine if some mechanical or programming mistake prompted some irreversible outcome. Organized therapeutic gadgets are powerless against something other than a criminal goal. It's anything but a consequence of inappropriate designing or terrible parts; it is an aftereffect of assembling defects, programming botches, or other situations influenced by a huge number of conditions. The multifaceted nature of gadgets and operational innovation that controls physical procedures, for example, siphons, opens the doors to imperfections in plan, execution, or activity, any of which can prompt sensational unplanned failure.

10.5.2 Protecting patient's confidentiality

Ensuring security of patient information is paramount. As indicated by PricewaterhouseCoopers (PwC) Global State of Information Security review 2015 [33], the quantity of security breaks revealed by social insurance suppliers significantly increased by 60% from 2013 to 2014. Vulnerabilities in organized medical gadgets create immense inconveniences for proprietors of these apparatuses, since they monitor, access, and store the user's personal data. To satisfy the IoT worldview, the majority of the gadgets must have remote correspondence modules that give them the capacity to exchange information at any time from any place. So the ability of remote systems administration is one of the fundamental qualities of savvy social insurance for accomplishing high proficiency.

Similarly for wireless arrangements, the client needs to make sure that there is no transmission of decoded individual information, or that there are no secondary passages to the gadget from the system. Furthermore some smart gadgets are proficient at wellbeing checking as well as on charging (making on the web budgetary tasks), utilized as a different element or as an expansion for therapeutic purposes (e.g., as a piece of medical services supplier arrangement or for simplicity of tasks). Vulnerabilities in such arrangements put patients in danger of losing therapeutic data as well as money. Since therapeutic IoT is still in its initiation, nobody knows the manners in which data could be utilized by those with bad intentions [13].

10.5.3 Deliberate disrupt

Purposeful disrupt [31] is another worry. Mobsters, psychological militants, and different lawbreakers try to abuse vulnerabilities in the IT foundation to perpetrate wrongdoings and cause turbulent circumstances. When it contacts banks or transport frameworks, the outcomes are reasonable by the administration. However, when a gadget is on a human or even under a person's skin, the results of the digital wrongdoing may be especially harmful. Take a pacemaker, for example. It is entirely feasible for someone to interfere with its operation, causing harm to the patient. The same could occur with insulin siphons or any medication-conveying frameworks.

10.5.4 Prevalent disrupt

Prevalent disturbance [32] is a significant problem, however, as medical gadgets are associated with the worldwide system, some focused on malware could spread over the Internet and act against just chosen therapeutic gadgets, so everybody with the defenseless gadget is influenced.

10.5.5 Threats of cyberattacks

Digital assaults [2] can infuse false information into a framework, causing basic harm in IoT applications. Therefor it is important to guard against digital assaults in home applications for patients with disabilities. However, the asset compelled nature of huge numbers of IoT gadgets existing in a smart home condition don't permit to execute the standard security arrangements.

10.5.6 Information eavesdropping and confidentiality

The wellbeing information of patients, including patients with disabilities, are alleged under the legitimate commitments of secrecy and made accessible only to those the patients have given approval for. It is essential to prevent stealing of information or snooping while patient's stream their data over remote connections [34].

Authors [35] propose moving code cryptographic conventions and body-coupled correspondence to moderate snooping on patient information. They [35] propose a bi-polar different base information-concealing procedure for pictures, where a pixel esteem contrast between a unique picture and its default JPEG lossy decompressed picture is taken as the number transformation base. The calculation permits to stow away, e.g., specialists' computerized seals and PHR inside a still picture. (The specialists' advanced seals are basic to the verification of PHR.) The still picture could be a log of a doctor's facility recognizing where the PHR originates from. An indicative report and a biomedical flag, for example, an electrocardiogram (ECG), can likewise be covered up in a picture. The proposed methodology permits concealing different information types in a similar picture.

Information confidentiality can be enhanced by utilizing Public Key Encryption (PKI). PKI is a compelling way to deal with information encryption as it can give an abnormal state of certainty to trading data in a shaky situation. Article [21] presents a theoretical plan and a model execution of a framework dependent on IoT doors that aggregate wellbeing sensor information and resolve security issues through advanced authentications and PKI information encryption.

10.5.7 Location privacy

Location protection [34] is concerned with area security dangers and snooping stealthily on a client's area. Location security in WSNs, that is, explicitly concealing the message sender's area, can be accomplished through direct to a haphazardly chosen middle of the road hub (RRIN) [36]. Snooping and following of parcels can be anticipated by the Location Privacy Routing (LPR) convention, which consistently conveys the bearings of approaching and active traffic at sensor hubs. Ghost single-way directing makes guarantees that parcels achieve the Base Station (BS) following diverse ways so that each bundle made by a source pursues an alternate irregular way toward the BS.

10.5.8 Privacy of IoT-based applications

Privacy is a major issue for all internet-based applications. In the smart healthcare environment, all personal health records are stored on the internet, which makes these records more vulnerable. Therefore, to ensure privacy, before redistributing and separating the healthcare framework, it is important to implement the following steps: encrypt the stored information, examine whether delicate information should be private or not, install cyber assault detection strategies, and provide a framework for recuperation [37].

10.5.9 Interoperability

Medical services interoperability portrays the capacity for heterogeneous data innovation frameworks and programming applications, for example, the EHR framework, to convey, trade, and utilize information [15]. Enabling data frameworks to cooperate together inside and across hierarchical boundaries is paramount for developing effective precaution delivery for people and communities [16]. For instance, interoperability empowers suppliers to safely share medical records with each other (given that the patients have consented to such), paying little respect to supplier area and trust connections between them. Secure and adaptable information sharing is fundamental for providing synergistic treatment and care choices for patients. Information sharing enhances indicative accuracy [18] by social occasion affirmations or suggestions from a gathering of therapeutic specialists, just as avoiding deficiencies [17] and blunders in the treatment plan and drug delivery [19, 20]. Likewise, amassed knowledge and bits of knowledge [21, 22, 24] enable clinicians to comprehend tolerant requirements and thusly apply for increasingly compelling medicines. For instance, gatherings of doctors with various claims to fame in malignant growth care shape tumor sheets that meet consistently to examine disease cases, share information, and decide powerful malignant growth treatment and care gets ready for patients [26].

10.5.10 Lack of trust between service providers

The incapable information sharing procedure [38] in social insurance results to some degree from the absence of trust among suppliers and the absence of interoperability between wellbeing IT frameworks and applications today. Social insurance interoperability includes three dimensions: (1) fundamental, (2) auxiliary, and (3) semantic [29]. Care suppliers must have the capacity to distinguish different suppliers and furthermore trust their personalities before any patient wellbeing-related correspondence happens [2]. Trust connections frequently exist between in-network suppliers as well as wellbeing associations, yet they are difficult to set up when the office that accepts the information for consideration does not utilize the same wellbeing framework with a mutual supplier registry, for example, in private practice or a clinic.

10.5.11 Scalability concerns

Restorative information may contain huge volumes of information, and big data are hard to share electronically because of confinements in transmission capacity or prohibitive firewall settings [35].

10.5.12 Identity threats and privacy of stored data

Loss of a patient's protection, particularly their personal information, may result in noteworthy physical and economic damage to the patient. This is called *Unlinkability*, *Anonymity*, and *Identity management*.

Secrecy and pseudonymity administrations can be utilized to conceal the genuine character that is fixing to put away information. Security administrations can ensure securing records by utilizing different protection methods that depend on adding clamor to patients' records [39]. This might be utilized to guarantee that a database permits recovering only factual information.

10.5.13 Privacy requirements for IoT applications for users with disabilities [40]

- **(i)** Disabled should realize who claims their medicinal information.
- **(ii)** Appropriate differently abled authorization for utilizing her wellbeing information (e.g., the intensity of lawyer for a guardian).
- **(iii)** Anonymity or pseudonymity shroud genuine characters of the crippled clients by methods for isolating the personality of an individual into sub-characters.
- **(iv)** Location protection to keep areas of clients, gadgets, and so forth, private.
- **(v)** Maximizing the area of data.
- **(vi)** Privacy for IoT gadgets.
- **(vii)** Emphasizing protection from the earliest starting point of the application structure.

There are some more security challenges in ubiquitous healthcare frameworks like corporeal attacks, assimilating RFID labels into IoT, assimilating WSNs into IoT, denial-of service (DoS) attacks, communication attacks (on layers), and so on.

10.6 Future concerning directions

The therapeutic sciences have made incredible steps in creating more secure and compelling treatments, like a negligibly obtrusive medical procedure and new medication to relieve the impacts of sickness like arthritis. Yet, ubiquitous healthcare is an evolving area of open research challenges.

10.6.1 Cybernetic precaution

This implies the utilization of electronics to reduce or supplant face-to-face connection of specialists and patients. There are, however, obstructions [41] to virtual correspondence among patients and doctors and among doctors and different suppliers, such as the following:

- administration of remuneration components for private division and virtual consideration
- administrative obstructions

- security of individual data
- advanced distribution and access to innovation

10.6.2 Insufficient information arrangements and machine learning prototypes

Medical information can be noisy, unstructured, time-associated, vast in volume, shifting in amplitudes, filled with missing qualities, and above all, without legitimate marks. These deficiencies may keep existing machines taking in calculations from conveying high arrangement exactness. For instance, SVM, tree-based design, and Bayesian systems miscarry to measure well when the feature measurement upsurges for additionally difficult tasks, for example, biomedical appearance arrangement and EHR examination [41].

10.6.3 Calibration procedure and organization provision

The calibration procedure is expected to encourage smooth progress between and synchronization of various ubiquitous medical service assets. Ubiquitous medical service systems comprise different treatment/checking procedures and heterogeneous devices for both individual and clinical operation. They have to perform smart medical services errands and work cooperatively in the social insurance circle to enhance the wellbeing of the client. For instance, on account of a crisis, individual human services frameworks ought to discuss proficiently with clinical social insurance frameworks. Then again, when the crisis dies down, clinical development and treatment/restoration should be well matched with individual medical services frameworks. In this way, smooth change and synchronization dependent on standard conventions are fundamental for the legitimate working of individual and clinical medical services assets [42].

10.6.4 Cloud computing alternative: "fog computing"

Distributed computing, with its versatile computational power and adaptable stockpiling, has greatly engaged medical services applications. In spite of its undeniable advantages, its pertinence is restricted in numerous medical services applications, specifically when the application is mission perilous, information-dominant, or latency-penetrating. Even a brief time period of cloud inaccessibility, which might be a consequence of a catastrophe in cloud servers or absence of web connectivity, can cause serious problems.

Furthermore, medical applications (e.g., seizure or arrhythmia recognition) may need to process an enormous amount of information consistently. For such information-predominant requests, sending the information to the cloud is not cost-effective, particularly if it depends on web network connectivity [42]. Also, due to the round-trip delay, raw information stored in the cloud creates implications and sends the resulting noise inferences back to the consumer side.

10.7 Conclusion: Trust in ubiquitous medicinal framework

Present-day media communications and machine learning open new ways for symptomatic processes and treatment strategies. The IoT offers the idea of the interconnected world, where medical administrations are upheld by each part of being, from nourishment to transportation. All things considered, the advantages of savvy benefits of medical services close to home outweigh the disadvantages. Nonetheless, security and protection, client experience, and selection are areas to focus on in the future [38].

This chapter focused on the use of ubiquitous healthcare monitoring and the IoT to perform health care close to patients' homes. IoT-based solutions are improving the Government Medical Services and also providing the clients to access the computerized medical services. The smart healthcare system is the revolution in the medical industry, now, patients can check their medical record. Hence, the system provides better healthcare to everyone and error-free and smooth communication to patients [40].

The pervasive registering condition expands the current healthcare administrations and restorative organizations to the individual and the home. While pervasive human health services present an incredible opportunity, they also present significant challenges and hazards. The move towards observing patients remotely or at their homes by means of body sensors is not only a gradual enhancement over existing practice, it is a subjective advance change.

References

[1] World Health Organization, Primary Health Care: Report of the International Conference 1 on Primary Health Care, Alma-Ata, USSR, 6–12 September 1978, Jointly Sponsored by the World Health Organization and the United Nations Children's Fund. Geneva, World Health Organization, 1978 (Health for All Series No. 1).

[2] C. Rotariu, V. Manta, Wireless system for remote monitoring of oxygen saturation and heart rate, in: Federated Conference on Computer Science and Information Systems (FedCSIS), Wroclaw, Poland, 2012, pp. 193–196.

[3] M. Abadi, A. Agarwal, P. Barham, E. Brevdo, Z. Chen, C. Citro, G.S. Corrado, A. Davis, J. Dean, M. Devin, et al., Tensorflow: Large-Scale Machine Learning on Heterogeneous Distributed Systems, arXiv preprint arXiv:1603.04467, 2016.

[4] World Health Statistics 2008. Geneva, World Health Organization, 2008.

[5] A.O. Akmandor, N.K. Jha, Keep the stress away with SoDA: stress detection and alleviation system, IEEE Trans. Multi-Scale Comput. Syst. 3 (4) 2017 269–282.

[6] B. Alipanahi, A. Delong, M.T. Weirauch, B.J. Frey, Predicting the sequence specificities of DNA-and RNA-binding proteins by deep learning, Nat. Biotechnol. 33 (8) 2015 831–838.

[7] A.O. Akmandor, N.K. Jha, Smart health care: an edge-side computing perspective, IEEE Consum. Electron. Mag. 7 (1) 2018 29–37.

[8] M. Arif, S. Basalamah, Similarity-dissimilarity plot for high dimensional data of different attribute types in biomedical datasets, Int. J. Innov. Comput. Inform. Control 8 (2) 2012 1275–1297.

[9] X.H. Cao, I. Stojkovic, Z. Obradovic, A robust data scaling algorithm to improve classification accuracies in biomedical data, BMC Bioinform. 17 2016 1–10.
[10] L. Atzori, A. Iera, G. Morabito, The Internet of things: a survey, Comput. Netw. 54 (15) 2010 2787–2805.
[11] Y. Cao, S. Chen, P. Hou, D. Brown, FAST: a fog computing assisted distributed analytics system to monitor fall for stroke mitigation, in: Proc. IEEE Int. Conf. Networking, Architecture and Storage, 2015, pp. 2–11.
[12] F.R. Cerqueira, T.G. Ferreira, A. de Paiva Oliveira, D.A. Augusto, E. Krempser, H.J. C. Barbosa, S. do Carmo Castro Franceschini, B.A.C. de Freitas, A.P. Gomes, R. Siqueira-Batista, NICeSim: an open-source simulator based on machine learning techniques to support medical research on prenatal and perinatal care decision making, Artif. Intell. Med. 62 (3) 2014 193–201.
[13] R. Chandrasekar, Elementary? Question answering, IBM's Watson, and the Jeopardy! Challenge, Resonance 19 (3) 2014 222–241.
[14] C. Chen, D. Haddad, J. Selsky, J.E. Hoffman, R.L. Kravitz, D. Estrin, I. Sim, Making sense of mobile health data: an open architecture to improve individual and population-level health, J. Med. Internet Res. 14 (4) 2012 e112.
[15] Y. Chen, E. Argentinis, G. Weber, IBM Watson: how cognitive computing can be applied to big data challenges in life sciences research, Clin. Therap. 38 (4) 2016 688–701.
[16] D.K. Das, M. Ghosh, M. Pal, A.K. Maiti, C. Chakraborty, Machine learning approach for automated screening of malaria parasite using light microscopic images, Micron 45 2013 97–106.
[17] C. Joo-Hak, The Benefits of Ubiquitous Healthcare, Korea IT Times, 2008.
[18] A.V. Dastjerdi, R. Buyya, Fog computing: helping the Internet of Things realize its potential, IEEE Comput. 49 (8) 2016 112–116.
[19] J. Nam, A Trust Framework of Ubiquitous Healthcare With Advanced Petri-Net Model, vol. 0001, 2009.
[20] X.H. Lee, S. Lee, T.K. Lee, H. Lee, M. Khalid, R. Sankar, Activity oriented access control to ubiquitous hospital information and services, Inform. Sci. 180 2010 2979–2990.
[21] A. Self, Measuring National Well-Being: Insights Across Society, the Economy and the Environment, Available at: http://webarchive.nationalarchives.gov.uk/20160105160709, 2014. http://www.ons.gov.uk/ons/dcp171766_371427.pdf.
[22] J. Thomas, S. Wise, Global Health Care Outlook: Reconciling Rapid Growth & Cost Consciousness, The Carlyle Group, 2015. Available at: https://www.carlyle.com/sites/default/files/market-commentary/october_2015_-_global_health_care_investment_outlook.pdf.
[23] N. Dey, A.S. Ashour, F. Shi, S.J. Fong, R.S. Sherratt, Developing residential wireless sensor networks for ECG healthcare monitoring, IEEE Trans. Consum. Electr. 63 (4) 2017 442449, https://doi.org/10.1109/TCE.2017.015063. ISSN 00983063. Available at http://centaur.reading.ac.uk/73898/.
[24] K.R. Darshan, K.R. Anandakumar, A comprehensive review on usage of Internet of things (IoT) in healthcare system, in: International Conference on Emerging Research in Electronics, Computer Science and Technology (ICERECT), Mandya, India, 132–136 2015, pp. 374–380.
[25] E. Ortiz, C.M. Clancy, Use of information technology to improve the quality of health care in the United States (AHRQ), Health Serv. Res. 38 (2) 2003 xi–xxii, https://doi.org/10.1111/1475-6773.00127.

[26] P. Sundaravadivel, E. Kougianos, S.P. Mohanty, M. Ganapathiraju, Everything You Wanted to Know about Smart Healthcare, https://www.researchgate.net/publication/322187294_Everything_You_Wanted_to_Know_about_Smart_Health_Care_Evaluating_the_Different_Technologies_and_Components_of_the_Internet_of_Things_for_Better_Health/stats, 2018.
[27] P. Gope, T. Hwang, BSN-care: a secure IoT-based modern healthcare system using body sensor network, IEEE Sens. J. 16 (5) 2016 1368–1376.
[28] R.K. Kodali, G. Swamy, B. Lakshmi, An implementation of IoT for healthcare, IEEE Recent Adv. Intell. Comput. Syst. (RAICS) 2015 411–416.
[29] S.K. Dhar, S.S. Bhunia, N. Mukherjee, Interference aware scheduling of sensors in IoT enabled health-care monitoring system, in: Fourth International Conference of Emerging Applications of Information Technology, Kolkata, India, 2014, pp. 152–157.
[30] Ambient Assisted Living Communications, Available online: http://www.comsoc.org/files/Publications/Magazines/ci/cfp/cfpcommag0115a.html.
[31] L. Catarinucci, et al., An IoT-aware architecture for smart healthcare systems, IEEE Internet Things J. 2 (6) 2015 515–526.
[32] S.M.R. Islam, D. Kwak, M.H. Kabir, M. Hossain, K.S. Kwak, The Internet of things for health care: a comprehensive survey, IEEE Access 3 2015 678–708.
[33] Y.J. Fan, Y.H. Yin, L.D. Xu, Y. Zeng, F. Wu, IoT-based smart rehabilitation system, IEEE Trans. Ind. Inform. 10 (2) 2014 1568–1577.
[34] S.-H. Cheng, An intelligent infant healthcare system of vital signs integrated by active RFID, in: International Conference on Machine Learning and Cybernetics, Tianjin, China, 2013, pp. 1157–1160.
[35] L. Yang, Y. Ge, W. Li, W. Rao, W. Shen, A home mobile healthcare system for wheelchair users, in: IEEE International Conference on Computer Supported Cooperative Work in Design (CSCWD), Hsinchu, China, 2014, pp. 609–614.
[36] H. Sue, J. Wan, C. Zou, J. Liu, Security in the Internet of things: a review, in: IEEE Inter. Conf. on Computer Science and Electronics Engineering, vol. 3, 2012, pp. 648–651.
[37] K. Patel, "Health and Medicine" IoT Can Help You Obtain Greater Efficiency Through Smarter Asset Management (Online). https://www.ibm.com/blogs/internet-ofthings/6-benefits-of-iot-for-healthcare/.
[38] L. Yang, P. Yu, W. Bailing, B. Xuefeng, Y. Xinling, L. Geng, IOT secure transmission based on integration of IBE and PKI/CA, Int. J. Control Autom. 6 (2) 2013 245–254.
[39] K.H. Yeh, A secure IoT-based healthcare system with body sensor networks, IEEE Access 4 2016 10288–10299.
[40] F. Jimenez, R. Torres, Building an IoT-aware healthcare monitoring system, in: International Conference of the Chilean Computer Science Society (SCCC), Santiago, Chile, 2015, pp. 1–4.
[41] White Paper on, The Future of Technology in Health and Health care: A Primer, Canadian Medical Association. https://cmahealthsummit.ca/app/uploads/2018/08/HS-Backgrounder_ENG_Final.pdf.
[42] A.M. Nia, S. Sur-Kolay, A. Raghunathan, N.K. Jha, Physiological information leakage: a new frontier in health information security, IEEE Trans. Emerg. Topics Comput. 4 (3) 2016 321–334.

PART

3

Applications of pattern recognition algorithms in U-healthcare

CHAPTER

PNN-based classification of retinal diseases using fundus images 11

Jitendra Virmani*, Gajendra Pratap Singh[†], Yashmeet Singh[†], Kriti[‡]

*CSIR-Central Scientific Instruments Organization (CSIR-CSIO), Chandigarh, India**

Punjab Engineering College, Chandigarh, India[†] Thapar Institute of Engineering and Technology, Patiala, India[‡]

11.1 Introduction

The two most common retinal diseases responsible for visual impairment are diabetic retinopathy and glaucoma. It is estimated that there are 60 million Glaucoma patients all over the world and this number will increase to 80 million by 2020. Among the 2.65% of world population aged 40 years and above, glaucoma is prevalent [1]. On the other hand, the number of diabetic retinopathy cases was found to be 126.6 million globally in the year 2010 and it is expected to increase to 191 million in 2030 [2].

Glaucoma is an eye disease in which intraocular pressure increases and damages the optic nerve. If damage to the optic nerve continues, then glaucoma may lead to permanent blindness [3].

Diabetic retinopathy is a disease, which is commonly found among diabetic patients. In this disease, retinal blood vessels become abnormal and may bleed or leak fluid. This results in distorted vision and swelling in the area of the retina called the macula. It is also a major cause of blindness worldwide [4].

Fundus of the eye refers to the interior surface opposite the lens. The photograph of this region of the eye is called a fundus image that shows the optic disc, macula, central and peripheral region of the retina. The camera used for fundus imaging is a low-power microscope attached with a camera.

The different retinal conditions as visible on a fundus image are described as:

(1) Normal fundus image: A normal fundus image has the following features: a pale pink disc, with sharp, flat margins; the central retinal artery travels

Sensors for Health Monitoring. https://doi.org/10.1016/B978-0-12-819361-7.00011-7

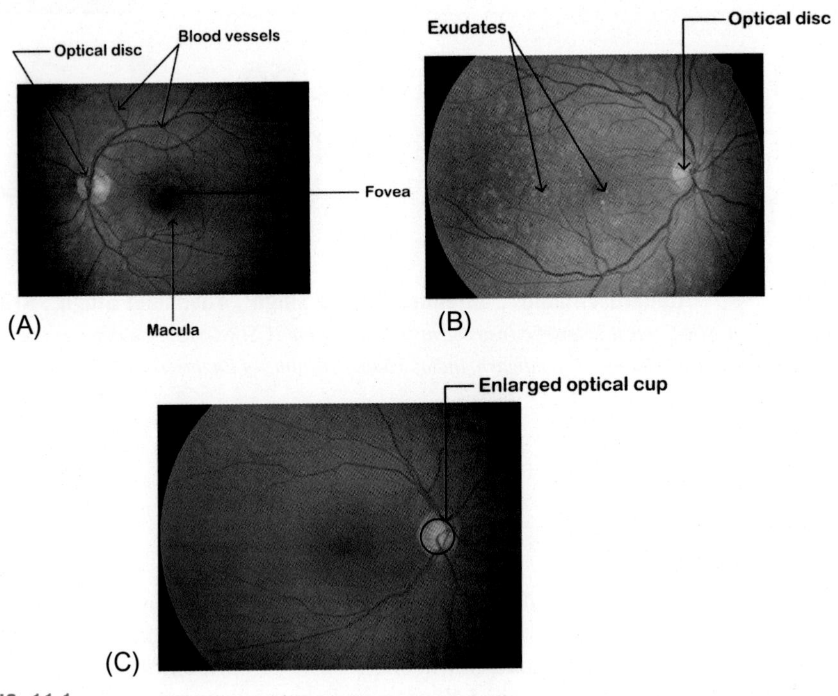

FIG. 11.1

(A) Normal fundus image. *Note: Pale pink* disc, optical nerve traveling through the optic disc, and *red* background is visible in the image. (B) Diabetic retinopathy-affected fundus image. *Note:* Exudates are visible in the image. (C) Glaucoma-affected fundus image. *Note*: Enlarged optic cup is visible.

within the optic nerve, branching near the surface; the background is red. A normal fundus image is shown in Fig. 11.1A.

(2) **Diabetic retinopathy-affected fundus image:** A retinal image affected by diabetic retinopathy is characterized with microaneurysms (little bulges from vessels), leaking fluid or blood (exudates). A diabetic retinopathy-affected fundus image is shown in Fig. 11.1B.

(3) **Glaucoma-affected fundus image:** Glaucoma is characterized by thinning of neuroretinal rim of the optical disc and enlarged optic cup size. A glaucoma-affected fundus image is shown in Fig. 11.1C.

The sample images randomly taken from the HRF dataset showing the normal, diabetic retinopathy-affected, and glaucoma-affected fundus images are represented in Fig. 11.2.

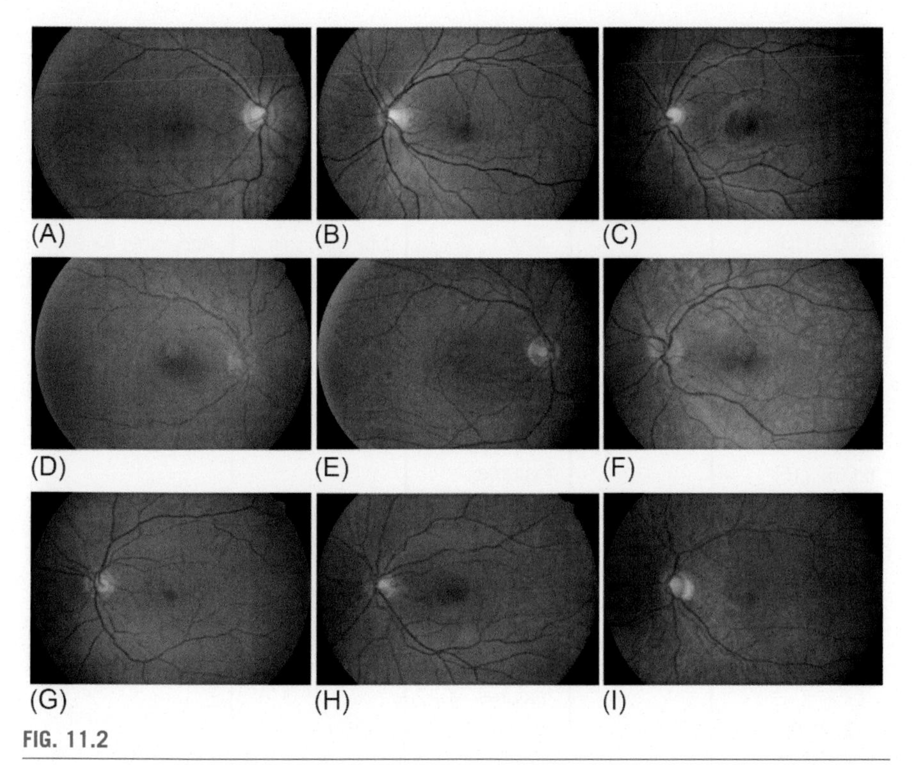

FIG. 11.2

Sample images of retina showing normal retinal images (A–C), retinal image affected with diabetic retinopathy (D–F), and retinal image affected with glaucoma (G–I).

11.2 Related work

The analysis of retinal diseases using Fundus images is a manual procedure requiring the expertise of a trained person that can clearly define and identify the margins of the optic nerve. This approach however results in errors during the subjective analysis due to misinterpretations on the observer's part. Therefore, the main challenge faced by the research community is to design a computer-based system that could automate the analysis of diagnosis of retinal diseases. Numerous attempts have been made in the recent times for the analysis of retinal fundus images like segmentation of optical disc or retinal vessels, diameter measurement of disc [5–11], classification of different retinal diseases [12–54]. A brief description of the studies carried out over the years for the diagnosis of different retinal diseases is given in Table 11.1.

From this literature review, it has been observed that very few attempts have been made to differentiate between diabetic retinopathy and glaucoma fundus images. Therefore, an attempt has been made in the present work to design a computer-aided

Table 11.1 Studies carried out for disease diagnosis of retina

Author (Year)	No. of images	Extracted features	Classifier	Performance
Gardner et al. (1996) [12]	NOR: 101 DR: 200	Retinal features	BPNN	Sensitivity: 88.4% Specificity: 83.5%
Ege et al. (2000) [13]	134	Shape and color features	Bayes Mahalanobis kNN	Accuracy: 86.0% Accuracy: 83.9% Accuracy: 83.7%
Bock et al. (2007) [14]	200	Intensity, texture, FFT coefficients, histogram features	NB kNN SVM	Accuracy: 86% (SVM)
Acharya et al. (2008) [15]	300	HOS-based features	SVM	Accuracy: 82.0%
Fink et al. (2008) [16]	NOR: 60 G: 60	–	kNN	Accuracy: 91.0%
Nayak et al. (2008) [17]	140	Blood vessel area, exudates area, texture features	ANN	Sensitivity: 90.0% Specificity: 100%
Yun et al. (2008) [18]	124	Area and perimeter features of RGB channels	BPNN	Sensitivity:91.7% Specificity: 100%
Acharya et al. (2009) [19]	331	HOS-based features	SVM	Sensitivity: 82.0% Specificity: 86.0%
Nayak et al. (2009) [20]	NOR: 24 G: 37	Cup-to-disc ratio, distance between optic disc center and optic nerve head, ISNT ratio	ANN	Sensitivity: 100% Specificity: 80.0%
Acharya et al. (2011) [21]	NOR: 30 G: 30	HOS and texture features	SVM SMO RF NB	Accuracy: 91.0% (RF)
Dua et al. (2012) [22]	NOR: 30 G: 30	DWT-based features	SVM SMO RF NB	Accuracy: 93.3% (SVM, SMO)
Mookiah et al. (2012) [23]	NOR: 30 G: 30	HOS and DWT features	SVM	Accuracy:95.0%

Table 11.1 Studies carried out for disease diagnosis of retina—cont'd

Author (Year)	No. of images	Extracted features	Classifier	Performance
Ramani et al. (2012) [24]	NOR: 15 DR:15 G: 15	Texture features	C4.5 DT	Accuracy: 100%
Adarsh et al. (2013) [25]	NOR: 20 DR: 110	Area and texture features	SVM	Accuracy: 95.3%
Annu et al. (2013) [26]	NOR: 10 G: 10	Wavelet energy features	PNN	Accuracy: 95%
Rama Krishnan et al. (2013) [27]	NOR: 30 G: 30	HOS, TT, DWT, and energy features	SVM	Accuracy: 91.6%
Mookiah et al. (2013) [28]	156	Texture features	C4.5 SVM PNN	Accuracy: 88.4% Accuracy: 77.5% (linear kernel) Accuracy: 96.1%
Wagle et al. (2013) [29]	60	Texture features	Modified kNN	Accuracy: 94.4%
Antal et al. (2014) [30]	1200	AM/FM-based features	Ensemble of classifiers	Accuracy: 90.0%
Gayathri et al. (2014) [31]	NOR: 15 G: 15	Texture energy features	MLP-BP ANN	Accuracy: 97.6%
Noronha et al. (2014) [32]	272	Texture features	SVM NB	Accuracy: 90.0% (RBF kernel) Accuracy: 92.6%
Simonthomas et al. (2014) [33]	NOR: 30 G: 30	Texture features	kNN	Accuracy: 98%
Acharya et al. (2015) [34]	510	Gabor transform-based features	PCA-SVM PCA-NB	Accuracy: 93.1% (polynomial kernel) Accuracy: 90.1%
Chinaret al. (2015) [35]	30	Curvic features	PNN	Accuracy:90%
Rao et al. (2015) [36]	300	Wavelet-based features	NB MLP-BP ANN	Accuracy: 89.6% Accuracy: 97.6%
Choudhary et al. (2015) [37]	90	Cup-to-disc ratio, ISNT quadrant features	ANN	Accuracy: 96%

Continued

Table 11.1 Studies carried out for disease diagnosis of retina—cont'd

Author (Year)	No. of images	Extracted features	Classifier	Performance
Chen et al. (2015) [38]	ORIGA SCES	–	CNN	AUC: 0.838 AUC: 0.898
Chen et al. (2015) [39]	ORIGA SCES	–	CNN	AUC: 0.831 AUC: 0.887
Pratt et al. (2016) [40]	80,000	–	CNN	Accuracy: 75.0%
Acharya et al. (2017) [41]	NOR: 143 G: 559	Texton and LCP features	kNN	Accuracy: 95.8%
Gargeya et al. (2017) [42]	75,137	CNN-based features	DT	AUC: 0.97
Al-Bander et al. (2017) [43]	NOR: 255 G: 200	AlexNet-based features	SVM	Accuracy: 88.2%
Choi et al. (2017) [44]	10,000	VGG-19-based features	RF	Accuracy: 72.8%
Maheshwari et al. (2017) [45]	NOR: 30 G: 30 NOR: 255 G: 250	2D-EWT, correntropy features	LS-SVM	Accuracy: 98.3%
Hemanth et al. (2018) [46]	540	Texture features	MHNN	Accuracy: 99.2%
Wang et al. (2018) [47]	35000	DenseNet features	LightGBM	Kappa score: 0.84
Norouzifard et al. (2018) [48]	NOR: 277 G: 170	–	VGG19 Inception-ResNet-v2	Specificity: 90.9% Sensitivity: 93.3%
Kwasigroch et al. (2018) [49]	88000	–	VGG-D	Accuracy: 81.7%
Raghavendra et al. (2018) [50]	NOR: 500 G: 500	Radon transform-based GIST features	SVM	Accuracy: 97.0%
Raghavendra et al. (2018) [51]	NOR: 589 G: 837	CNN-based features	LDA	Accuracy: 98.1%
Wang et al. (2018) [52]	166	–	AlexNet VGG16 Inception-V3	Accuracy: 37.4% Accuracy: 50.0% Accuracy: 63.2%

Table 11.1 Studies carried out for disease diagnosis of retina—cont'd

Author (Year)	No. of images	Extracted features	Classifier	Performance
Zhen (2018) [53]	5978	–	VGG16 VGG19 ResNet DenseNet InveptionV3 Inception ResNet Xception NasNet Mobile	Accuracy: 75.5% (DenseNet)
Kim et al. (2018) [54]	1080	–	VGG16	Accuracy: 91.0%

Note: ANN, *artificial neural network;* BPNN, *backpropagation neural network;* CNN, *convolutional neural network;* DR, *diabetic retinopathy;* DT, *decision tree;* G, *glaucoma;* HOS, *higher-order statistics;* kNN, *k-nearest neighbor;* MHNN, *modified hop neural network;* NB, *naïve Bayes;* NOR, *normal;* PCA, *principal component analysis;* PNN, *probabilistic neural network;* RF, *random forest;* SVM, *support vector machine.*

classification (CAC) system that can help in the clear differentiation between the diabetic retinopathy and glaucoma images based on the textural features

11.3 Methodology

For the differentiation between the diabetic retinopathy and glaucoma images, the workflow of the computer-aided classification (CAC) system design is shown in Fig. 11.3.

11.3.1 Database description

The present database corresponds to HRF database, which is freely available gold standard, high-resolution set of retinal fundus images. The HRF database contains 15 healthy, 15 glaucomatous, and 15 diabetic retinopathy images [55].

From this database, two data subsets, namely, training and testing datasets have been obtained. The training dataset has 8 images belonging to diabetic retinopathy and 8 images belonging to glaucoma along with their vasculature (vessel). The remaining 7 images of diabetic retinopathy and 7 images of glaucoma along with their vasculature (vessel) are used in forming the testing dataset. The bifurcation of dataset has been shown in Fig. 11.4.

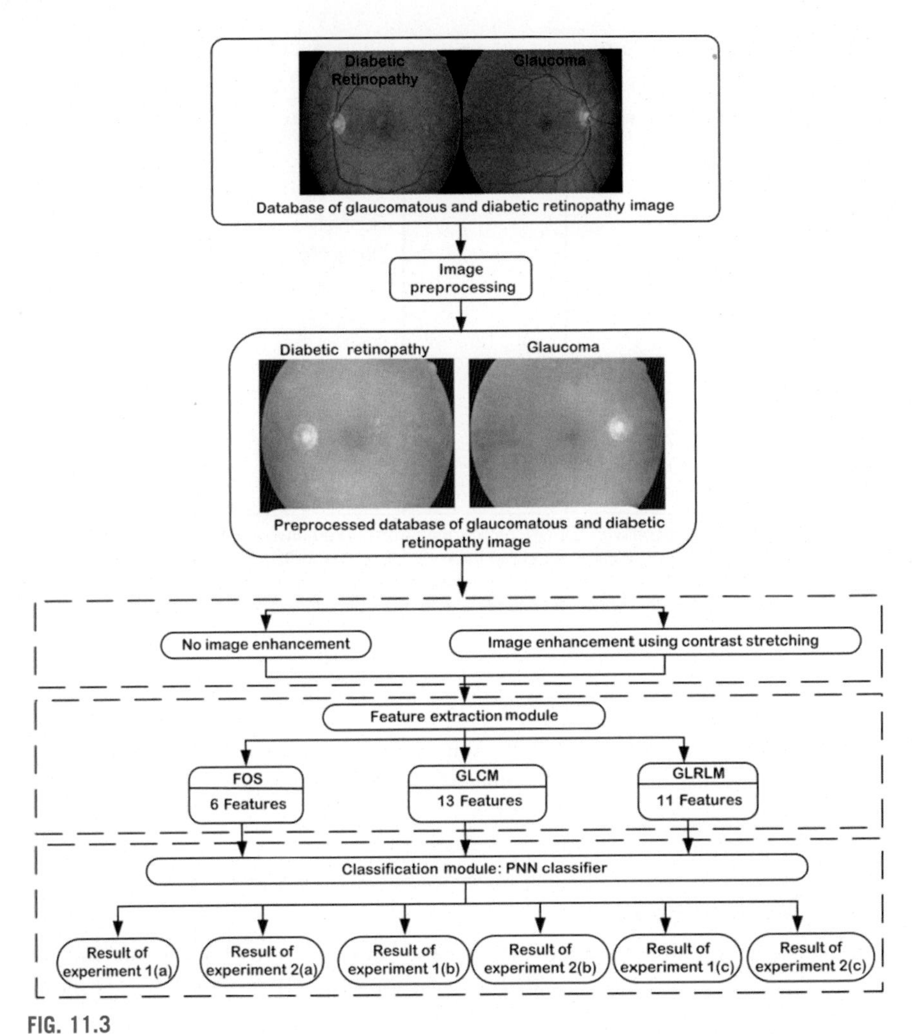

FIG. 11.3

Workflow of the CAC system design.

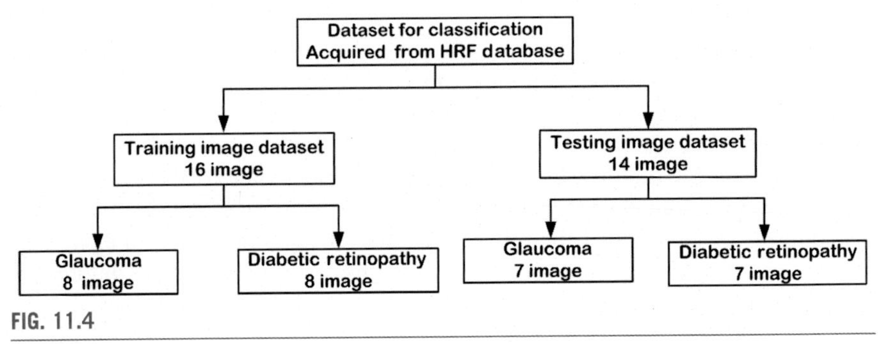

FIG. 11.4

Description of dataset.

The size of each image in the dataset is 3504 × 2336 pixels with a resolution of 72 dpi.

11.3.2 Image preprocessing

Preprocessing is crucial part for differentiating diabetic retinopathy and Glaucoma. In this step, retinal blood vessels are segmented from raw image and an enhanced vessel-free image is obtained. The steps undertaken for this purpose are shown in Fig. 11.5.

11.3.2.1 Grayscale conversion

The raw images presented in the database are RGB color images. For making images more informative and their computations in the later stages less intensive, these images have been converted into grayscale images. The sample images are shown in Fig. 11.6.

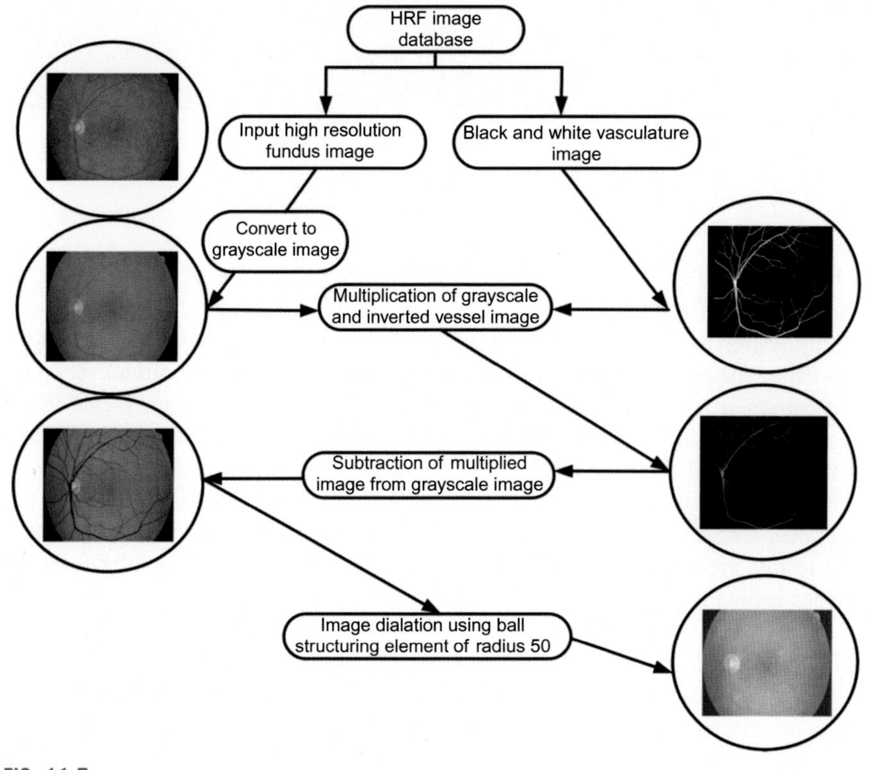

FIG. 11.5

Workflow of preprocessing.

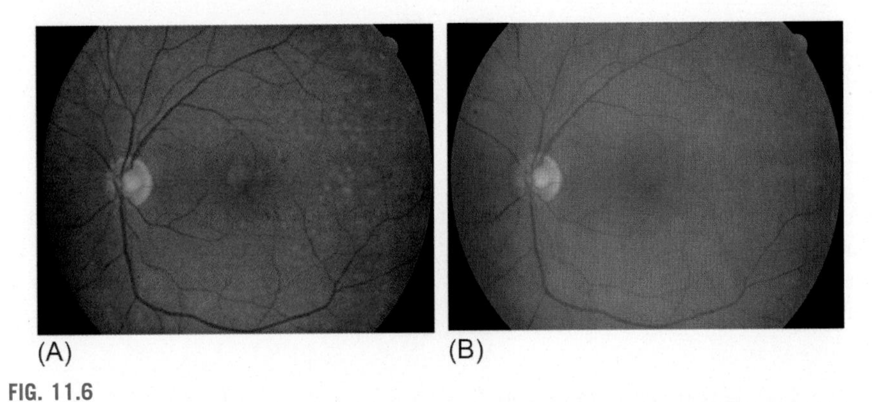

FIG. 11.6

(A) Original RGB image, and (B) original grayscale image.

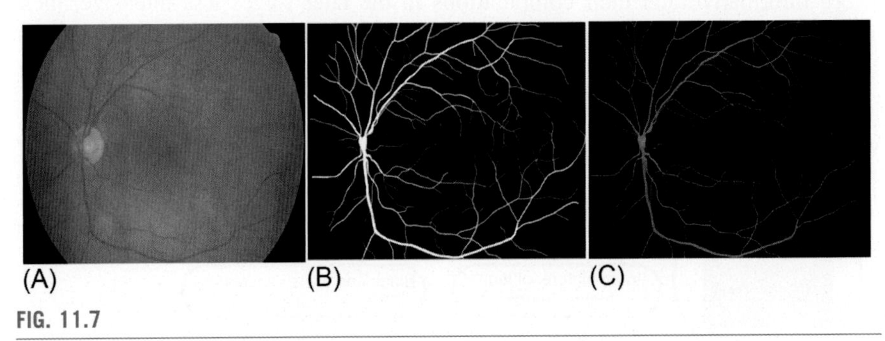

FIG. 11.7

(A) Grayscale image, (B) retinal vessel mask, and (C) resultant multiplied image.

11.3.2.2 Multiplication of gray image with vasculature of raw image (retinal vessel mask)

Multiplication of grayscale image showing whole fundus, with its vasculature image also called the retinal vessel mask (which is an image just showing retinal vessels of that particular fundus image), has been done in order to obtain a grayscale image consisting only of retinal vessels present in our original grayscale fundus image. Every element of the gray scale image is multiplied by the corresponding element in the retinal vessel mask and in the resultant image only those parts are retained that fall under the white region (value = 1) of the mask, in this case the retinal vessels. The results are represented in Fig. 11.7. The vascular retinal mask of the fundus images is obtained from the HRF dataset.

11.3.2.3 Subtraction of multiplied image from gray image

After the previous step, in order to obtain a fundus image free from these vessels, the resultant image of the multiplication operation is subtracted from the grayscale fundus image. This results in the grayscale fundus image without the retinal vessels as shown in Fig. 11.8 (seen as black, 0 intensity areas in the fundus image).

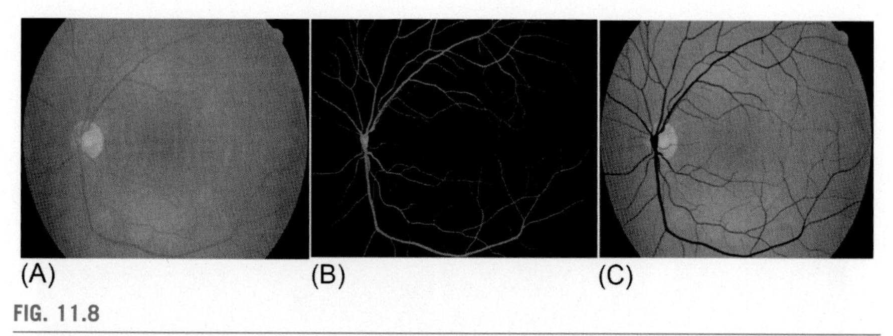

FIG. 11.8

(A) Grayscale image, (B) multiplied image, and (C) subtracted image.

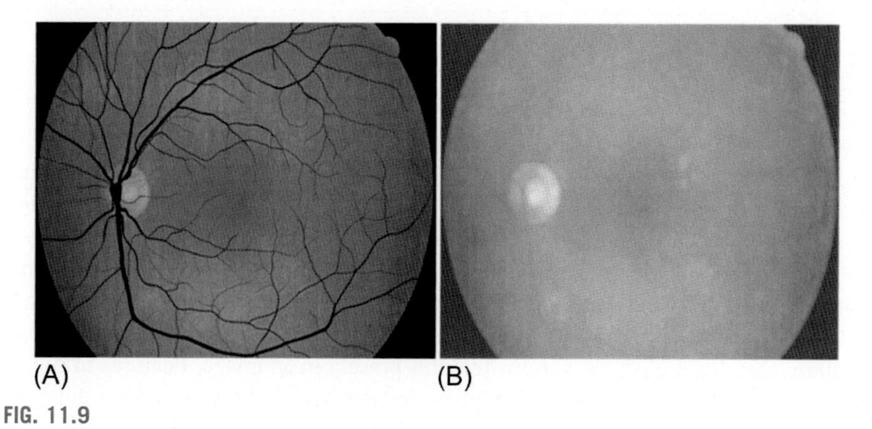

FIG. 11.9

(A) Subtracted image, and (B) dilated image.

11.3.2.4 Image dilation

The subtracted image contains black trails of removed retinal vessels. In order to fill this trail, image dilation has been done using ball-type structuring element of radius 50.This results in a grayscale fundus image, free from retinal vessels and their black trails. The resultant images are shown in Fig. 11.9.

11.3.2.5 Image enhancement by contrast stretching

Image enhancement is a technique by which manipulation of pixel value of an image is done and, due to this, there is an increment in visual interpretation and understanding of image.

Contrast stretching, an image enhancement method also known as normalization, is used to improve the image contrast by stretching the range of intensity values that a particular image contains. The resultant enhanced images are shown in Fig. 11.10.

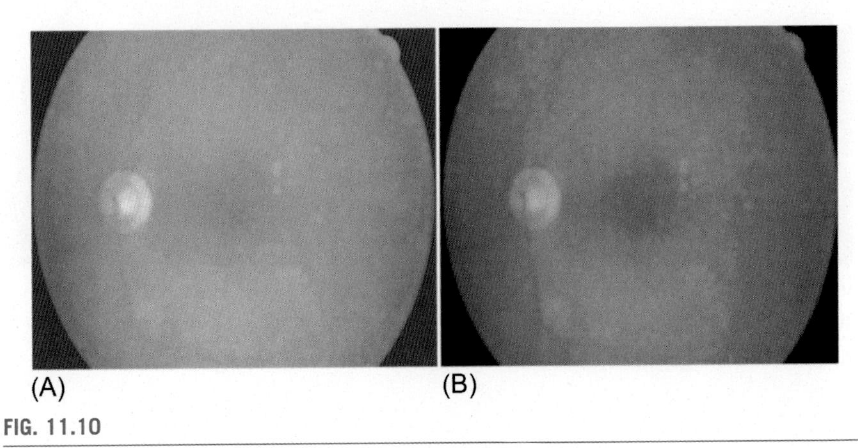

FIG. 11.10

(A) Original image and (b) enhanced image by contrast stretching.

11.3.3 Feature extraction

Feature extraction in image processing is the process of obtaining relevant and useful information from the whole image and it is a special form of dimensionality reduction. The features of an image can be extracted by its content, i.e., color, position, texture, shape, dominant edge of an image item and region, etc. Texture analysis is a challenging job due to the presence of complex patterns of the texture and infinitely different lighting conditions that must be taken into consideration during analysis. It is possible to perform texture segmentation using the knowledge of dominant features that distinguish the various features present in an image. Features in fundus images are difficult to extract visually and redundant in nature, and hence there is a need to transform them into mathematical set of features. Out of all the features for present work, textural features are considered [56, 57].

Statistical-based method determines the spatial distribution of pixel intensity (gray level) values by calculating local feature on each point and calculating set of statistical data from the distribution of local features. The different feature extraction methods used in the present work are shown in Fig. 11.11.

11.3.3.1 First-order statistics

First-order statistics (FOS) is a well-established tool for describing texture features by the help of intensity histogram of an image in the field of medical image processing. The parameters determined by FOS depend upon single individual pixel rather than neighboring pixel of image. From these pixel intensity values, various parameters like standard deviation, entropy, average gray level (mean), smoothness, uniformity, etc. are computed [58–62]. A summary of algorithm used in formation of FOS feature vector is shown in Fig. 11.12.

The FOS-based features are computed considering the spatial distribution of various intensity levels in the region of interest (ROI) using the following equations.

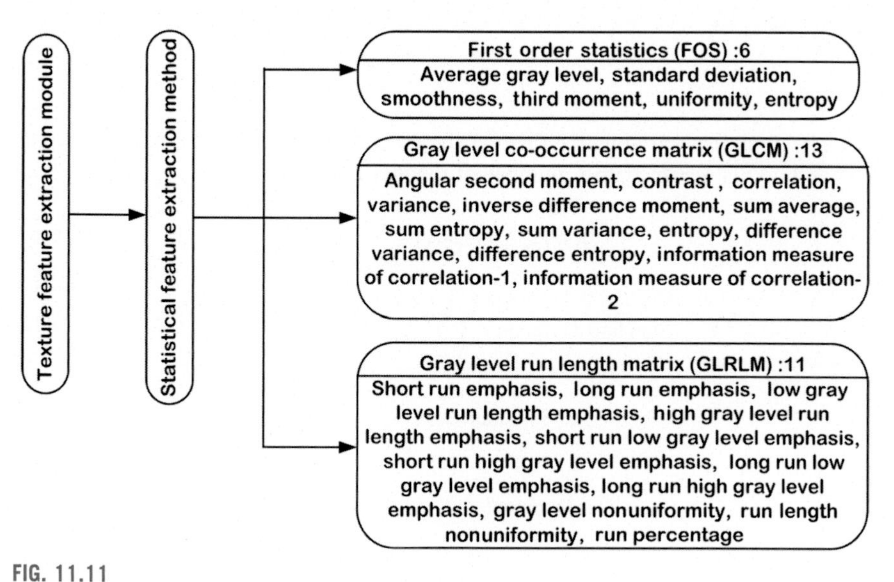

FIG. 11.11

Different methods of feature extraction.

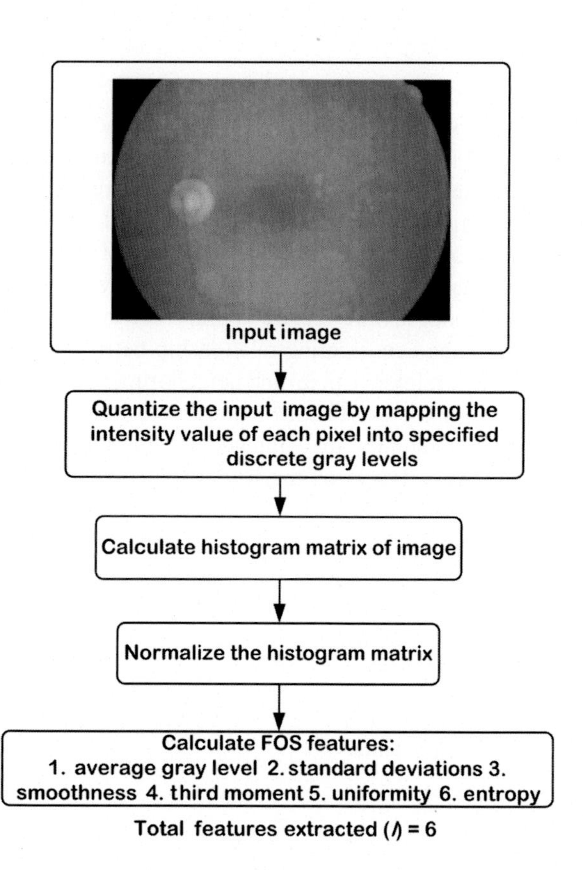

FIG. 11.12

Flowchart of FOS.

Let x be a random variable that denotes the gray levels of an image, $p(x_i)$ is the corresponding histogram with L number of distinct gray levels

$$\text{Average gray level} = \sum_{i=0}^{L-1} x_i p(x_i) \tag{11.1}$$

$$\text{Standard deviation} = \left(\sum_{i=0}^{L-1} (x_i - \text{mean})^2 p(x_i) \right)^{1/2} \tag{11.2}$$

$$\text{Smoothness} = 1 - \frac{1}{1+\sigma^2} \tag{11.3}$$

$$\text{Third moment} = \sum_{i=0}^{L-1} (x_i - \text{mean})^3 p(x_i) \tag{11.4}$$

$$\text{Uniformity} = \sum_{i} p(i)^2 \tag{11.5}$$

$$\text{Entropy} = -\sum_{i} p(x_i) \log_2 p(x_i) \tag{11.6}$$

11.3.3.2 Gray-level co-occurrence matrix

Gray-level co-occurrence matrix (GLCM) or Co-occurrence distribution is a matrix showing different combination of gray levels found within the image [63, 64]. The textural features extracted from the images by GLCM were helpful in identification of different regions in the images. In GLCM, the pixel values of neighboring pixels are compared and a co-occurrence matrix is developed [60, 65, 66]. The number of rows and columns is equal to number of pixel brightness value (gray level). GLCM provides information about texture feature and is a second-order statistical texture feature computation method considering relationship between two pixels for mathematical calculation, i.e., reference pixel and neighboring pixel [61, 67]. A summary of algorithm used in formation of GLCM feature vector is shown in Fig. 11.13.

The 2nd-order GLCM features are computed considering the spatial relationship between any two intensity levels of the ROI using the following mathematical equations.

$$\text{Angular second moment} = \sum_{i,j} P_{ij}^2 \tag{11.7}$$

$$\text{Contrast} = \sum_{i,j} P_{i,j} (i-j)^2 \tag{11.8}$$

$$\text{Correlation} = \sum_{i,j} P_{i,j} \left[\frac{(i-\mu_i)(j-\mu_j)}{\sigma_i \sigma_j} \right] \tag{11.9}$$

$$\text{Variance} = \sum_{i,j} P_{i,j} (i-\mu_i)^2 \tag{11.10}$$

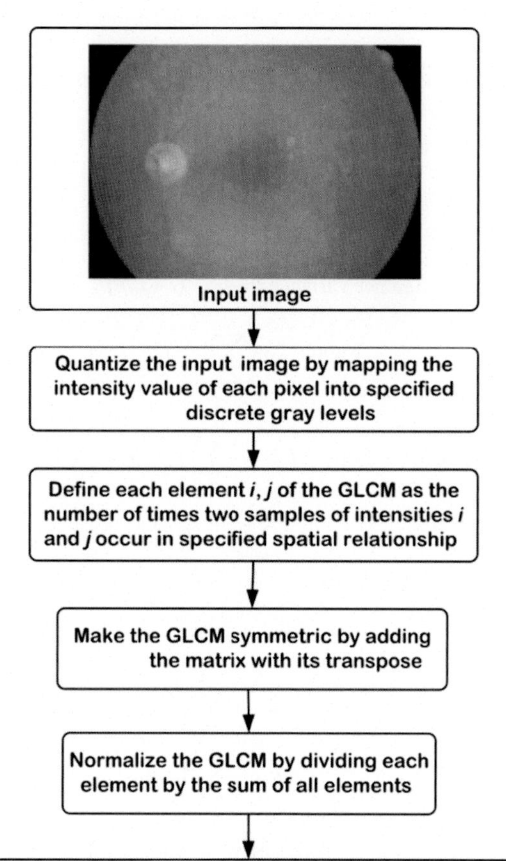

FIG. 11.13

Flowchart of GLCM.

$$\text{Inverse difference moment} = \sum_{i,j} \frac{P_{i,j}}{1+(i-j)^2} \tag{11.11}$$

$$\text{Sum average} = f_{12} = \sum_{i=2}^{2N_g} i p_{x+y}(i) \tag{11.12}$$

$$\text{Sum entropy} = f_{14} = -\sum_{i=2}^{2N_g} p_{x+y}(i) \log\left(p_{x+y}(i)\right) \tag{11.13}$$

$$\text{Sum variance} = f_{13} = \sum_{i=2}^{2N_g} (i - f_{14})^2 p_{x+y}(i) \tag{11.14}$$

$$\text{Entropy} = -\sum_{i,j} p_{i,j} \log (p_{i,j}) \tag{11.15}$$

$$\text{Difference variance} = -\sum_{i=0}^{N_g-1} (i - f_6)^2 p_{x-y}(i)$$
$$\text{where, } f_6 = \sum_{i,j} |i-j| p_{ij} \tag{11.16}$$

$$\text{Difference entropy} = f_{16} = -\sum_{i=0}^{N_g-1} p_{x-y}(i) \log (p_{x-y}(i)) \tag{11.17}$$

$$\text{Information measure of correlation1} = \frac{-\sum_{i,j} p_{i,j}((\log (p_{i,j})) - \log (p_x(i)p_y(i)))}{\max (HX, HY)} \tag{11.18}$$

$$\text{Information measure of correlation2} = f_{18} = \sqrt{1 - e^{-2(a-b)}}$$
$$\text{where, } a = -\sum_{i,j} p_x(i)p_y(i) \log (p_x(i)p_y(i)) \tag{11.19}$$
$$b = -\sum_{i,j} p_{i,j} \log (p_x(i)p_y(i))$$

where $P(i, j)$ gives the statistical probability values for changes between gray levels i and j at a given distance d and angle θ.

11.3.3.3 Gray-level run length matrix

Gray-level run length matrix (GLRLM) is higher-order statistical feature extraction method [60, 68]. It considers three or more pixel intensity values and determines the size of homogenous run for each pixel brightness value (gray level). The consecutive pixel having same gray level is known as gray-level run and run length value is number of times such a run occurs in image and this compute matrix in 13 different directions in three-dimensional space and 4 directions in two-dimensional space from a given image. Each element of GLRLM represents number of homogenous runs j and number of same gray level [61, 62, 69, 70]. A summary of algorithm used in formation of GLRL-based feature vector is shown in Fig. 11.14.

The higher-order GLRLM features are computed considering the spatial relationship between more than two intensity levels of the ROI using the following mathematical equations.

$$\text{Short run emphasis} = \sum_{i=1}^{G}\sum_{j=1}^{R} \frac{p(i,j|\theta)}{j^2} / \sum_{i=1}^{G}\sum_{j=1}^{R} p(i,j|\theta) \tag{11.20}$$

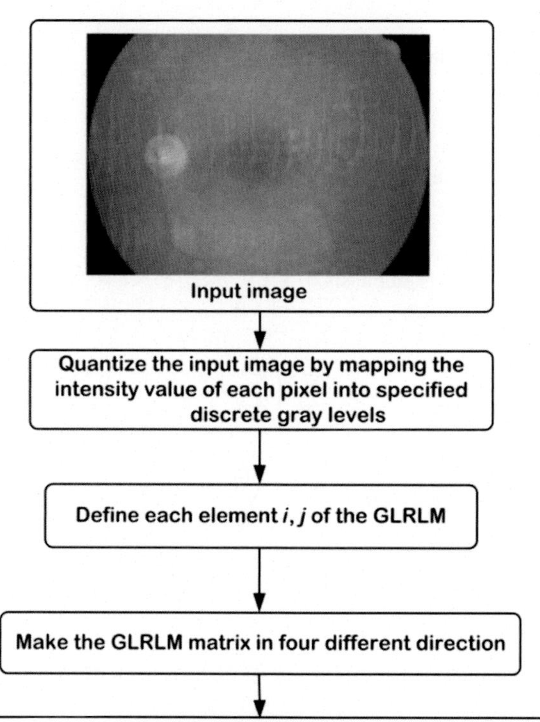

FIG. 11.14

Flowchart of GLRLM.

$$\text{Long run emphasis} = \sum_{i=1}^{G}\sum_{j=1}^{R} j^2 p(i,j|\theta) \Big/ \sum_{i=1}^{G}\sum_{j=1}^{R} p(i,j|\theta) \tag{11.21}$$

$$\text{Low gray-level run emphasis} = \sum_{i=1}^{G}\sum_{j=1}^{R} \frac{p(i,j|\theta)}{i^2} \Big/ \sum_{i=1}^{G}\sum_{j=1}^{R} p(i,j|\theta) \tag{11.22}$$

$$\text{High gray-level run emphasis} = \sum_{i=1}^{G}\sum_{j=1}^{R} i^2 p(i,j|\theta) \Big/ \sum_{i=1}^{G}\sum_{j=1}^{R} p(i,j|\theta) \tag{11.23}$$

$$\text{Short run low gray-level emphasis} = \sum_{i=1}^{G}\sum_{j=1}^{R} \frac{p(i,j|\theta)}{i^2 \times j^2} \Big/ \sum_{i=1}^{G}\sum_{j=1}^{R} p(i,j|\theta) \tag{11.24}$$

$$\text{Short run high gray-level emphasis} = \sum_{i=1}^{G}\sum_{j=1}^{R}\frac{p(i,j|\theta)\times i^2}{j^2} / \sum_{i=1}^{G}\sum_{j=1}^{R}p(i,j|\theta) \tag{11.25}$$

$$\text{Long run low gray-level emphasis} = \sum_{i=1}^{G}\sum_{j=1}^{R}\frac{p(i,j|\theta)\times j^2}{i^2} / \sum_{i=1}^{G}\sum_{j=1}^{R}p(i,j|\theta) \tag{11.26}$$

$$\text{Long run high gray-level emphasis} = \sum_{i=1}^{G}\sum_{j=1}^{R}p(i,j|\theta)\times i^2 \times j^2 / \sum_{i=1}^{G}\sum_{j=1}^{R}p(i,j|\theta) \tag{11.27}$$

$$\text{Gray-level nonuniformity} = \sum_{i=1}^{G}\left(\sum_{j=1}^{R}p(i,j|\theta)\right)^2 / \sum_{i=1}^{G}\sum_{j=1}^{R}p(i,j|\theta) \tag{11.28}$$

$$\text{Run length nonuniformity} = \sum_{i=1}^{R}\left(\sum_{j=1}^{G}p(i,j|\theta)\right)^2 / \sum_{i=1}^{G}\sum_{j=1}^{R}p(i,j|\theta) \tag{11.29}$$

$$\text{Run percentage} = \frac{1}{n}\sum_{i=1}^{G}\sum_{j=1}^{R}p(i,j|\theta) \tag{11.30}$$

where $p(i,j|\theta)$ represents total number of occurrences of the runs (with length j) at gray level i in a direction θ.

11.3.4 Classification module

Neural network is most widely used technique in the field of medical image processing. Artificial neural network is information processing paradigm similar to biological neural network in which neuron collects signal through dendrites and axon serves as sender of that signal and then synapse converts those signals into activity [71–74]. In the following work, PNN has been used as a classifier in order to perform the required differential medical diagnosis.

PNN, which stands for Probabilistic Neural Network, is a feed forward neural network used for classification and pattern recognition. PNN is formed by replacing the sigmoid activation function with an exponential function that can compute non-linear decision boundaries approaching the Bayes optimal. A PNN classifier, having four layers of neural network, can be used to map any input pattern to any number of classification [75–78]. These four layers are:

(1) Input Layer: The neurons making up input layer are simply responsible for supplying input units to all the neurons in the hidden layer.
(2) Hidden Layer: This layer is responsible for computing the distance between the test instance and the training instances and then applies PDF function having sigma values during training of the network.

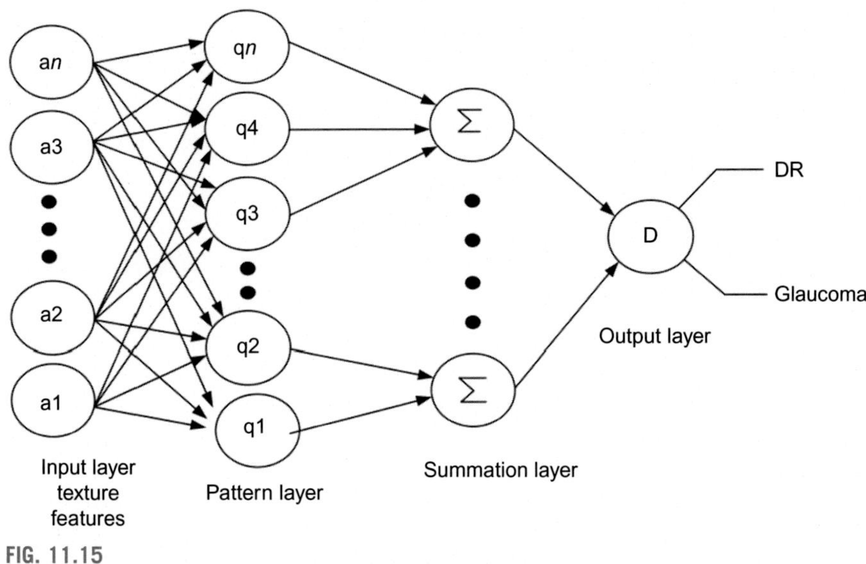

FIG. 11.15

PNN classifier.

(3) Summation Layer: The output of the hidden layer is given to the pattern neuron corresponding to the hidden neuron's category. These pattern neurons then add the values for the class they represent and give out the result to output layer.

(4) Output Layer: The final layer in the PNN and is responsible for comparing the weighted votes for each target class (i.e., outputs the final judgment of the network on the basis of comparing activation in previous layer).

A build up of a typical PNN classifier is shown in Fig 11.15.

11.4 Results and discussion

In the present work, different techniques for differentiating diabetic retinopathy and Glaucoma have been studied. These two diseases contribute mainly in the vision loss all around the world. With the help of CAC model developed in the work, bifurcation of diabetic retinopathy and Glaucoma can be done more efficiently and cheaply. Two experiments have been carried out in this work with the description given in Table 11.2.

11.4.1 Experiment 1: Classification of diabetic retinopathy and glaucoma without image enhancement

The workflow adopted for carrying out this experiment is shown in Fig. 11.16.

Table 11.2 Description of experiments

Experiment	Description
Experiment 1:	Classification of diabetic retinopathy and glaucoma without image enhancement **(a)** FOS feature extraction method **(b)** GLCM feature extraction method **(c)** GLRLM feature extraction method
Experiment 2:	Classification of diabetic retinopathy and glaucoma using contrast stretching image enhancement **(a)** FOS feature extraction method **(b)** GLCM feature extraction method **(c)** GLRLM feature extraction method

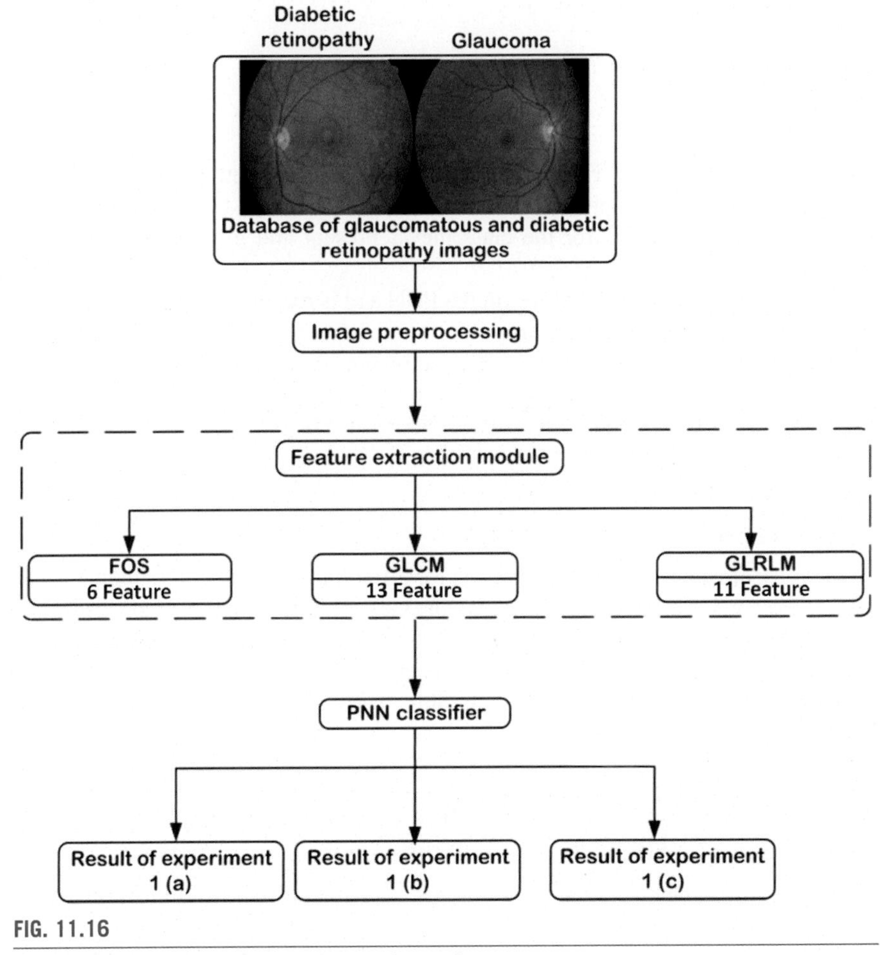

FIG. 11.16

Workflow adopted for carrying out experiment 1.

Table 11.3 Classification performance without image enhancement

FV(*l*)	CM			Accuracy %	Sensitivity (DR) %	Sensitivity (G) %
		DR	G			
FOS (6)	DR	7	0	71.43	100	42.85
	G	4	3			
GLCM (13)	DR	2	5	57.14	28.57	85.71
	G	1	6			
GLRLM (11)	DR	6	1	78.57	85.71	71.43
	G	2	5			

Note: CM, *confusion matrix;* DR, *diabetic retinopathy;* FV, *feature vector;* G, *glaucoma;* I, *length of feature vector.*

In this experiment from the preprocessed images, different feature vectors have been extracted using FOS, GLCM, and GLRLM methods of feature extraction. The results of the experiment are shown in Table 11.3.

From the results obtained in Table 11.3, it can be observed that highest accuracy for correct classification of DR and G images is 78.57 % obtained using the feature vector formed by applying GLRLM method of feature extraction. The sensitivity values for DR and G classes are 85.71 % and 71.43 %, respectively. Out of 14 testing instances, 3 have been misclassified.

11.4.2 Experiment 2: Classification of diabetic retinopathy and glaucoma using contrast stretching image enhancement

The workflow adopted for carrying out this experiment is shown in Fig. 11.17.

In this experiment from the enhanced images, different feature vectors have been extracted using FOS, GLCM, and GLRLM methods of feature extraction. The results of the experiment are shown in Table 11.4.

From the results obtained in Table 11.4, it can be observed that highest accuracy for correct classification of DR and G images is 85.71 % obtained using the feature vector formed by applying GLRLM method of feature extraction. The sensitivity values for DR and G classes are 71.43 % and 100 %, respectively. Out of 14 testing instances, only 2 have been misclassified.

A CAC system has been proposed in Fig. 11.18 on the basis of the results of the experimentation and can be used in differential diagnosis between Diabetic Retinopathy and Glaucoma. First of all, the acquired images of eyes affected with Glaucoma or Diabetic Retinopathy are converted to grayscale images. Then through preprocessing techniques explained in the paper, the retinal blood vessels are removed. The obtained images are then enhanced using contrast stretching. GLRLM features are extracted from these images and are used as input to the PNN classifier model,

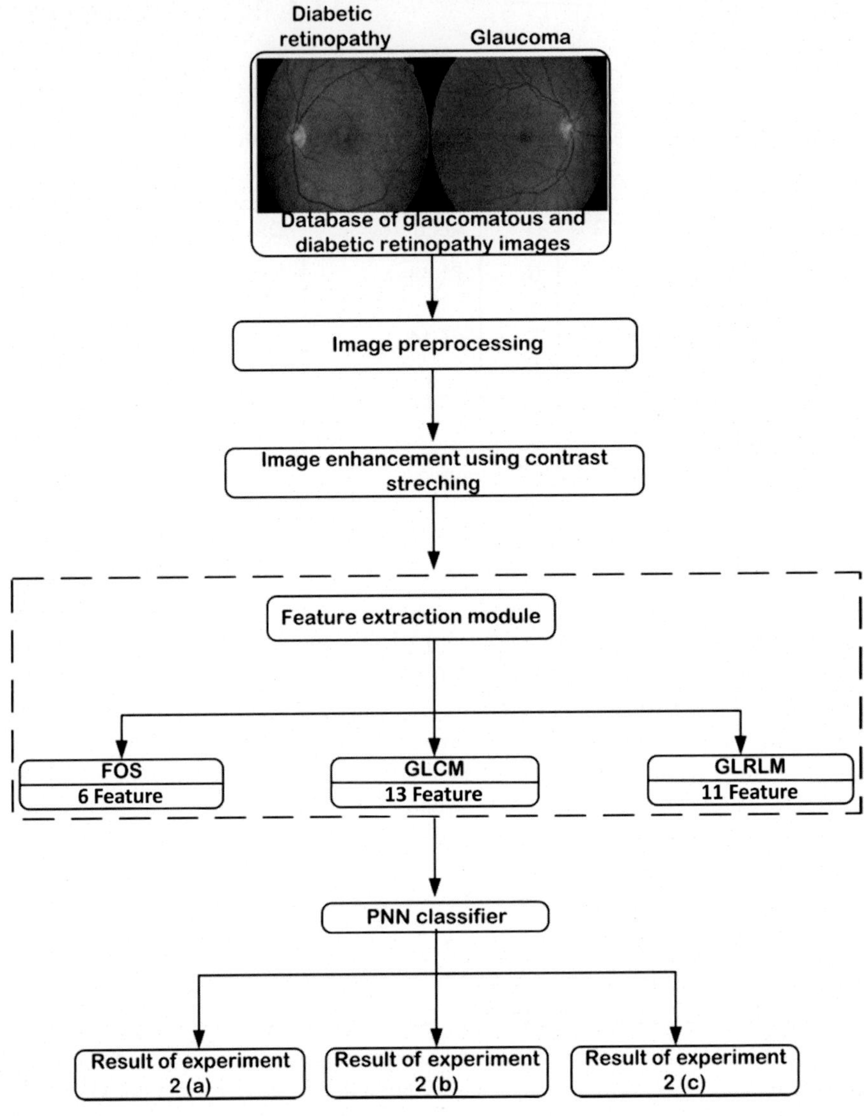

FIG. 11.17

Workflow adopted for carrying out experiment 2.

Table 11.4 Classification performance with image enhancement

FV(*l*)	CM			Accuracy %	Sensitivity (DR) %	Sensitivity (G)%
		DR	G			
FOS (6)	DR	4	3	71.43	57.14	85.71
	G	1	6			
GLCM (13)	DR	5	2	78.57	71.43	85.71
	G	1	6			
GLRLM (11)	DR	5	2	85.71	71.43	100
	G	0	7			

Note: CM, *confusion matrix;* DR, *diabetic retinopathy;* FV, *feature vector;* G, *glaucoma;* l, *length of feature vector.*

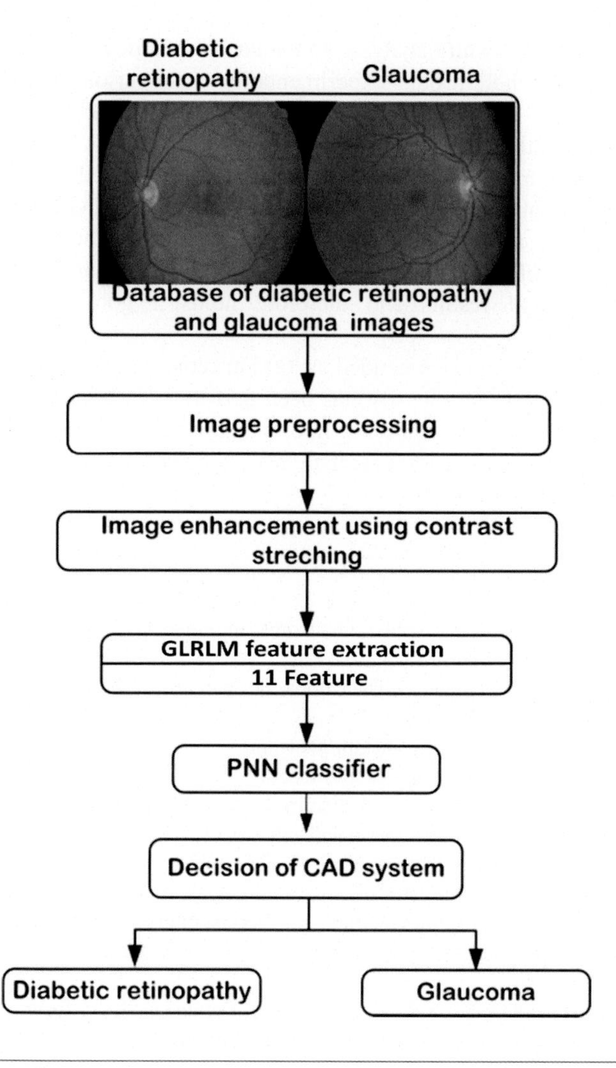

FIG. 11.18

Proposed CAC system.

outputting the decision of whether the eye is affected with Glaucoma or Diabetic Retinopathy.

11.5 Conclusion and future scope

The texture analysis of fundus image of eyes has been shown to give considerably good results for development of a differential diagnosis system between Diabetic Retinopathy and Glaucoma. There are some preprocessing techniques that play an important role as classification analysis is dependent upon them and hence are to be performed before texture analysis. In the course of study, these techniques have been explained. Further, through experimentation using various feature extractors, it was found that the best results (classification accuracy of 85.71%) were obtained using GLRLM on the "enhanced image dataset."

Overall, the proposed algorithm, so developed, can be used in medical diagnosis of eyes for differentiating between Diabetic Retinopathy and Glaucoma with confidence. It can even assist ophthalmologist as a second opinion tool. The computational efficiency of this proposed CAC system is high enough to be used for quick and practical results and therefore holds a good future prospect in medical field.

The present work can be extended as: (a) For removal of retinal vessels, dilation with ball-type structuring element has been performed. The performance of other structuring elements, e.g., diamond, line, octagon, square, rectangle, etc. can be tested. (b) Different deep learning techniques can be employed for the classification of different retinal diseases.

References

[1] R. Saxena, D. Singh, P. Vashist, Glaucoma: an emerging peril, Indian J. Community Med. 38 (3) (2013) 135–137.

[2] Y. Zheng, M. He, N. Congdon, The worldwide epidemic of diabetic retinopathy, Indian J. Ophthalmol. 60 (5) (2012) 428–431.

[3] https://www.webmd.com/eye-health/glaucoma-eyes#1.

[4] https://nei.nih.gov/health/diabetic/retinopathy.

[5] D. Shriranjani, S.G. Tebby, S.C. Satapathy, N. Dey, V. Rajinikanth, Kapur's entropy and active contour-based segmentation and analysis of retinal optic disc, in: Computational Signal Processing and Analysis, 2018, pp. 287–295.

[6] S. Chakraborty, S. Chatterjee, N. Dey, A.S. Ashour, F. Shi, Gradient approximation in retinal blood vessel segmentation, in: Proceedings of 4th IEEE Uttar Pradesh Section International Conference on Electrical, Computer and Electronics (UPCON), 2017, pp. 618–623.

[7] L. Moraru, C.D. Obreja, N. Dey, A.S. Ashour, Dempster-Shafer fusion for effective retinal vessels' diameter measurement, in: Soft Computing Based Medical Image Analysis, 2018, pp. 149–160.

[8] S.E. Zohora, S. Chakraborty, A.M. Khan, N. Dey, Detection of exudates in diabetic retinopathy: a review, in: Proceedings of International Conference on Electrical, Electronics, and Optimization Techniques (ICEEOT), 2016, pp. 2063–2068.

[9] L. Cao, N. Dey, A.S. Ashour, S. Fong, R.S. Sherratt, L. Wu, F. Shi, Diabetic plantar pressure analysis using image fusion, Multimed. Tools Appl. (2018), https://doi.org/10.1007/s11042-018-6269-x.

[10] Z. Li, N. Dey, A.S. Ashour, L. Cao, Y. Wang, D. Wang, P. McCauley, V.E. Balas, K. Shi, F. Shi, Convolutional neural network based clustering and manifold learning method for diabetic plantar pressure imaging dataset, J. Med. Imaging Health Inform. 7 (3) (2017) 639–652.

[11] A. Pachiyappan, U.N. Das, T.V.S.P. Murthy, R. Tatavarti, Automated diagnosis of diabetic retinopathy and glaucoma using fundus and OCT images, Lipids Health Dis. 11 (2012), https://doi.org/10.1186/1476-511X-11-73.

[12] G.G. Gardner, D. Keating, T.H. Williamson, A.T. Elliott, Automatic detection of diabetic retinopathy using an artificial neural network: a screening tool, Br. J. Ophthalmol. 80 (11) (1996) 940–944.

[13] B.M. Ege, O.K. Hejlesen, O.V. Larsen, K. Moller, B. Jennings, D. Kerr, D.A. Cavan, Screening for diabetic retinopathy using computer based image analysis and statistical classification, Comput. Methods Prog. Biomed. 62 (3) (2000) 165–175.

[14] R. Bock, J. Meier, G. Michelson, L.G. Nyul, J. Hornegger, Classifying glaucoma with image-based features from fundus photographs, in: Joint Pattern Recognition Symposium, 2007, pp. 355–364.

[15] U.R. Acharya, C.K. Chua, E.Y.K. Ng, W. Yu, C. Chee, Application of higher order spectra for the identification of diabetes retinopathy stages, J. Med. Syst. 32 (2008) 481–488.

[16] F. Fink, K. Worle, P. Gruber, A.M. Tome, J.M. Gorriz-Saez, C.G. Puntonet, E.W. Lang, ICA analysis of retina images for glaucoma classification, in: Proceedings of 30th Annual International IEEE EMBS Conference, 2008, pp. 4664–4667.

[17] J. Nayak, P.S. Bhat, U.R. Acharya, C.M. Lim, M. Kagathi, Automated identification of diabetic retinopathy stages using digital fundus images, J. Med. Syst. 32 (2) (2008) 107–115.

[18] W.L. Yun, U.R. Acharya, Y.V. Venkatesh, C. Chee, L.C. Min, E.Y.K. Ng, Identification of different stages of diabetic retinopathy using retinal optical images, Inf. Sci. 178 (1) (2008) 106–121.

[19] U.R. Acharya, C.M. Lim, E.Y. Ng, C. Chee, T. Tamura, Computer-based detection of diabetes retinopathy stages using digital fundus images, Proc. Inst. Mech. Eng. H 223 (5) (2009) 545–553.

[20] J. Nayak, U.R. Acharya, P.S. Bhat, N. Shetty, T.C. Lim, Automated diagnosis of glaucoma using digital fundus images, J. Med. Syst. 33 (5) (2009) 337.

[21] U.R. Acharya, U. Rajendra, et al., Automated diagnosis of glaucoma using texture and higher order spectra features, IEEE Trans. Inf. Technol. Biomed. 15 (3) (2011) 449–455.

[22] S. Dua, U.R. Acharya, P. Chowriappa, S. Vinitha Sree, Wavelet-based energy features for glaucomatus image classification, IEEE Trans. Inf. Technol. Biomed. 16 (1) (2012) 80–87.

[23] R.K. Mookiah, U.R. Acharya, C.M. Lim, A. Petznick, J.S. Suri, Data mining technique for automated diagnosis of glaucoma using higher order spectra and wavelet energy features, Knowl.-Based Syst. 33 (2012) 73–82.

[24] R.G. Ramani, L. Balasubramanian, S.G. Jacob, Automatic prediction of diabetic retinopathy and glaucoma through retinal image analysis and data mining techniques, in: Proceedings of International Conference on Machine Vision and Image Processing (MVIP), 2012, https://doi.org/10.1109/MVIP.2012.6428782.

[25] P. Adarsh, D. Jeyakumari, Multiclass SVM-based automated diagnosis of diabetic retinopathy, in: Proceedings of International Conference on Communication and Signal Processing, 2013, , pp. 206–210.

[26] N. Annu, J. Judith, Automated classification of glaucoma images by wavelet energy features, Int. J. Eng. Technol. 5 (2) (2013) 1716–1721.

[27] M.M. Rama Krishnan, O. Faust, Automated glaucoma detection using hybrid feature extraction in retinal fundus images, J. Mech. Med. Biol. 13 (1) (2013) 1350011.

[28] M.R.K. Mookiah, U.R. Acharya, R.J. Martis, C.K. Chua, C.M. Lim, E.Y.K. Ng, A. Laude, Evolutionary algorithm based classifier parameter tuning for automatic diabetic retinopathy grading: a hybrid feature extraction approach, Knowl.-Based Syst. 39 (2013) 9–22.

[29] S. Wagle, J.A. Mangai, V. Santosh Kumar, An improved medical image classification model using data mining techniques, in: Proceedings of GCC Conference and Exhibition, 2013, https://doi.org/10.1109/IEEEGCC.2013.6705760.
[30] B. Antal, A. Hajdu, An ensemble-based system for automatic screening of diabetic retinopathy, Knowl.-Based Syst. 60 (2014) 20–27.
[31] R. Gayathri, P.V. Rao, A. S, Automated glaucoma detection system based on wavelet energy features and ANN, in: Proceedings of International Conference on Advances in Computing, Communications and Informatics, 2014, pp. 2808–2812.
[32] K.P. Noronha, U.R. Acharya, K.P. Nayak, R.J. Martis, S.V. Bhandary, Automated classification of glaucoma stages using higher order cumulant features, Biomed. Signal Process. Control 10 (2014) 174–183.
[33] S. Simonthomas, N. Thulasi, P. Ashraf, Automated diagnosis of glaucoma using Haralick texture features, in: Proceedings of International Conference on Information Communication and Embedded Systems, 2014, https://doi.org/10.1109/ICICES.2014.7033743.
[34] U.R. Acharya, E.Y.K. Ng, L.W.J. Eugene, K.P. Noronha, L.C. Min, K.P. Nayak, S. V. Bhandary, Decision support system for the glaucoma using Gabor transformation, Biomed. Signal Process. Control 15 (2015) 18–26.
[35] D.M. Chinar, A radial segmented feature based PNN model for retinal disease, Int. J. Comput. Sci. Mobile Comput. 4 (11) (2015) 102–107.
[36] P.V. Rao, R. Gayathri, R. Sunitha, A novel approach for design and analysis of diabetic retinopathy glaucoma detection using cup to disk ration and ANN, in: Proceedings of 2nd International Conference on Nanomaterials and Technologies, vol. 10, 2015, pp. 446–454.
[37] K. Choudhary, S. Tiwari, ANN glaucoma detection using cup-to-disk ratio and neuroretinal rim, Int. J. Comput. Appl. 111 (11) (2015) 8–14.
[38] X. Chen, Y. Xu, S. Yan, D.W.K. Wong, T.Y. Wong, J. Liu, Automatic feature learning for glaucoma detection based on deep learning, in: Navab, et al. (Eds.), MICCAI-2015, vol. 9351, 2015, pp. 669–677.
[39] X. Chen, Y. Xu, D.W.K. Wong, T.Y. Wong, J. Liu, Glaucoma detection based on deep convolutional neural network, in: Proceedings of 37th Annual International Conference of the IEEE Engineering in Medicine and Biology Society (EMBC), 2015, pp. 715–718.
[40] H. Pratt, F. Coenen, D.M. Broadbent, S.P. Harding, Y. Zheng, Convolutional neural networks for diabetic retinopathy, Procedia Comput. Sci. 90 (7) (2016) 200–205.
[41] U.R. Acharya, S. Bhat, J.E.W. Koh, S.V. Bhandary, H. Adeli, A novel algorithm to detect glaucoma risk using texton and local configuration pattern features extracted form fundus images, Comput. Biol. Med. 88 (2017) 72–83.
[42] R. Gargeya, T. Leng, Automated identification of diabetic retinopathy using deep learning, Ophthalmology 124 (7) (2017) 962–969.
[43] B. Al-Bander, W. Al-Nuaimy, M.A. Al-Taee, Y. Zheng, Automated glaucoma diagnosis using deep learning approach, in: Proceedings of 14th International Multi-Conference on Systems, Signals and Devices, 2017, pp. 207–210.
[44] J.Y. Choi, T.K. Yoo, J.G. Seo, J. Kwak, T.T. Um, T.H. Rim, Multi-categorical deep learning neural network to classify retinal images: a pilot study employing small database, PLoS One 12 (11) (2017), https://doi.org/10.1371/journal.pone.0187336.
[45] S. Maheshwari, R.B. Pachori, U.R. Acharya, Automated diagnosis of glaucoma using empirical wavelet transform and correntropy features extracted from fundus images, IEEE J. Biomed. Health Inform. 21 (3) (2017) 803–813.

[46] D.J. Hemanth, J. Anitha, L.H. Son, M. Mittal, Diabetic retinopathy diagnosis from retinal images using modified Hopfield neural network, J. Med. Syst. 42 (12) (2018), https://doi.org/10.1007/s10916-018-1111-6.
[47] Y. Wang, G.A. Wang, W. Fan, J. Li, A deep learning based pipeline for image grading of diabetic retinopathy', in: Proceedings of International Conference on Smart Health, 2018, pp. 240–248.
[48] M. Norouzifard, A. Nemati, H. Gholamhosseini, R. Klette, Automated glaucoma diagnosis using deep and transfer learning: proposal of a system for clinical testing, in: IEEE Image and Vision Computing, 2018, pp. 1–6.
[49] A. Kwasigroch, B. Jarzembinski, M. Grochowski, Deep CNN based decision support system for detection and assessing the stage of diabetic retinopathy, in: International Interdisciplinary PhD Workshop (IIPhDW), 2018, pp. 111–116.
[50] U. Raghavendra, S.V. Bhandary, A. Gudigar, U.R. Acharya, Novel expert system for glaucoma identification using non-parametric spatial envelope energy spectrum with fundus images, Biocybern. Biomed. Eng. 38 (2018) 170–180.
[51] U. Raghavendra, H. Fujita, S.V. Bhandary, A. Gudigar, J.H. Tan, U.R. Acharya, Deep convolution neural network for accurate diagnosis of glaucoma using digital fundus images, Inf. Sci. 441 (2018) 41–49.
[52] X. Wang, Y. Lu, Y. Wang, W.B. Chen, Diabetic retinopathy stage classification using convolutional neural networks, in: Proceedings of IEEE International Conference on Information Reuse and Integration, 2018, pp. 465–471.
[53] Y. Zhen, Performance assessment of the deep learning technologies in grading glaucoma severity, in: Proceedings of IEEE Conference on Computer Vision and Pattern Recognition, 2018, pp. 1–12.
[54] M. Kim, J. Zuallaert, O. Janssens, S. Van Hoecke, H. Park, W. De Neve, Web applicable computer-aided diagnosis of glaucoma using deep learning, in: Machine Learning for Health (ML4H) Workshop, 2018, , pp. 1–5.
[55] https://www5.cs.fau.de/research/data/fundus-images/.
[56] B.S. Manjunath, W.Y. Ma, Texture features for browsing and retrieval of image data, IEEE Trans. Pattern Anal. Mach. Intell. 18 (8) (1996) 837–842.
[57] P. Howarth, S. Rüger, Evaluation of texture features for content-based image retrieval, in: Proceedings of International Conference on Image and Video Retrieval, 2004, pp. 326–333.
[58] J.J. Jaime-Rodriguez, C.A. Gutierrez, D.U. Campos-Delgado, J.M. Luna-Rivera, First-order statistics analysis of two new geometrical models for non-WSSUS mobile-to-mobile channels, in: Proceedings of 2016 IEEE 12th International Conference on Wireless and Mobile Computing, Networking and Communications (WiMob), 2016, https://doi.org/10.1109/WiMOB.2016.7763232.
[59] B. Julesz, A theory of preattentive texture discrimination based on first-order statistics of textons, Biol. Cybern. 41 (2) (1998) 131–138.
[60] I. Kumar, J. Virmani, H.S. Bhadauria, Optimization of ROI size and development of computer assisted framework for breast tissue pattern characterization using digitized screen film mammograms, in: Machine Learning in Bio-Signal Analysis and Diagnostic Imaging, 2019, pp. 127–157.
[61] M.B. Subramanya, J. Virmani, Kriti, A DEFS based system for differential diagnosis between severe fatty liver and cirrhotic liver using ultrasound images, in: Machine Learning in Bio-Signal Analysis and Diagnostic Imaging, 2019, pp. 53–72.

[62] N. Manth, J.V. Kriti, Comparison of multiclass and hierarchical CAC design for benign and malignant hepatic tumors, in: U-Healthcare Monitoring Systems, 2019, pp. 119–146.

[63] D.J. Marceau, P.J. Howarth, J.M. Dubois, D.J. Gratton, Evaluation of the grey-level co-occurrence matrix method for land-cover classification using SPOT imagery, IEEE Trans. Geosci. Remote Sens. 28 (4) (1990) 513–519.

[64] V. Sebastian, A. Unnikrishnan, K. Balakrishnan, Gray Level Co-Occurrence Matrices: Generalisation and Some New Features, 2012. arXivpreprint arXiv:1205.4831.

[65] A. Gebejes, R. Huertas, Texture characterization based on grey-level co-occurrence matrix, in: Proceedings of Conference of Informatics and Management Sciences, 2013.

[66] L.S. Davis, S.A. Johns, J.K. Aggarwal, Texture analysis using generalized co-occurrence matrices, IEEE Trans. Pattern Anal. Mach. Intell. PAMI-3 (1) (1979) 251–259.

[67] S.A. Shearer, R.G. Holmes, Plant identification using color co-occurrence matrices, Trans. ASAE 33 (6) (1990) 1237–1244.

[68] A.S.M. Sohail, P. Bhattacharya, S.P. Mudur, S. Krishnamurthy, Local relative GLRLM-based texture feature extraction for classifying ultrasound medical images, in: Proceedings of 24th Canadian Conference on Electrical and Computer Engineering (CCECE), 2011, https://doi.org/10.1109/CCECE.2011.6030630.

[69] H.H. Loh, J.G. Leu, R.C. Luo, The analysis of natural textures using run length features, IEEE Trans. Ind. Electron. 35 (2) (1988) 323–328.

[70] M.M. Galloway, Texture analysis using grey level run lengths, Comput. Graph. Image Process. 4 (2) (1975) 172–179.

[71] D.F. Specht, Probabilistic neural networks for classification, mapping, or associative memory, in: Proceedings of IEEE International Conference on Neural Networks, 1988, pp. 525–532.

[72] L. Rutkowski, Adaptive probabilistic neural networks for pattern classification in time-varying environment, IEEE Trans. Neural Netw. 15 (4) (2004) 811–827.

[73] L. Ning, Network intrusion classification based on probabilistic neural network, in: Proceedings of 2013 5th International Conference on Computational and Information Sciences (ICCIS), 2013, pp. 57–59.

[74] S. Bansal, G. Chhabra, B.S. Chandra, J.V. Kriti, A hybrid CAD system design for liver diseases using clinical and radiological data, in: U-Healthcare Monitoring Systems, 2019, pp. 289–314.

[75] J.C. Luo, Y. Leung, J. Zheng, J.H. Ma, An elliptical basis function network for classification of remote-sensing images, J. Geogr. Syst. 6 (3) (2004) 219–236.

[76] S. Khoshnoud, M. Teshnehlab, M.A. Shoorehdeli, Probabilistic neural network oriented classification methodology for Ischemic Beat detection using multi resolution wavelet analysis, in: Proceedings of 17th Iranian Conference of Biomedical Engineering (ICBME), 2010, pp. 1–4.

[77] M. Kusy, R. Zajdel, Probabilistic neural network training procedure based on Q (0)-learning algorithm in medical data classification, Appl. Intell. 41 (3) (2014) 837–854.

[78] D.J. Jwo, C.C. Lai, Neural network-based GPS GDOP approximation and classification, *GPS* Solutions 11 (1) (2007) 51–60.

CHAPTER

12 Performance comparison analysis of different classifier for early detection of knee osteoarthritis

Ankit Vijayvargiya*, Puru Lokendra Singh*, Samidha Mridul Verma*, Rajesh Kumar*, Sanjiv Bansal†

**Department of Electrical Engineering, Malviya National Institute of Technology, Jaipur, India*

†Department of Physical Medicine and Rehabilitation, Sawai Man Singh Hospital, Jaipur, India

12.1 Introduction

Knee osteoarthritis (OA) is very common joint disease in the elderly. About 10% of males and 13% of females older than 60 years have symptomatic knee OA [1]. It can cause pain and may affect a person's daily activities, like playing sports, climbing stairs, getting up from a sitting position, and so on. In the United States, more than 1 in every 10 adults are suffering from knee OA [2]. A survey conducted by the National Library of Medicine (NLM) showed that OA symptoms are most common at the age of 70 years.

The knee joint is a synovial joint that contains a fluid-filled capsule. The prominent function of the knee joint is to provide stability for the body, act as a shock absorber, and allow movement of the leg. The main parts of knee joints are bones, cartilages, ligaments, fluids, muscles, and tendons. If any of these structures is defective it may lead to a knee problem. Knee OA is the most common problem in the knee. The knee joint is formed by three major bones, the femur (thigh bone), tibia (shin bone), and the patella (knee cap). In addition, there are four ligaments: anterior cruciate ligament (ACL), posterior cruciate ligament (PCL), medial collateral ligament (MCL), and lateral collateral ligament (LCL). These ligaments are used to connect bones to other bones. The femur is the strongest bone of the human body. It runs from the hip to the knee joint. The tibia is the longest bone in human body and runs from the knee joint to the ankle. There are two types of cartilage: meniscus and articular cartilage. Meniscus cartilage reduces the friction at the knee joint between the femur and tibia as well as disperses the weight of the body. Articular cartilage is thin cartilage that exists between the femur and tibia and provides smooth movement of these bones [3] (Fig. 12.1).

Sensors for Health Monitoring. https://doi.org/10.1016/B978-0-12-819361-7.00012-9

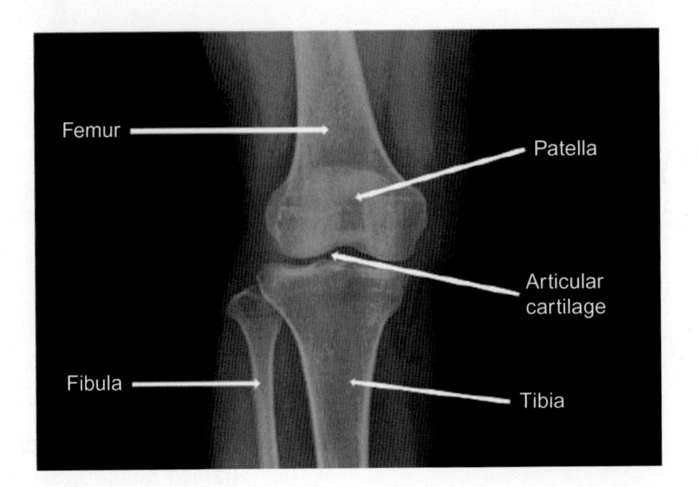

FIG. 12.1

Anatomy of the knee.

The degeneration and deterioration of the articular cartilage of the knee leads to knee OA. There are many reasons for degeneration of articular cartilage like aging, heavy weight, heredity, repetitive stress injuries, and so on. These reasons cause our lower limb muscles to become weak and hence knee OA occurs [4]. The foremost symptom of knee OA is pain at the knee joint that increases when a person moves or runs but that feels better with rest. There are also other symptoms like swelling, stiffness at the knee, cracking sounds at the joint when the knee moves, and so on. Degeneration of cartilage is a very slow process. Therefore if we can find lower-limb muscle weakness at an early stage, then we can predict knee OA at an early stage as well. If knee OA can be predicted early, healthcare professionals may be able to provide some precautions or remedies to the patient. The main cause of knee OA is not clear and there is no modification once it occurs. Only joint replacement surgery can help at this stage [5] (Fig. 12.2).

Clinically, knee OA is diagnosed by X-ray and magnetic resonance imaging (MRI) techniques [6–8]. These methods are efficient ways to diagnose the structure of the bones, but they are very expensive. Not everyone may have the money to spend for early detection of knee OA.

In this chapter, we propose an automated diagnostic technique for knee OA by using surface electromyography (sEMG) electrodes and AI classification techniques. The chapter is divided into five sections. Section 12.1 provides an introduction. In Section 12.2, we provide an overview of EMG and analysis techniques. In Section 12.3, we discuss different kernels of a SVM classifier. We provide our results in Section 12.4. Finally, in Section 12.5 we conclude our research and make recommendations for future work.

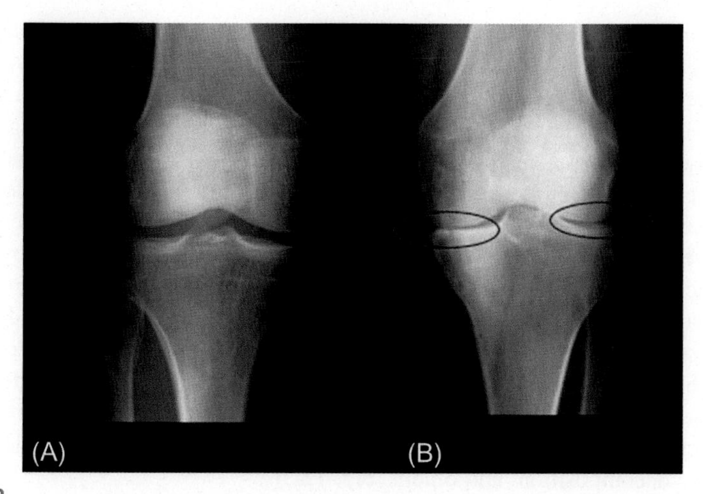

FIG. 12.2

(A) No OA and (B) with OA.

12.2 Analysis of EMG signal

This section provides an overview of EMG and various analysis techniques. Information for becoming a proficient electromyographer is beyond the scope of this chapter, but the content provided gives a basic knowledge of EMG signal analysis.

EMG records the movement of our muscles. It is based on the simple fact that whenever a muscle contracts, a burst of electrical activity is generated that propagates through the adjacent tissue and bone and can be recorded from neighboring skin areas. EMG signals can be recorded by using two techniques: sEMG and intramuscular EMG.

sEMG is a noninvasive technology in which the EMG electrodes can be easily placed on the surface of the skin. The signals recorded during a voluntary muscle action refer to the contributions of all active Muscle Units (MUs). Myoelectric activity detected with surface electrodes above a given muscle can be considered as an aggregate of tissue-filtered signals generated by a number of concurrently active MUs. sEMG signals are dependent on the properties of the contributing MUs, their firing patterns, and their interdependence. In this technique we measure the electrical potential difference in muscles, so we require at least two electrodes.

Surface detection of myoelectric signals is preferred for a variety of reasons, as it provides a number of advantages over needle detection or invasive techniques. The electrodes can be applied without any discomfort or any medical supervision and there are no risks of infection. Long-term monitoring is also easier when using surface electrodes than when using needles. A number of reports have also shown that surface potentials contain much more information than is commonly detected by other non-invasive techniques, making it a very effective and efficient tool to record EMG signals [9].

In the intramuscular method, EMG signals can be recorded by using a fine wire inserted into a muscle with a surface electrode as a reference, or by using two fine wires inserted into muscle referenced to each other. This method is very painful and skin preparation is very important before insertion of the needle electrodes.

While recording the signal from muscles, mainly four types of noises may be introduced in the EMG signal. These noises are: ambient noise produced by electromagnetic devices, inherent noise due to electronic components, motion artifacts noise due to motion of electrodes during the exercise or movement, and inherent noise instability due to firing rate of the motor units [10]. Due to these noises analyzing the EMG signals is very complicated. Therefore we have to reduce these noises before using the recorded EMG signal. There are lots of different methods for reducing these noises, including wavelet denoising, higher order statistics, empirical mode decomposition, and others (Fig. 12.3).

For the classification task in this chapter we took EMG data from the UCI Machine Learning Repository [11]. For collecting the sEMG data, we may in the future design an IoT-based wireless sEMG sensor [12, 13]. We obtained 22 samples; 11 subjects with knee abnormalities confirmed by MRI or X-ray (unhealthy) and 11

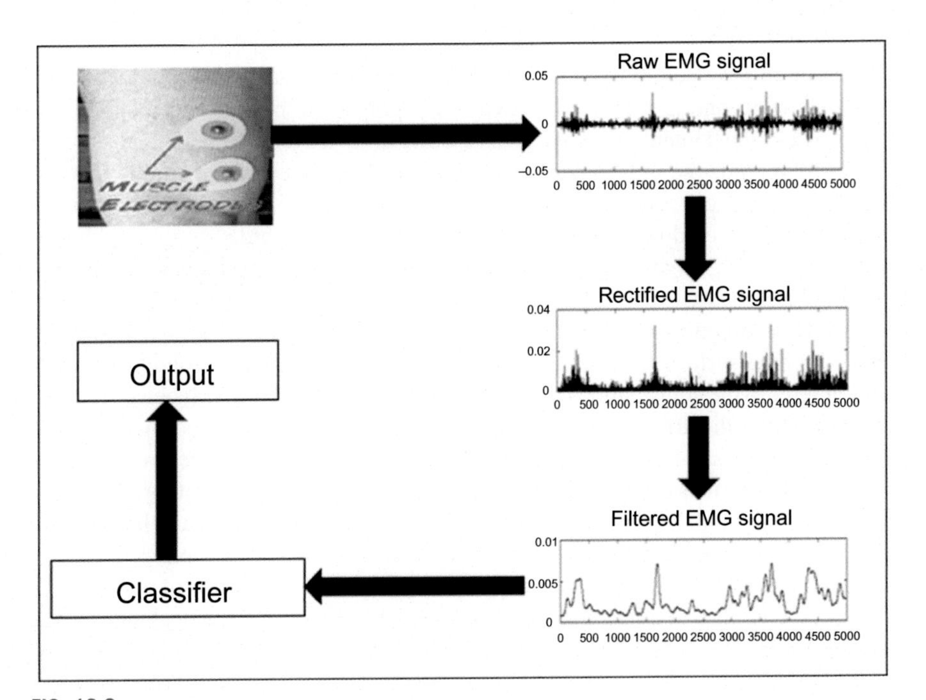

FIG. 12.3

Functional block diagram.

subjects without knee problems (healthy). We assumed knee OA as the type of knee abnormality in the 11 subjects. We took three different exercises for data collection: gait, flexion of the leg up, and leg extension from a sitting position.

We recorded data around four muscles (vastus medialis, semitendinosus, biceps femoris, and rectus femoris) and stored them in a computer using Bluetooth and 14-bit resolution and fs = 1000 Hz.

The frequency range of a sEMG signal is between 50 and 500 Hz. These signals contain DC bias. This DC bias can be removed by applying a high-pass filter with a cut-off frequency of 50 Hz. After removing the DC offset, we used a rectification process for the signal. This process is very common in EMG signal analysis. Both positive and negative components are available in the raw EMG signal, so the mean value of the raw EMG signal may be zero. Due to this reason a rectification process is used in EMG analysis. After this process, only positive components are present in the raw signal of EMG signal. Two types of rectification processes can be used for the EMG signal: full wave rectification and half wave rectification. Full wave rectification adds negative components of the raw EMG signal to the positive. Half wave rectification removes negative components of the raw EMG signal. Full wave rectification is preferred because it conserves both positive and negative components of the EMG signal. After the rectification process, filtration is used for reducing the different types of noises. According to the literature, a fourth-order Butterworth filter is suitable for this purpose.

After the filtration of the EMG signal, we used the mean absolute value technique for feature extraction. We used the principal component analysis (PCA) technique for feature reduction. Then we calculated the mean value of the integrated EMG (IEMG) signal. IEMG is the simplest feature of the EMG signal. It can be calculated by summing of all the absolute values of the EMG signal.

$$\text{Mean absolute value} = \frac{1}{N} * \text{IEMG} \tag{12.1}$$

$$\text{IEMG} = \sum \overline{X}_i \tag{12.2}$$

Figs. 12.4A and 12.5A show the raw data of EMG signals for healthy and unhealthy subjects. This data is very noisy due to different types of noise like ambient noise, inherent noise, movement artifact noise, and so on. Both positive and negative components are available in this data. So a rectification process has been used. Figs. 12.4B and 12.5B show the rectified data for the healthy and unhealthy subjects. The entire negative component adds in positive then only positive component is present. Figs. 12.4C and 12.5C show the filtered data for the healthy and unhealthy subjects. With the help of this figure we may be able to say that the magnitude of the voltage of the healthy subject is higher than that from the unhealthy subject. We may also conclude that the muscle strength of the healthy subject is higher than that of the unhealthy subject.

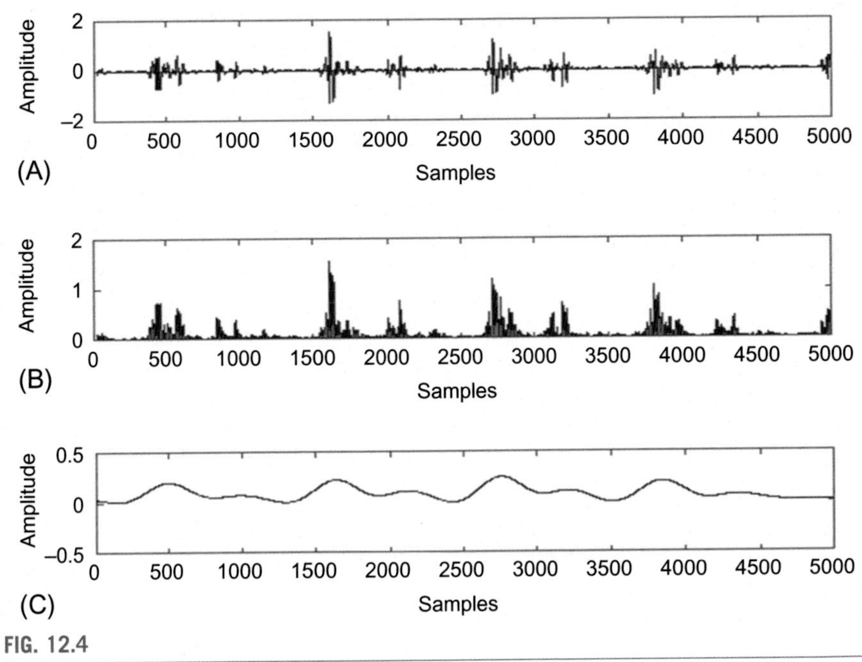

FIG. 12.4

(A) Raw signal of healthy subject. (B) Rectified signal of healthy subject. (C) Filtered signal of healthy subject.

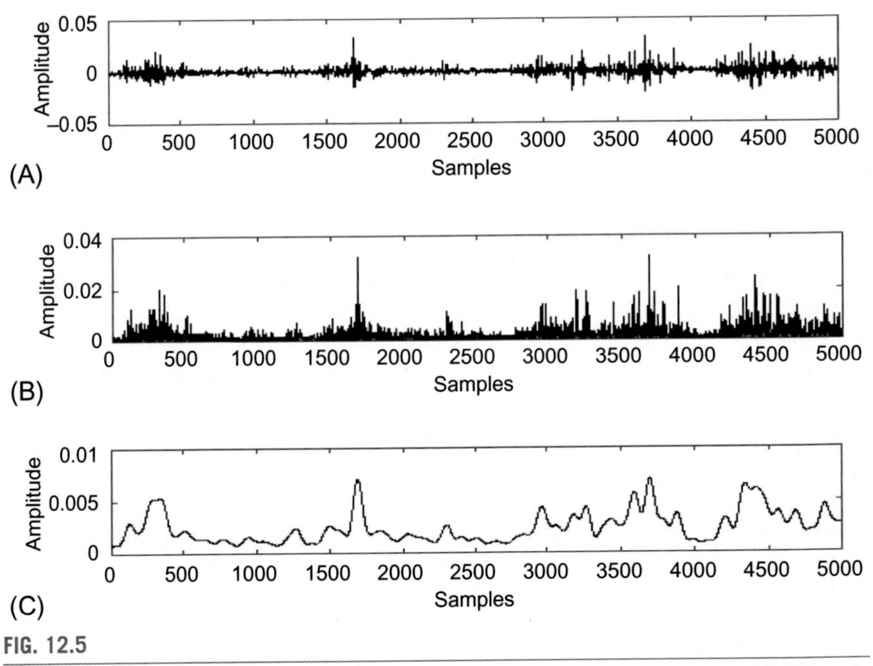

FIG. 12.5

(A) Raw signal of unhealthy subject. (B) Rectified signal of unhealthy subject. (C) Filtered signal of unhealthy subject.

12.3 Principal component analysis

In this modern age, the large size of data is not only a problem for computer hardware but it also affects the performance of many machine learning algorithms. After the feature extraction process, a dimensionality reduction technique is generally used due to the following reasons [14, 15]:

a. It decreases the amount of computation required by the machine learning algorithms
b. Certain features can be correlated and hence may convey redundant information
c. Some of the features might be insignificant and can influence the execution of the machine learning algorithm

PCA, also known as Karhunen-Loeve transformation, is the most popular and useful technique that aims to detect correlation between variables and hence can be used for feature reduction.

The goal is to reduce the dimensions of an $\boldsymbol{m}$-dimensional feature-set by projecting it into a $\boldsymbol{k}$-dimensional subspace where $\boldsymbol{k}<\boldsymbol{m}$. The PCA method is as follows:

1. The data is standardized.
2. The eigenvectors and eigenvalues are then calculated from the covariance matrix. The covariance matrix $\sum$, which is a $\boldsymbol{m}\times\boldsymbol{m}$, is obtained by Eq. (12.3).

$$\sum \text{ln} - 1((\boldsymbol{X}-\boldsymbol{x})\boldsymbol{T}(\boldsymbol{X}-\boldsymbol{x})) \tag{12.3}$$

where $\boldsymbol{X}$ is the original feature matrix with dimensions $\boldsymbol{p}\times\boldsymbol{m}$, $\boldsymbol{x}$ is the mean vector, which represents the mean of the feature columns, $\boldsymbol{m}$ is the dimension of the original feature matrix, and $\boldsymbol{p}$ is the number of data samples.

3. The eigenvectors corresponding to the lowest eigenvalues contain the least information about the data and hence can be dropped without any significant loss of information. Hence the eigenvalues are ranked from highest to lowest and the top $\boldsymbol{k}$ eigenvectors are then selected for projection corresponding to the $\boldsymbol{k}$ eigenvalues.
4. In the last step, the projection matrix $\boldsymbol{W}$ is constructed from the $\boldsymbol{k}$ eigenvectors selected in the previous step, and the original feature-set $\boldsymbol{X}$ ($\boldsymbol{p}\times\boldsymbol{m}$) is then transformed via the projection matrix $\boldsymbol{W}$ ($\boldsymbol{m}\times\boldsymbol{k}$) in order to obtain the samples onto a new subspace $\boldsymbol{Y}$ ($\boldsymbol{p}\times\boldsymbol{k}$).

$$\mathrm{Y}=\mathrm{X}\cdot\mathrm{W}$$

In this research, a total of 12 features for 22 subjects (11 for healthy and 11 for unhealthy) are available in the given dataset. Total dataset size is 22×12. If this size is directly used for the classification then it will take high computation time. PCA is used to reduce the features and retain most important components that contribute the variation [11]. It reduced the dimensionality by taking six principal components, and projecting the original dataset (22×12) to a new sub-space with 22×6 dimension.

12.4 SVM classifier

Classification is the process of predicting the class of given data points. There are two types of classification. The first is supervised learning wherein a training set of correctly identified observations is available. The second is unsupervised learning wherein data is grouped into categories based on some measure of inherent similarity. Binary classification and regression analysis can be performed via the supervised learning method.

In machine learning, SVM comes under the classification of supervised learning. With the help of the different kernel tricks (linear, Gaussian, polynomial, etc.), SVM can be used for linear classification as well as nonlinear classification.

In linear classification a hyperplane is constructed. After plotting the data, we can use different hyperplanes to classify it. Fig. 12.6A shows three hyperplanes (H1, H2, and H3) that can easily classify this data. So which hyperplane is better?

There are different hyperplanes that can classify the data, but only one gives maximum separation and that hyperplane is the better one compared to the others. So we can say that the margin width (shown in Fig. 12.6B) should be high.

Let us assume that $(X_1, Y_1), (X_2, Y_2), (X_3, Y_3), \ldots, (X_n, Y_n)$ are the training dataset where $X_1, X_2, X_3, \ldots, X_n$ are the training inputs and $Y_1, Y_2, Y_3, \ldots, Y_n$ are the outputs, which are either +1 or −1.

Classified as:	+1	if	$w{\cdot}X+b \geq 1$
	−1	if	$w{\cdot}X+b \leq -1$
	Universe explodes	if	$-1 < w{\cdot}X+b < 1$

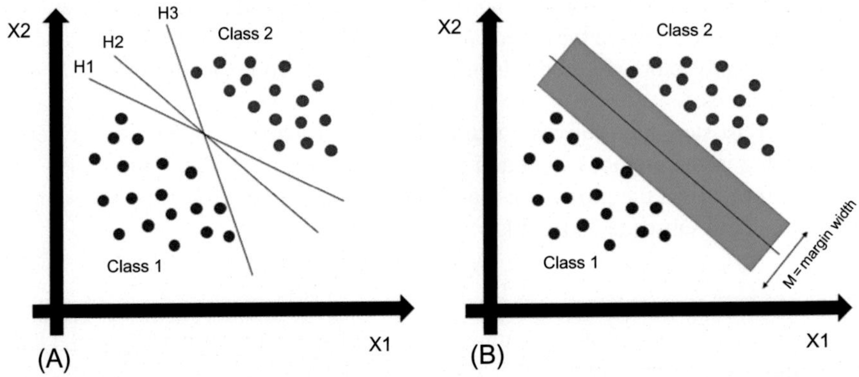

FIG. 12.6

(A) Three different hyperplanes that can classify the data. (B) Illustration of Linear SVM.

Let X^- be any point on the class 1 plane and X^+ be any point on the class 2 plane. Then

$$X^+ = X^- + \lambda w \tag{12.4}$$

$$\text{Plus-plane}: w \cdot X + b = +1 \tag{12.5}$$

$$\text{Minus-plane}: w \cdot X + b = -1 \tag{12.6}$$

$$\text{So at } X = X^+ : w \cdot X^+ + b = +1 \tag{12.7}$$

$$\text{at } X = X^- : w \cdot X^- + b = -1 \tag{12.8}$$

$$|X^+ - X^-| = M = \lambda w \tag{12.9}$$

From Eqs. (12.4), (12.7)

$$w \cdot (X^- + \lambda w) + b = 1$$

$$w \cdot X^- + b + \lambda w \cdot w = 1$$

From Eq. (12.8)

$$-1 + \lambda w \cdot w = 1$$

$$\lambda = 2/w \cdot w \tag{12.10}$$

From Eqs. (12.9), (12.10)

$$M = (2/w \cdot w) \times w \tag{12.11}$$

$$M = 2/\sqrt{w \cdot w} \tag{12.12}$$

So margin width M should be maximum or we can maximize the M by the minimization of $w \cdot w$.

Vapnik [16] proposed the maximum margin hyperplane algorithm in 1963 for linear classification. A linear classifier can be used only for linear data. If the data is nonlinear in nature then a linear classifier will fail or give very low accuracy. In 1992, Bernhard E. Boser, Isabelle M. Guyon, and Vladimir N. Vapnik [16] proposed a method to create a nonlinear classifier by using different kernel tricks (Fig. 12.7).

A kernel trick is a way to make optimization efficient when there are lots of features. A kernel or a window function can be defined as:

$$K(\bar{x}) = \begin{cases} 1; & \text{if } ||\bar{x}|| \leq 1 \\ 0; & \text{Otherwise} \end{cases}$$

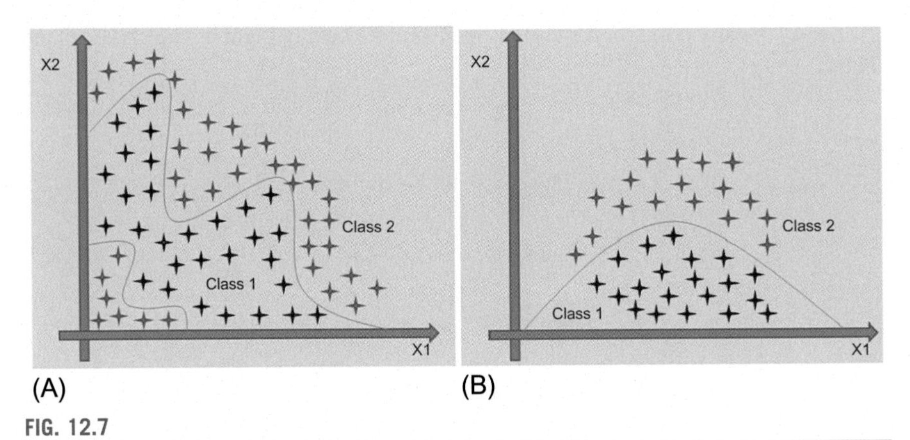

FIG. 12.7

(A) SVM with fine Gaussian kernel and (B) SVM with quadratic kernel.

12.5 Performance analysis

This section discusses the performance analysis of the different kernels (linear, quadratic, and fine Gaussian) of the SVM classifier. A SVM classifier is a binary classifier and it predicts the dataset as either positive or negative. It mainly produces four outcomes: (1) true positive (TP), which means correct positive prediction; (2) true negative (TN), which means correct negative prediction; (3) false positive (FP), which means incorrect positive prediction, and (4) false negative (FN), which means incorrect negative prediction. A confusion matrix is formed from these four outcomes of the classification (Fig. 12.8).

From the UCI dataset, a total of 22 subjects were used for this study. Of these subjects, 16 (around 70%) were used for training the SVM and the remaining six (around 30%) were used for testing the SVM. There are three exercises and at each exercise we recorded data from the four muscles listed earlier for a total of 12 features obtained. Features are reduced by using the PCA technique. Three different types of kernels were used under this study. The confusion matrix for different cross-validation ($K=3$, 5, 8, 10) is shown in Tables 12.1–12.4.

With the help of this confusion matrix we can derive the following four measures: Accuracy, Error Rate, Sensitivity, and Selectivity.

Accuracy is defined as the ratio of all correct predictions to the total number of dataset. Table 12.5 shows the accuracy of the different kernels.

$$\text{Accuracy}\ (\%) = \left(\frac{TP+TN}{N}\right) \times 100\% \tag{12.13}$$

Error rate is defined as the ratio of all incorrect predictions to the total number of dataset. Table 12.6 shows the error rate of the different kernels.

$$\text{Error rate}\ (\%) = \left(\frac{FP+FN}{N}\right) \times 100\% \tag{12.14}$$

or

$$\text{Error rate}\ (\%) = 1 - \text{Accuracy}\% \tag{12.15}$$

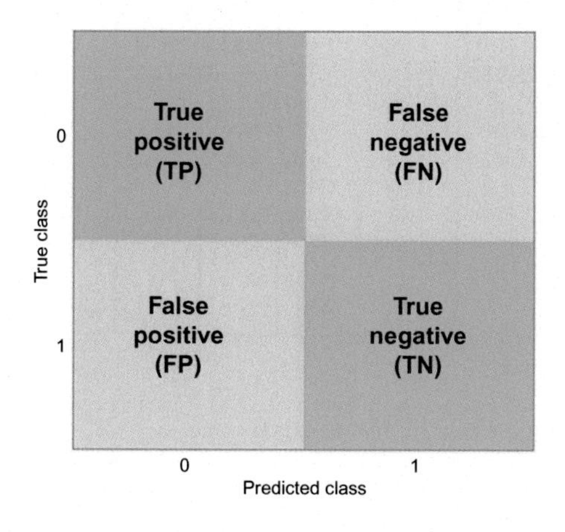

FIG. 12.8

Confusion matrix.

Table 12.1 Confusion matrix for $K=3$

		Estimated value by linear SVM				**Estimated value by quadratic SVM**				**Estimated value by fine Gaussian SVM**	
		No OA	**Has OA**			**No OA**	**Has OA**			**No OA**	**Has OA**
True value	No OA	4	7	True value	No OA	5	6	True value	No OA	11	0
	Has OA	4	7		Has OA	2	9		Has OA	3	8

Table 12.2 Confusion matrix for $K=5$

		Estimated value by linear SVM				**Estimated value by quadratic SVM**				**Estimated value by fine Gaussian SVM**	
		No OA	**Has OA**			**No OA**	**Has OA**			**No OA**	**Has OA**
True value	No OA	6	5	True value	No OA	4	7	True value	No OA	10	1
	Has OA	1	10		Has OA	3	8		Has OA	3	8

Table 12.3 Confusion matrix for $K=8$

		Estimated value by linear SVM				**Estimated value by quadratic SVM**				**Estimated value by fine Gaussian SVM**	
		No OA	**Has OA**			**No OA**	**Has OA**			**No OA**	**Has OA**
True value	No OA	4	7	True value	No OA	2	9	True value	No OA	9	2
	Has OA	1	10		Has OA	4	7		Has OA	4	7

Table 12.4 Confusion matrix for $K=10$

		Estimated value by linear SVM				**Estimated value by quadratic SVM**				**Estimated value by fine Gaussian SVM**	
		No OA	**Has OA**			**No OA**	**Has OA**			**No OA**	**Has OA**
True value	No OA	4	7	True value	No OA	3	8	True value	No OA	11	0
	Has OA	3	8		Has OA	3	8		Has OA	4	7

Table 12.5 Accuracy of SVM with different kernels for different values of K-fold cross-validation sets

		Accuracy			
Sr. no.	**Classifiers**	$K=3$	$K=5$	$K=8$	$K=10$
1	SVM with linear kernel	50%	72.70%	63.60%	54.50%
2	SVM with quadratic kernel	63.60%	54.50%	40.90%	50.00%
3	SVM with fine Gaussian kernel	86.40%	81.80%	72.70%	81.80%

Table 12.6 Error rate of SVM with different kernels for different values of K-fold cross-validation sets

		Error rate			
Sr. no.	**Classifiers**	$K=3$	$K=5$	$K=8$	$K=10$
1	SVM with linear kernel	50%	27.3%	36.4%	45.5%
2	SVM with quadratic kernel	36.4%	45.5%	59.1%	50%
3	SVM with fine Gaussian kernel	13.6%	18.2%	27.3%	18.2%

Sensitivity is defined as the ratio of correct positive predictions to the total number of positive dataset. Table 12.7 shows the sensitivity of the different kernels.

$$\text{Sensitivity}\ (\%) = \left(\frac{\text{TP}}{\text{TP}+\text{FN}}\right) \times 100\% \tag{12.16}$$

Specificity is defined as the ratio of correct negative predictions to the total number of negative dataset. Table 12.8 shows the specificity of the different kernels.

$$\text{Specificity}\ (\%) = \left(\frac{TN}{TN+FP}\right) \times 100\% \tag{12.17}$$

From Table 12.5, we can say that the SVM with fine Gaussian kernel gives the best results for all the different *K*-fold cross-validation sets. The fine Gaussian kernel with a threefold cross-validation set gives the highest accuracy at 86.4%. Table 12.6 shows the error rate, which refers to the ratio of incorrect predictions to the total number of dataset. Those classifiers that give a higher accuracy will give a minimum error rate and vice versa. In Table 12.6 SVM with fine Gaussian kernel with a threefold cross-validation set gives the lowest error rate at 13.6%. SVM with quadratic kernel with an eightfold cross-validation set gives the highest error rate at 59.1%

Table 12.7 shows the sensitivity of SVM with different kernels for different values of *K*-fold cross-validation sets. Sensitivity gives the correct positive prediction to the total number of positive dataset. From Table 12.7, we can say that the SVM with fine Gaussian kernel with 3 or 10 *K*-fold cross-validation set gives the highest sensitivity at 100%. SVM with quadratic kernel with 8 *K*-fold cross-validation gives the minimum sensitivity at 18.2%.

Table 12.7 Sensitivity of SVM with different kernels for different values of *K*-fold cross-validation sets

		Sensitivity			
Sr. no.	**Classifiers**	***K*=3**	***K*=5**	***K*=8**	***K*=10**
1	SVM with linear kernel	36.3%	54.5%	36.3%	36.3%
2	SVM with quadratic kernel	45.4%	36.3%	18.2%	27.3%
3	SVM with fine Gaussian kernel	100%	90.9	81.8%	100%

Table 12.8 Specificity of SVM with different kernels for different values of *K*-fold cross-validation sets

		Specificity			
Sr. no.	**Classifiers**	***K*=3**	***K*=5**	***K*=8**	***K*=10**
1	SVM with linear kernel	63.6%	90.9%	90.9%	72.7%
2	SVM with quadratic kernel	81.8%	72.7%	63.6%	72.7%
3	SVM with fine Gaussian kernel	72.7%	72.7%	63.6%	63.6%

Table 12.8 shows the specificity of SVM with different kernels for different values of K-fold cross-validation sets. Specificity gives the correct negative prediction to the total number of negative dataset. From Table 12.8, we can say that the SVM with linear kernel with 5 or 8 K-fold cross-validation set gives the highest selectivity at 90.9%.

12.6 Conclusion and future scope

Our results support the hypothesis that muscle weakness of the lower limb is the cause of knee OA. We studied performance of different kernels of SVM for different K-fold cross-validation sets in the results section.

The overall accuracy of SVM classifier with fine Gaussian kernel at different cross-validation gives the best results. One limitation of this research is that only a small amount of data are available. In the end, we concluded that SVM with fine Gaussian kernel with threefold cross-validation gives maximum accuracy of 86.4%.

The linear, quadratic, and Gaussian kernels are completely different in cases of constructing the hyperplane or boundary between the categories. The linear kernel classifies the dataset linearly. The polynomial (quadratic) kernel is also the same as the linear classifier, with the only difference being the shape of the boundary line. Fine Gaussian kernel uses normal curves around the data points and sums these so that the decision boundary can be defined. It gives more prediction accuracy, fast evaluation of the learned target function, and consumes less time than the linear or quadratic kernel. In future work, other classifiers may be proposed.

Involving a greater number of subjects may increase the efficacy of the algorithm used to classify the subjects. Subjects involved in the study were taken from the UCI dataset. In this dataset, data for only 22 subjects was available. In future work, we suggest collecting and incorporating sEMG data for more subjects. For collecting the sEMG, we will try to design a IoT-based wireless sEMG sensor.

References

[1] Y. Zhang, J.M. Jordan, Epidemiology of osteoarthritis, Clin. Geriatr. Med. 26 (3) (2010) 355–369.

[2] D.K. White, C. Tudor-Locke, D.T. Felson, K.D. Gross, J. Niu, M. Nevitt, C.E. Lewis, J. Torner, T. Neogi, Do radiographic disease and pain account for why people with or at high risk of knee osteoarthritis do not meet physical activity guidelines? Arthr. Rheumat. 65 (1) (2013) 139–147.

[3] A. Sakamoto, M.T. Khan, T. Kurita, EMG signals in co-activations of lower limb muscles for knee joint analysis, in: 2015 International Conference on Informatics, Electronics & Vision (ICIEV), IEEE, 2015, pp. 1–5.

[4] N.A. Segal, J.C. Torner, D. Felson, J. Niu, L. Sharma, C.E. Lewis, M. Nevitt, Effect of thigh strength on incident radiographic and symptomatic knee osteoarthritis in a longitudinal cohort, Arthr. Care Res.: Off. J. Am. Coll. Rheumatol. 61 (9) (2009) 1210–1217.

[5] National Collaborating Centre for Chronic Conditions (Great Britain) and National Institute for Clinical Excellence (Great Britain), Osteoarthritis: National Clinical Guidelines for Care and Management in Adults, Royal College of Physicians, 2008.

[6] J. Bedson, K. Jordan, P. Croft, How do gps use x rays to manage chronic knee pain in the elderly? A case study, Ann. Rheum. Dis. 62 (5) (2003) 450–454.

[7] C. Peterfy, M. Kothari, Imaging osteoarthritis: magnetic resonance imaging versus x-ray, Curr. Rheumatol. Rep. 8 (1) (2006) 16.

[8] F. Eckstein, D. Burstein, T.M. Link, Quantitative MRI of cartilage and bone: degenerative changes in osteoarthritis, NMR Biomed.: Int. J. Devot. Dev. Appl. Magnet. Reson. In Vivo 19 (7) (2006) 822–854.

[9] R.O. Merletti, C.J. De Luca, New techniques in surface electromyography, Comput. Aided Electromyogr. Expert Syst. 2 (1989) 115–124.

[10] M.B.I. Reaz, M.S. Hussain, F. Mohd-Yasin, Techniques of EMG signal analysis: detection, processing, classification and applications, Biol. Proced. 8 (1) (2006) 11.

[11] O.F.A. Sanchez, J.L.R. Sotelo, M.H. Gonzales, G.A.M. Hernandez, EMG dataset in lower limb data set, in: UCI Machine Learning Repository: 2014-02-05, 2014.

[12] G. Elhayatmy, N. Dey, A.S. Ashour, Internet of things based wireless body area network in healthcare, in: Internet of Things and Big Data Analytics Toward Next-Generation Intelligence, Springer, Cham, 2018, pp. 3–20.

[13] N. Dey, A.S. Ashour, F. Shi, S.J. Fong, R.S. Sherratt, Developing residential wireless sensor networks for ECG healthcare monitoring, IEEE Trans. Consum. Electr. 63 (4) (2017) 442–449.

[14] G.R. Naik, S.E. Selvan, M. Gobbo, A. Acharyya, H.T. Nguyen, Principle component analysis applied to surface electromyography: a comprehensive review, IEEE Access (2016).

[15] C. Sapsanis, G. Georgoulas, A. Tzes, EMG based classification of basic hand movements based on time-frequency features, in: 2013 21st Mediterranean Conference on Control & Automation (MED), IEEE, 2013, pp. 716–722.

[16] C. Cortes, V. Vapnik, Support-vector networks, Mach. Learn. 20 (3) (1995) 273–297.

CHAPTER

A comparative study for brain tumor detection in MRI images using texture features

13

P.K. Bhagat, Prakash Choudhary, Kh. Manglem Singh

Department of Computer Science and Engineering, NIT Manipur, Imphal, India

13.1 Introduction

The texture is one of the important characteristics in MRI images. The texture information present in the medical image is vital for automatic processing of the image. The texture pattern of pathological tissues is quite different from normal tissues and contains a unique intensity pattern. These changes can be exploited to classify MRI images. The application domain plays a critical role in the selection of types of textural properties to be computed.

Brain cancer is a disease in which abnormal cells grow uncontrollably in the brain. The incidence of brain tumors is increasing rapidly, particularly in the young generation. Tumors can directly destroy all healthy brain cells. Curing cancer has been a major goal of medical researchers for decades, but the development of new treatments takes time and money. Since the advent of MRI, there has been a significant improvement in the quality of images produced by it. Also, there are different types of MR techniques for different body parts and different diseases. Still, the improved MRI images do not guarantee that visual interpretation of an MRI image by all the experts will be the same. When a human inspects an MRI image, it is time consuming and sometimes leads to misclassification. The subjectivity and time-consuming nature of human inspection are not suitable for treatment. For accurate treatment planning, the brain tumor must be identified accurately as early as possible. To minimize human intervention, a method for classification should be developed that can automatically classify whether the images contain a tumor or not. The aid of a computer in the interpretation of brain MRI images has helped a lot. The interpretation by the computer must be correct; otherwise, it would be fatal for treatment planning. There have been many kinds of research in this area and studies are still ongoing to achieve 100% accuracy. Chakraborty et al. provide a detailed discussion on the impact of intelligent computing methods for biomedical image analysis and health care [1]. Chaki et al. and Chaki and Dey present an excellent review of sensor-based biometric recognition and genetics and genomics, respectively [2, 3]. Chaki et al. [2] present a detailed discussion on the state-of-the-art sensor-based

Sensors for Health Monitoring. https://doi.org/10.1016/B978-0-12-819361-7.00013-0

biometric recognition techniques including preprocessing to model design, and the various issues faced during data acquisition. The numerous steps associated with the processing of gene data are explained by Chaki and Dey [3].

The automatic detection of a tumor requires having a set of distinct features that accurately represent the image followed by a classification model to detect the pathological image. The features may be of different types such as region-based features, shape-based features, texture features, and so on. The accuracy of feature extraction is the basis of accurate image classification. Feature extraction is a type of dimensionality reduction where a large number of pixels of the image are efficiently represented in such a way that interesting parts of the image are captured effectively. Feature extraction techniques are applied to get features that will be useful for the classification of the image. The types of features to be extracted from the image vary from application to application.

The shape is one of the important visual features of an image. Shape-based features play a critical role in picture characterization and they are important clues for human beings to identify objects. Color and shape of the brain cells play a very critical role for brain tumor detection. The shape and size of a tumor determine the gravity of the tumor and next course of action [4]. The shape can be detected through edge detection methods or region determination methods. Shape-based features are a numerical representation of the geometrical data and are independent of the position and rotation. Edge-based methods calculate shape features from the boundary of the shape. They first detect the edges in the shape, and detected edges are exploited to generate shape features. The region-based methods exploit the entire regions of the shape to generate features. The analysis of the interior part of the shape starts from a given initial point and expands slowly toward the edges of shape and likewise covers the whole area of the shape. The problems with shape-based features extraction methods are that they are usually computationally complex and sensitive to noise. If the edges of the shape are not detected accurately, the method will encounter a problem and will generate the wrong shape. This will ultimately result in a wrongly shaped feature and hence poor performance of the algorithm. Also, it is very difficult to determine the quality of the object through shape-based features.

Texture plays an important role in the human visual perception system [5]. In the absence of any single definition of texture, surface characteristics and appearance of an object can be considered as the texture of that object [6]. While describing the texture, the computer vision researchers are describing it as everything of an image after extracting the color components and local shapes [7]. The various methods for extraction of texture features include histogram-based first-order texture features [8], gray tone spatial dependence matrix [9] and its variants [10, 11], local binary pattern (LBP) [12] and its variants [13], local directional pattern [14] and its variants [15], wavelet features [16], and so on.

A combination of various types of feature extraction methods is known as hybrid features. In hybrid feature representation, various types of features, such as shape features, texture features, and others, are combined systematically and represented together as a single feature. Hybrid features are suitable to represent complex properties of the image and can be applied when the choice of any single feature extraction method is not specified. However, hybrid features are complex in nature, and the selection of appropriate features for the fusion is difficult. Also, it poses additional challenges and may produce worse results if the features are not well chosen and may require high-dimensional features, which imply high computation cost [17]. However, the dimensions can be reduced through proper feature selection methods [18, 19], although even that will increase computational cost.

The texture is a key component of how a human visually perceives any image. The traditional approach for texture feature extraction for medical images has been used extensively to describe different image textures by unique features and has found application in identifying normal and cancerous pathology [20, 21]. The importance of texture-based features can be comprehended by the fact that various types of texture features have been used for the classification of brain tumors [22, 23]. According to Lowitz [24], if the same histogram is obtained twice, then the same class of spatial stimulus has been encountered. For a 512×512 image, there are 26×10^6 possible permutations of pixels, so the probability of getting the same histogram twice for different texture is almost impossible [24]. Although it has been asserted that texture pair agreeing in their second-order statistics cannot be discriminated [25]. The advantages of texture feature over others are that it is easy to extract with less computational complexity. Brain cells in MRI images have very powerful texture characteristics. Hence, texture-based feature extraction methods play a very important role in characterizing MRI images.

The aims of this study are twofold. First, this chapter organizes the rich literature in taxonomy to highlight the ingredients of the principal works in the literature and recognize their advantages and limitations. Notably, our focus is on a detailed explanation of all the texture-based feature extraction methods and their significance in brain tumor detection. The chapter aims to explain how a specific method extracts texture features and constructs a feature vector. Second, this chapter has designed a common experimental protocol to establish and evaluate the efficiency of key feature extraction methods, because in the literature there does not exist any thorough empirical comparison. Our proposed protocol contains a standard dataset and a range of classifiers to assess the effectiveness of specific feature extraction methods over MRI images. The various existing methods for image preprocessing to classification are shown in Fig. 13.1.

The rest of the chapter is organized as follows: Section 13.2 presents a taxonomy to structure the literature on texture feature extraction for brain tumor detection. Section 13.3 introduces a common experimental protocol to evaluate the feature extraction method for detection of a brain tumor. Section 13.4 describes the implementation of feature extraction methods and presents extensive comparison of their results for the detection of a tumor, followed by a conclusion in Section 13.5.

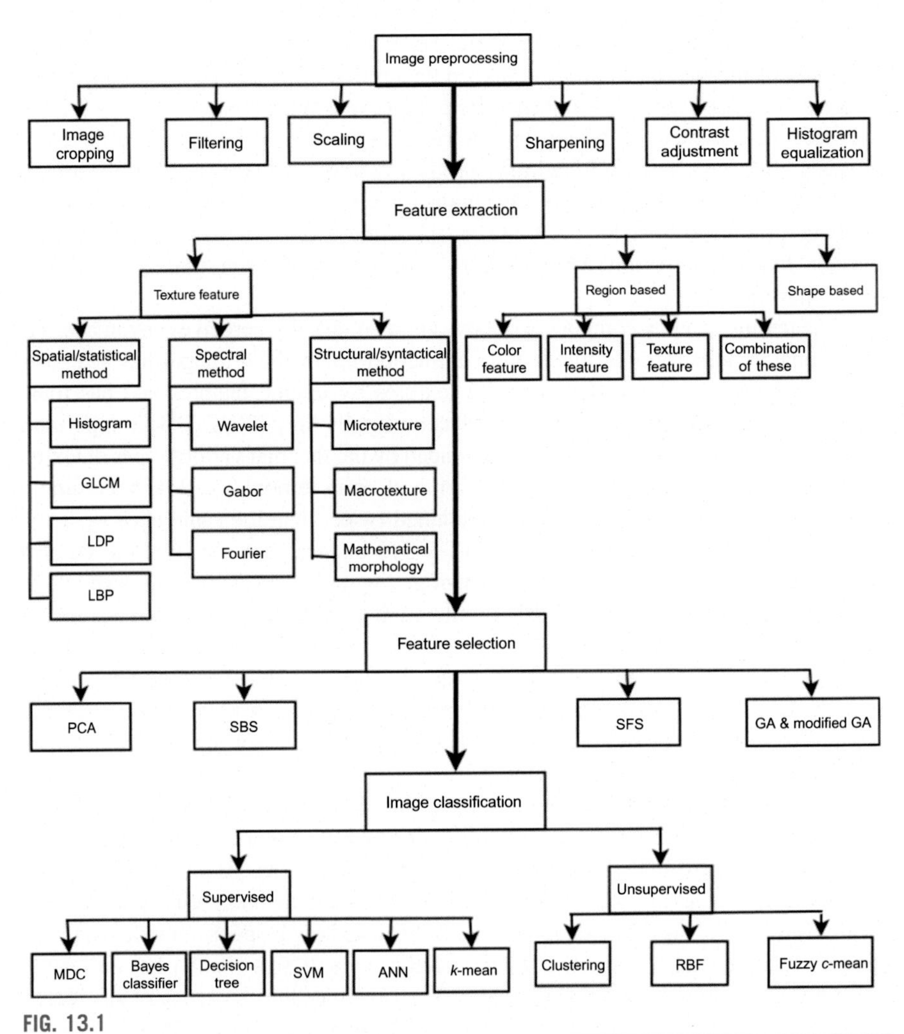

FIG. 13.1

Various existing methods from image preprocessing to classification.

13.2 Taxonomy and reviews

A feature can be defined as a piece of information that is the abstraction of the image to describe or represent an image uniquely. Each image contains multiple features. A single feature is not sufficient to describe any image uniquely hence this chapter extracts various features from an image and puts it together in a feature vector. The types of features that should be extracted depend on the problem domain. Hence, feature extraction techniques vary from domain to domain (e.g., radius would be sufficient to describe a circle, but for eclipse the radius is not enough). Also from a given pattern, a unique feature vector can be generated, but from a feature vector, a unique pattern may not always be generated.

Color-based features of the image are one of the most powerful representations of the contents of the image. An image is composed of RGB colors, hence extracting RGB-based features has been a very active area of image content representation. Color plays an indispensable role for humans in the recognition of objects in an image, as human eyes are sensitive to colors. Color features are invariant to the rotation and transformation, which makes them some of the most dominant visual features. If normalization is used then color features are also invariant to scaling. Kodituwakku and Selvarajah present a comparative analysis of color features [26]. The color space can be represented using RGB, HSL, HSI, HSV, and so on, and are invariant to rotation and translation. No matter what color space representation is used, color information can be represented in 3D space (one for each color) [27]. A small variation in color value is not noticeable by the human eye as compared to the variation in gray-level values.

Recognition of objects present in the image is one of the most important aspects of characterization of the image. The best way to recognize objects is to segment the objects present in the image and then extract features of those segmented regions. However, segmentation of objects is itself a complex task. Moreover, an unsupervised segmentation algorithm is a brittle task. Semantic segmentation requires each pixel to be labeled with an object class [28], which can be implemented in a supervised manner [29, 30] or weakly supervised manner [31, 32] or unsupervised manner [33]. The deep neural network is also being used extensively for the segmentation of objects [34, 35]. When one segmented region contains all the pixels of one object, and no pixels of the other object it is considered as robust segmentation. Therefore researchers tried to implement weak segmentation. Although accurate, robust semantic segmentation is difficult to achieve, segmented regions are useful and powerful features to describe an image.

Salient attributes present in the image, usually identified by color, texture, or local shapes, are used to produce salient features. Although color and texture features are commonly used to represent the contents of the image. But, salient points deliver more discriminative features. In the absence of object-level segmentation, salient points act as weak segmentation and play an eternal role in the representation of an image. Salient points may be present at the different locations in the image and need not be corners, that is, they can be smooth lines as well. Sebe et al. [36] compared the salient points extracted using a wavelet with corner detection algorithm. A salient point detection and scale estimation method are proposed using the local minima based on fractional Brownian [37]. Although salient points are the substitution for the segmentation (as segmentation is a brittle task), if salient points are used along with segmentation it gives much more discriminative features.

Recently, the scale invariant feature transform (SIFT)-based features have become much more popular [38]. SIFT is a scale and rotation invariant local feature descriptor based on the edge-oriented histogram that extracts vital points (interest points) and its descriptors. A similar method that computes a histogram of the direction of gradients in a localized portion of an image called histogram of oriented gradients (HOG) [39] is a robust feature descriptor. Later, a speed-up version of SIFT called speeded-up robust features (SURF) was introduced [40]. For real-time applications, a machine learning-based corner detection algorithm called features from accelerated segment test (FAST) was introduced [41]. SIFT and SURF descriptors are usually converted into binary strings to speed up the matching process. However, binary robust independent elementary features (BRIEF) [42] provide a shortcut for finding binary strings directly without computing the descriptors. It is worth mentioning that BRIEF are a feature descriptor, not a feature detector. SIFT uses a 128-dimensional descriptor, and SURF has a 64-dimensional descriptor (i.e., a high computational complexity). An alternative of SIFT and SURF called oriented FAST and rotated BRIEF (ORB) have been proposed [43]. SIFT and SURF are patented algorithms; however, ORB is not patented and has a comparable performance with low computational complexity compared to SIFT and SURF.

Visual features have a significant role in the identification and recognition of objects and image content representation. Different types of features (texture, color, SIFT, etc.) have different characteristics. The local features (SIFT, SURF, shape, etc.) describe image patches, whereas global features represent an image as a whole. So, local features are a specification of an image and used for object recognition, while global features are a generalization of an image and suitable for object detection. The recent feature extraction methods (SIFT, SURF, HOG, etc.) are compelling object representation methods and have become standard feature representation methods. However, global features like texture, color, and so on have their significance and will continue to be in practice. The object-based feature extraction methods are usually not suitable for MRI images. Also MRI images are grayscale images, hence color-based feature extraction methods cannot be directly applied to MRI images. However, texture feature extraction methods are suitable for gray scale images and can be directly applied to MRI images.

Analysis of various aspects of a signal [44] or image describes various characteristics of a signal or image. Dimensionality reduction is one of the most popular techniques to remove redundant features before classification [19]. Feature extraction methods project features into a new feature space with lower dimensionality [45]. The different set of feature extraction methods exploits different characteristics of the image. The feature selection methods assist to select a small subset of features that minimize the redundancy and maximize significance to the target. The motive is to identify and select a useful subset of features to be used to represent patterns from a larger set of often mutually redundant, possibly irrelevant, features with different associated measurement risks [46].

There is no unilateral definition of texture, but texture can be described as the surface characteristic and appearance of an object. Texture explains about the spatial arrangements and pixels intensities in an image. Brodatz described the details of various types of texture. Texture analysis is the process of describing the distinctive nature of image regions based on their texture contents. Texture classification is an essential problem domain in the field of texture analysis where the task is to assign a set of predefined texture classes to an unknown image. To describe texture analysis methods and their components in a more formal way, some notations are introduced. Let there be an $N \times N$ image with G number of gray levels. The chapter uses x and $\overline{x}$ to represent an image and its transformed version, respectively. $x(i, j)$ refers to an intensity value of x at pixel position $\{(i,j)|0 \leq i \leq N-1 \text{ and } 0 \leq j \leq N-1\}$. Features from an image x form a vector of features called feature vector $f(x)$.

Texture analysis methods can be categorized in one of four classes: (1) statistical analysis, (2) geometrical analysis, (3) model-based analysis, and (4) signal processing analysis [47]. Xie and Mirmehdi [48] present a more detailed categorization and explanation of texture analysis methods and features extraction from these methods. The statistical texture analysis methods analyze the distribution of pixels intensities around every pixel of the image x and extract a local intensity for every pixel from the distribution and construct a transformed image $\overline{x}$ having locally transformed intensity value. Then, statistical information are extracted from $\overline{x}$, which acts as texture-based feature vector for x. The geometrical texture analysis methods use texture elements (edges, shapes, etc.) to find the geometric structure image (primitive) $\overline{x}$, and $\overline{x}$ is analyzed to extract statistical information to construct a feature vector. Construction of an image model using the intensity distribution is the primary goal of the model-based texture analysis methods. The signal processing texture analysis methods create a filtered image $\overline{x}$ by analyzing the frequency content of the image x. Then, certain features are computed from $\overline{x}$. The basic idea of texture analysis methods is shown in Eq. (13.1). The transformed image $\overline{x}$ is obtained by imposition of transformation function tf with image x. Features are obtained by performing statistical function sf over $\overline{x}$.

$$f(x) = \overline{x} \oplus sf, \text{ where } \overline{x} = x(i,j) \oplus tf \tag{13.1}$$

Most textures are defined as contextual properties involving the spatial distributions of gray levels. Therefore the statistical approaches based on spatial distribution of gray values are generally applicable. Geometrical texture analysis methods represent texture by a well-defined primitive, which is essentially a micro texture, and its governing rule are not as widely used as statistical methods, because the rule for building texture from primitive is a rigorous formula, and placement rules are based on the idea that primitives have regular spatial relationships. It is unlikely to describe textures in certainty, which is usually irregular, distorted, or variant in a structure using a geometrical approach. Moreover, geometrical methods are very susceptible to local noise and structural errors, distortions, or variations. In texture classification problems, parameters need to be estimated first. However, if the texture structure is unknown, the estimation of the parameters is hard to achieve. Thus, model-based texture analysis methods are intractable because the issue of estimating model parameters is difficult. Furthermore, model-based approaches only capture micro textures well, since the practical considerations limit the order of the model. Model-based approaches also fail with inhomogeneous textures. It has been noted that the statistical methods and wavelet transform methods are superior to others from the above comparison of properties of each texture description method.

In the next four sections, the four most important texture feature analysis methods are explained with their application in MRI images for brain tumor detection.

13.2.1 First-order histogram-based texture analysis

A histogram shows the distribution of pixels value (intensity) in an image. It does not show any interrelationship among pixels. Histogram-based feature extraction methods use the histogram of the image to extract texture information. Let there be N rows and N columns and G gray level in an image, then the probability of occurrence of a gray value i in an image is given by Eq. (13.2).

$$P(i) = \frac{n(i)}{N*N}, \quad G-1 \geq i \geq 0 \tag{13.2}$$

Here $n(i)$ is total number of occurrence of gray value (i) in the image.

Histogram-based methods are very simple to calculate. However, they are less practical in some domains due to the absence of correlation between pixels. Through an experiment, Lowitz [24] asserted that if the same histogram is encountered twice, then the same class of spatial stimulus has been encountered. The author used local image histogram as a feature for Landsat, Daedalus, and Portraits images quantified at 16 gray levels. Leboucher and Lowitz [8] presented a clustering technique based on the partitioning of a data histogram. They used a matrix based on curvature applied to the self-information function attached to the Gaussian distribution. According to Pratt [49], the shape of an image histogram provides many clues about the image characteristic. The author suggested six measures (Eqs. 13.3–13.8) for quantitative shape description of a first-order histogram.

$$\text{Mean}: \ S_M = \bar{b} = \Sigma_{b=0}^{G-1} bP(b) \tag{13.3}$$

$$\text{Standard deviation}: \ S_D = \sigma_b = [\Sigma_{b=0}^{G-1}(b-\bar{b})P(b)]^{1/2} \tag{13.4}$$

$$\text{Skewness}: \ S_S = \frac{1}{\sigma_b^3}\Sigma_{b=0}^{G-1}(b-\bar{b})^3 P(b) \tag{13.5}$$

$$\text{Kurtosis}: \ S_K = \left[\frac{1}{\sigma_b^4}\Sigma_{b=0}^{G-1}(b-\bar{b})^4 P(b)\right] - 3 \tag{13.6}$$

$$\text{Energy}: \ S_N = \Sigma_{b=0}^{G-1}[P(b)]^2 \tag{13.7}$$

$$\text{Entropy}: \ S_E = -\Sigma_{b=0}^{G-1} P(b)\log_2\{P(b)\} \tag{13.8}$$

Authors reported works related to first-order histogram-based features [50–52]. As has already been stated, due to no correlation with neighbor pixels, first-order histogram-based features are not widely used. We have not seen any experimental paper on MRI brain images using the first-order histogram-based feature method.

13.2.2 Gray-level cooccurrence matrix

As in first-order histogram-based feature, there does not exist any interrelationship among pixels. However, in the second-order histogram-based approach, the aim is to establish a correlation among pixels. According to Julesz [25], no texture pair can be visually distinguished if they agree in their second-order statistics. The second-order histogram features are based on the definition of the joint probability distribution of a pair of pixels. Haralick et al. [9] defined second-order histogram as a cooccurrence matrix. The authors called this cooccurrence matrix the gray-tone spatial dependence matrix. The matrix is calculated as $P(i, j, d, \Theta)$ where Θ can be 0, 45, 90, and 135 degrees, and $d = 1, 2, 3, \ldots, G$, where G is the maximum gray value in the image and (i, j) indicates an entry in a cell. A detailed explanation of gray-level cooccurrence matrix (GLCM) is shown in Figs. 13.2–13.4.

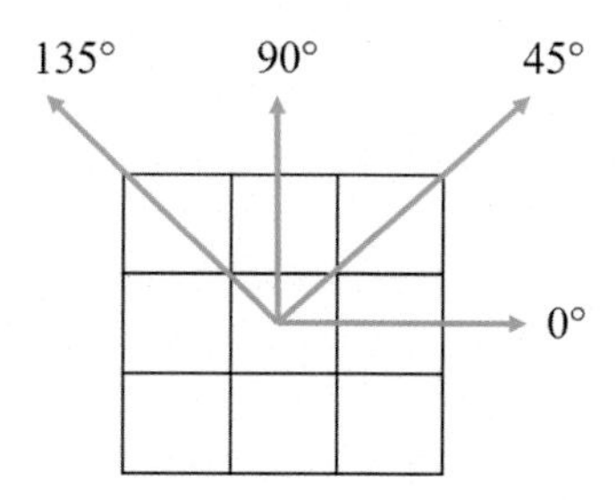

FIG. 13.2

All four directions in GLCM.

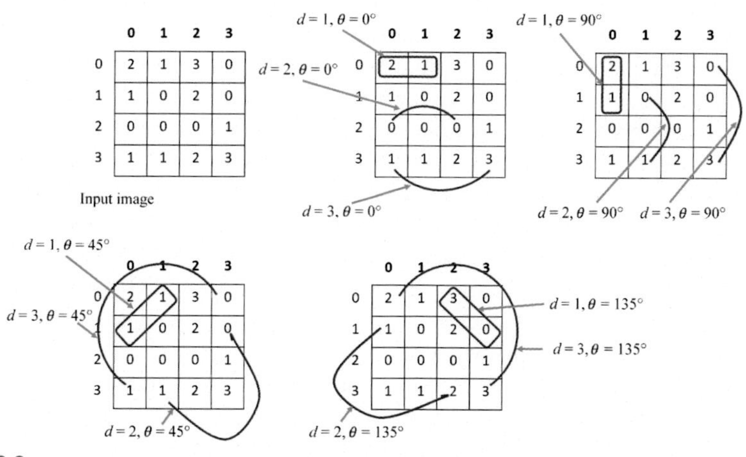

FIG. 13.3

Measuring the distance d in all four directions ($d = 1, 2, 3$).

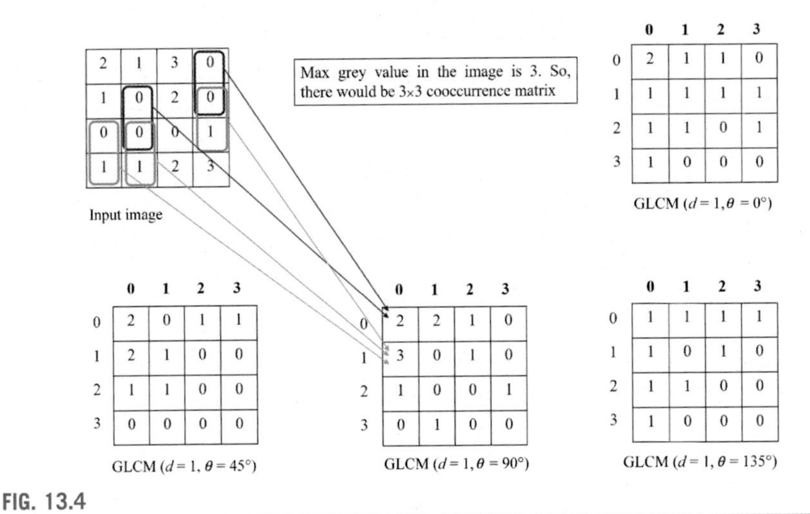

FIG. 13.4

Calculation of cooccurrence matrices for $d = 1$ in all four directions.

Also, the matrix is symmetric, that is, $P(i, j, d, \Theta) = P(j, i, d, \Theta)$. For each image, there are four matrices in $d = 1$, four matrices in $d = 2$, and so on. These four matrices are $P_{d,H}$, $P_{d,V}$, $P_{d,RD}$, and $P_{d,LD}$ (where H stands for horizontal or 0 degrees, V stands for vertical or 90 degrees, RD stands for right diagonal, or 45 degrees, and LD stands for left diagonal or 135 degrees). Hence, for one image there exist various cooccurrence matrices. Choosing an appropriate matrix with an appropriate distance is domain dependent. The 14 features that can be extracted from these matrices are explained by Haralick et al. [9]. If the image has a maximum gray value of 255, then the cooccurrence matrix will be very large, in this case, the original image can be quantized in either 4, 8, 16, 32, or 64. gray level and then on the quantized image the cooccurrence matrix can be calculated. Lerski et al. [53] discussed a total of 22 features (including 14 features from Haralick et al. [9]) derived from a cooccurrence matrix and used these features for tissue characterization of MRI-based images of the selected brain tumor. Alparone et al. [54] discussed a fast method for calculation of cooccurrence matrix parameters for image segmentation and presented an algorithm for the calculation of these parameters for windows centered on each pixel of the image. The algorithm is based on the fact that windows centered on adjacent pixels are mostly overlapped. Kovalev and Petrou [10] presented multidimensional cooccurrence matrices for feature extraction. Valkealahti and Oja [11] proposed a reduced multidimensional cooccurrence matrix using linear compression, dimension optimization, and vector quantization.

GLCM was used for texture feature extraction from MRI brain images for tumor classification [55, 56]. For the classification purpose, authors used a neuro-fuzzy classifier [56]. The paper satisfactorily classified the glioblastoma type of astrocytoma, which is most common glioma. Kadam et al. [57] also present an extensive lab work that used GLCM for texture feature extraction of MRI brain images. For the classification purpose, it used an artificial neural network (ANN) classifier, which uses backpropagation for learning. The use of GLCM and lifting discrete wavelet transform (LDWT) for feature extraction from MRI images can be seen in Ref. [58], and for the classification purpose, radial basis function network (RBFN) and feedforward neural network (FFNN) are used. The authors claimed better performance of RBFN over FFNN. The uses of GLCM for feature extraction from brain MRI images for tumor classification can also be seen in other papers as well [59, 60]. Fahrurozi et al. [61] used GLCM to extract features from wood images and used a naive-Bayes classifier for the classification of wood. The work carried out by Fahrurozi et al. [62] is based on a cooccurrence matrix where authors used GLCM-based features for retrieval of batik motif image. Sharma and Virmani performed an exhaustive experiment of GLCM features for ultrasound images [63]. The authors experimented with several feature extraction techniques such as Gabor, Log-Gabor, GLCM, and LBP, individually and in combination, and concluded that optimal feature fusion outperforms individual features for the retrieval of batik motif images.

13.2.3 Local binary pattern

LBP [64] is a recently developed texture feature extraction method having less computational complexity. Authors experimented with LBP over nine classes of texture: grass, paper, waves, raffia, sand, wood, calf, herringbone, and wool taken from Brodatz's album. The authors derived LBP from a model of texture unit developed by He and Wang [65, 66]. It can be said that LBP is a simplification of the method [65, 66]. Let X_c be the center pixel and P is number of pixels in the neighbor of X_c with a radius of R then LBP code is given in Eq. (13.9).

$$LBP_{P,R} = \Sigma_{i=0}^{P-1} s(X_i - X_c)2^i \tag{13.9}$$

$$s(X_i - X_c) = \begin{cases} 1 & \text{if } X_i - X_c \geq 0 \\ 0 & \text{if } X_i - X_c < 0 \end{cases} \tag{13.10}$$

The algorithm for the calculation of LBP value for any center pixel is given in Steps I–IV. Fig. 13.5 explains the steps of the algorithm through the diagram.

Step I. Select a window and choose the neighbors of center pixel X_c with respect to radius R. For $R = 1$ there are eight neighbors and for $R = 2$ there are 16 neighbors.

Step II. Find the binary bit value for each neighbor using Eq. (13.10).

Step III. Multiply each binary bit value with corresponding weight mask value.

Step IV. Calculate the sum.

In LBP, for a 3×3 neighborhood, there can be $2^8 = 256$ possible units, whereas in the original model [65, 66] it is $3^8 = 6561$ texture units. This LBP is gray scale invariant as the sign of the differences $s(X_i - X_c)$ has been considered instead of their exact value.

Ojala et al. [12] introduced rotation invariant ($LBP^{ri}_{P,R}$) and rotation invariant with uniform patterns ($LBP^{riu2}_{P,R}$) LBP operator. For rotation invariant Eq. (13.11),

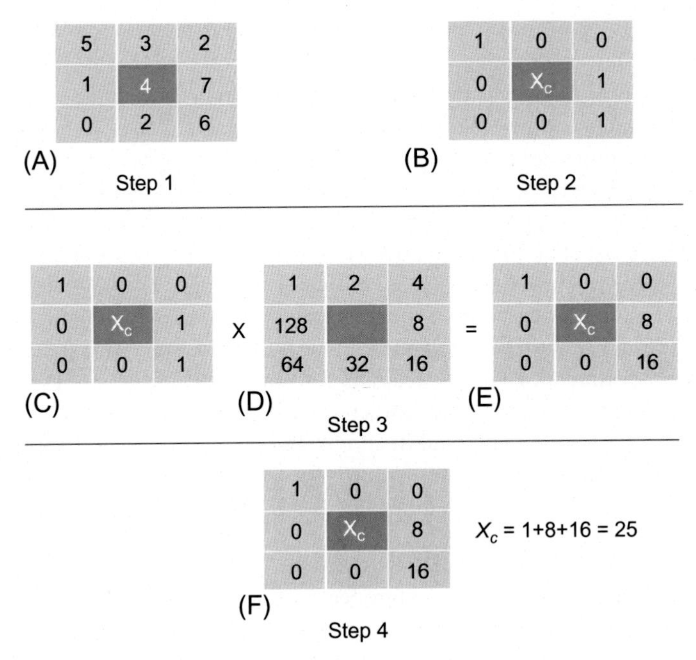

FIG. 13.5

Panel (A) indicates the neighbors of center pixel ($X_c = 4$) for $R = 1$. Panel (B) shows the binary value for each neighbor of X_c calculated using Eq. (13.10). Panels (C–E) show evaluation of step 3 of LBP algorithm where (D) is the weight mask. Panel (F) shows the LBP value of center pixel X_c.

the pixels in the neighbor set are indexed so that they form a circular chain, and the gray values of the diagonal pixels are determined by interpolation. $LBP_{8,\ 1}$ operator in 3 × 3 window gives 36 unique rotations for each pattern, except for the patterns having all zeros $(00000000)_2$ and all ones $(11111111)_2$, which remain constant at all rotation angles, all the 36 rotations provide a unique value.

$$LBP^{ri}_{P,R} = \min\{ROR(LBP_{P,R}, i)\}, \quad \text{for } i = 0, 1, 2, \ldots, P-1 \tag{13.11}$$

where $ROR(LBP_{P,\ R}, i)$ performs bitwise circular right shift.

Uniformity is measured by number of transition in binary value $s(X_i - X_c)$. If there are at most two transitions in a binary pattern, zero to one or one to zero, then it is uniform, otherwise it is not, as in Eq. (13.12). The uniformity was introduced in the observation that in significant image area some LBPs appear more frequently than others that are having less number of spatial transitions.

$$LBP^{riu2}_{P,R} = \begin{cases} \Sigma_{i=0}^{P-1} s(X_i - X_c) & \text{if pattern is uniform} \\ P+1 & \text{if pattern is nonuniform} \end{cases} \tag{13.12}$$

LBP, a easy-to-use and robust local texture descriptor, has shown promising results in industrial inspection, motion analysis, and face recognition [67]. Authors tested the robustness of rotation invariant LBP code for MRI brain image with bias field. Authors used three rectangular neighborhoods with three different LBP features $LBP_{8,1}$, $LBP_{16,2}$, and $LBP_{24,3}$ with simple $LBP_{P,R}$, rotation invariant ($LBP^{ri}_{P,R}$) and rotation invariant and uniform ($LBP^{riu2}_{P,R}$) and claimed that dissimilarity values decreases with (A) the complexity of the interpolation method increases (bicubic interpolation) with (B) rotation invariant with uniformity is introduced to LBP. Although there have been various experiments of LBP and its modification over a wide range of applications, a complete review of LBP and its applications can be seen in work by Pietikainen and Zhao [68]; any experiments of LBP over brain MRI images for classification purpose have not come under our notice. The use of rotation invariant LBP along with some other types of features (GLCM, etc.) for the classification of different types of a tumor can be seen in work by Sachdeva et al. [55].

13.2.4 Discrete wavelet transform

The wavelet is a mathematical tool to study the characteristics of a signal. The whole objective of wavelet analysis is the localization. The wavelet function is used to localize a signal in both time and frequency space. The wavelet functions are scaled and shifted version of some mother wavelets. There are two forms of wavelet transform [16]: the continuous wavelet transform and the discrete wavelet transform (DWT).

The continuous wavelet transform is used to analyze the continuous signal, which means the dilation and translation parameters vary continuously in the input signal. On the other hand, the DWTs analyze the discrete signal (i.e., dilation and translation parameters both take only discrete values).

Let $s(n)$ be a DWF, where $n = 0, 1, \ldots, M - 1$, having M number of samples. Then, the scaling coefficient is given by Eq. (13.13) and wavelet function coefficient is given by Eq. (13.14) where $j \geq j_0$

$$W_\varphi(j_0,k) = \frac{1}{\sqrt{M}} \Sigma_n s(n)\varphi_{j_0,k}(n) \tag{13.13}$$

$$W_\psi(j,k) = \frac{1}{\sqrt{M}} \Sigma_n s(n)\psi_{j,k}(n) \tag{13.14}$$

Every transformation has the original signal, the transformation signal, and the transformation kernel. With the help of the transformation kernel and the scaling coefficient function and discrete wavelet coefficient function, the original signal can be obtained again. This is called inverse DWT (see Eq. 13.15).

$$s(n) = \frac{1}{\sqrt{M}} \Sigma_k W_\varphi(j_0,k)\varphi_{j_0,k}(n) + \frac{1}{\sqrt{M}} \Sigma_{j=j_0}^{\infty} \Sigma_k W_\psi(j,k) \times \psi_{j,k}(n) \tag{13.15}$$

where $\varphi_{j0,k}(n)$ and $psi_{j,k}(n)$ are scaling and wavelet transformation kernel, respectively.

In case of images, which are two-dimensional, this chapter applies signal transformation in both directions, so $s(n)$ will be $x(n_1, n_2)$ [69]. Let the size of image be $N \times N$, then the intensity function is indicated by $x(n_1, n_2)$ where n_1 goes from 0 to $N-1$ and n_2 goes from 0 to $N-1$. Then the scaling coefficient function is given by Eq. (13.16) and wavelet transform coefficient is given by Eq. (13.17) and the inverse DWT is given by Eq. (13.18).

$$W_\varphi(j_0, k_1, k_2) = \frac{1}{\sqrt{N \times N}} \Sigma_{n_1=0}^{N-1} \Sigma_{n_2=0}^{N-1} x(n_1, n_2) \times \varphi_{j_0,k_1,k_2}(n_1, n_2) \tag{13.16}$$

$$W_\psi^i(j_0, k_1, k_2) = \frac{1}{\sqrt{N \times N}} \Sigma_{n_1=0}^{N-1} \Sigma_{n_2=0}^{N-1} x(n_1, n_2) \times \psi_{j_0,k_1,k_2}^i(n_1, n_2) \tag{13.17}$$

Here $i = \{H, V, D\}$. The explanation of H, V, and D is given in Eqs. (13.20)–(13.22), respectively.

$$\begin{aligned} x(n_1, n_2) &= \frac{1}{\sqrt{N \times N}} \Sigma_{k_1} \Sigma_{k_2} W_\varphi(j_0, k_1, k_2) \varphi_{j_0,k_1,k_2}(n_1, n_2) \\ &+ \frac{1}{\sqrt{N \times N}} \Sigma_{i=H,V,D} \Sigma_{j=j_0}^{\infty} \Sigma_{k_1} \Sigma_{k_2} W_\psi^i(j, k_1, k_2) \\ &\times \psi_{j,k_1,k_2}^i(n_1, n_2) \end{aligned} \tag{13.18}$$

When a scale analysis filter (also known as low-pass filter) and wavelet analysis filter (also known as high-pass filter) are applied on any signal (with P number of samples), there are P number of samples after low-pass filter and P number of samples after high-pass filter. In this way, the number of samples is increasing and bandwidth gets half. So, to accommodate this situation the signal needs to be subsampled after filtering by a factor of 2. The same needs to be done with the image. The original signal is filtered along the row (m) and column (n) directions by low-pass filter (h_φ) and high-pass filter (h_ψ). This results in four different filters (*HH, HL, LH, LL*). These four filters can be defined using Eqs. (13.19)–(13.22). The approximation coefficient (*LL*) is given in Eq. (13.19). The horizontal coefficient (*HL*) is obtained using Eq. (13.20), the vertical coefficient (*LH*) is obtained using Eq. (13.21), and the diagonal coefficient (*HH*) is obtained using Eq. (13.22).

$$\varphi(n_1, n_2) = \varphi(n_1)\varphi(n_2) \tag{13.19}$$

$$\psi^H(n_1, n_2) = \psi(n_1)\varphi(n_2) \tag{13.20}$$

$$\psi^V(n_1, n_2) = \varphi(n_1)\psi(n_2) \tag{13.21}$$

$$\psi^D(n_1, n_2) = \psi(n_1)\psi(n_2) \tag{13.22}$$

Images are generally very rich in low-frequency content. So, whenever signal analysis is performed on the image, further analysis is usually performed on the low-pass filtered (*LL*) version of the image. Fig. 13.6 shows the three levels of

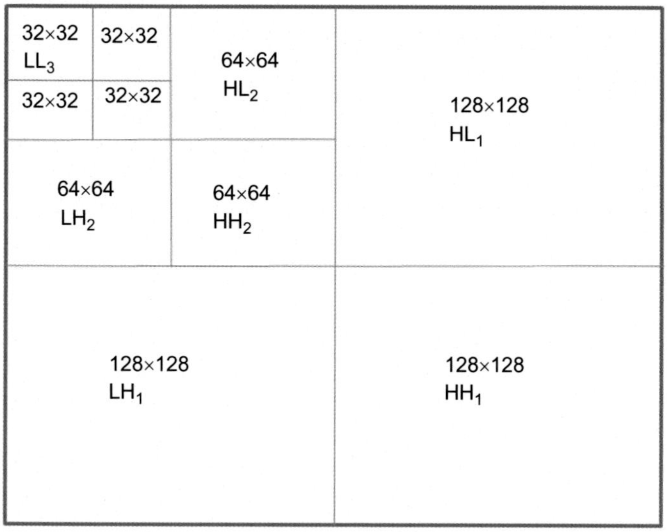

32×32 LL$_3$
32×32
32×32
32×32
64×64 HL$_2$
128×128 HL$_1$
64×64 LH$_2$
64×64 HH$_2$
128×128 LH$_1$
128×128 HH$_1$

FIG. 13.6

The three levels of decomposition of an input image of size 256 × 256.

decomposition performed on an image of size 256 × 256. The second and the third levels of decompositions are performed on the obtained *LL* filtered version only. In our experiment, the third-level *LL* coefficient is taken as the feature vector, which contains 1024 features. The highest scale was the original image, which contains the maximum resolution. After the first level of decomposition, every subband contains 1/4 of the original resolution. So, there is a bit of coarseness that has been incorporated after the first level of decomposition. Then, after the second-level, further coarseness is introduced and so on. In this way, from the finer domain, the coarser and coarser domain of analysis can be performed.

El-Dahshan et al. used a hybrid technique is for classification purpose [70]. The proposed system consists of three stages: feature extraction, feature reduction, and classification. For feature extraction, the authors used DWT with three levels of decomposition. The authors took third-level approximation coefficient (*LL*) as a feature vector. The principal component analysis (PCA) is used for dimensionality reduction. PCA reduces the size of the feature vector from 1024 to 7. Then, for classification purpose, they used a backpropagation artificial neural network (FP-ANN) classifier and *k*-nearest neighbor (*k*-NN) classifier. The authors claimed accuracy of 97% and 98.6% over a dataset of 70 images with FP-ANN and *k*-NN classifiers, respectively. The dataset used in the experiment was obtained from the Harvard Medical School website (http://www.med.harvard.edu/AANLIB/home.html). Apart from a tumor, the dataset used for this work also includes other types of brain abnormalities like Alzheimer's and multiple sclerosis. Furthermore the dimension of the dataset is very low to benchmark viability of the experiment results.

13.3 A common experimental protocol

To evaluate the performance of texture analysis methods, this chapter proposes a new protocol that is common for all the analysis methods. Existing works lack in the consensus on the performance, as different works in the literature use different datasets. Most of the works use privately obtained datasets from hospitals. As a result, the reported performance scores is not comparable across the chapters.

13.3.1 The dataset

The dataset used in this experiment was obtained from the MNI BITE database [71]. We used Group3 preoperative axial MRI brain images. Each image was resized into 256×256. Image resizing either enlarges or shrinks the image based on the image aspect ratio. Sometimes, MRI images may be affected by noise, which needs to be removed before feature extraction. A novel method for noise removal can be seen in Refs. [72, 73]. To maintain the quality of the image, various novel methods have been proposed [74]. In this experiment, a total of 819 images (535 normal images and 284 images with a tumor) have been used. The images were randomly shuffled, and out of 819 images, 546 images were used for training and 273 images used for testing. The dataset used in this experiment is shown in Table 13.1. Every feature extraction method used in this chapter uses the same dataset. This was done to compare and evaluate the efficiency of these feature extraction methods over the same dataset.

13.3.2 The classifiers

Once feature extraction using texture analysis methods is achieved, the next step is to classify the texture features using a classifier. An overview of the various steps involved in the image classification procedure are shown in Fig. 13.7. The proposed work used three classifiers: *k*-NN, support vector machine (SVM), and ANN classifier. To avoid the overfitting of the classifier estimator, the 10-fold cross-validation over the randomized feature set is performed for all the classifiers. *k*-NN classifier is theoretically a very simple but very efficient method having excellent classification accuracy. *k*-NN is an extension of the nearest neighbor (NN) classifier [75].

Table 13.1 Experimental images (size of dataset)

Image type	No. of images
Total images	819
Normal images	535
Pathological images	284
Training set	546
Test set	273

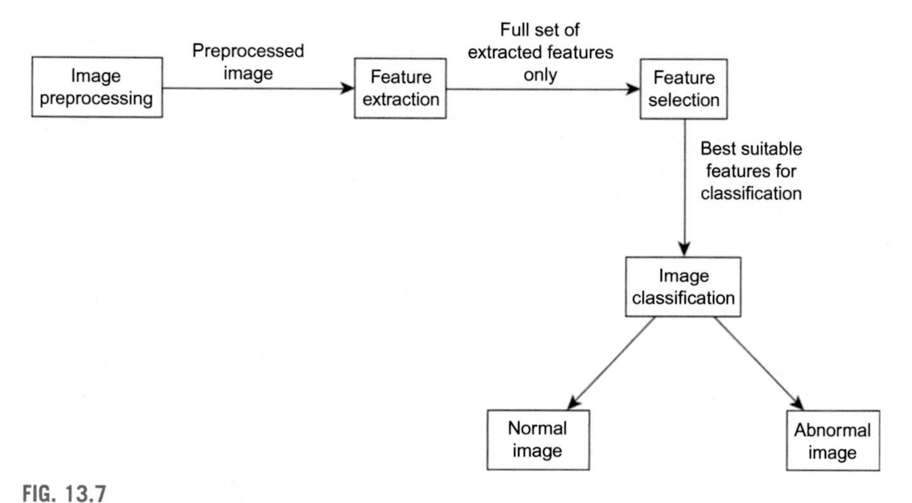

FIG. 13.7

Image classification in a compact form.

The Euclidean distance is used with the k-NN classifier. SVM is a powerful machine-learning classifier for binary classification and can be extended for multilabel classification [76]. The radical basis function kernel is used with SVM, and the penalty parameter (C) of the error is fixed. ANN is a mathematical model that simulates the functioning of the bioneurons [77]. An ANN is an adaptive, most often nonlinear system that learns to perform a function (an input/output map) from data. Adaptive means that the system parameters are changed during the learning phase. After the training phase, the ANN parameters are fixed, and the system is deployed to solve the problem at hand (the Recognition/Testing phase). The neural network having 500 hidden layers and optimized quasi-Newton solver is used with a back-propagation algorithm.

13.3.3 Evaluation measures

Analysis of the performance is critical to make a judgment about a method and to improve its effectiveness. Evaluation can help us to identify the areas of improvement and ultimately help us to realize the goals more effectively. Evaluation measures use some techniques for measuring the effectiveness of the methods. The confusion matrix is exploited for evaluation purpose. Our task to identify the MRI images containing the tumor is simply a binary classification task. Our confusion matrix is a 2×2 matrix containing four elements from Eqs. (13.23) to (13.26) [78]. a and p are actual and predicted class, respectively, and value 0 and 1 stand for negative and positive, respectively.

$$TN = \sum (a,p) = 1 \mid a = 0 \text{ and } p = 0 \tag{13.23}$$

$$FP = \sum (a,p) = 1 \mid a = 0 \text{ and } p = 1 \tag{13.24}$$

$$FN = \sum (a,p) = 1 \mid a = 1 \text{ and } p = 0 \tag{13.25}$$

$$TP = \sum (a,p) = 1 \mid a = 1 \text{ and } p = 1 \tag{13.26}$$

where

TP(true positive) = Both actual and obtained results are the same and indicate the presence of a tumor.
TN(true negative) = Both actual and obtained results are the same and indicate the absence of a tumor.
FN(false negative) = Although the tumor is present in actual image (ground truth), the obtained result shows absence of the tumor.
FP(false positive) = The obtained results show presence of a tumor even though a tumor is absent in ground truth.

Confusion matrix allows the visualization of the performance of a supervised classification model. The confusion matrix can be utilized directly or indirectly to present the various facets of a classification model. The precision, recall, and F1 score are computed from the confusion matrix [79] using Eqs. (13.27)–(13.29). Precision indicates how often a positive class is predicted correctly without making any wrong decision. Given a positive instance, recall suggests how sensitive the classifier is for detecting positive classes without making any wrong decision in identifying it as a negative class. The F1 score is a weighted average of the precision and recall and is useful for the imbalance class in the test dataset. Along with this, receiver operating characteristic (ROC) and area under curve (AUC) with the percentage of ROC curve that is the underneath the curve are also calculated. The AUC is useful for the imbalance class in the database and can be used as a single number summary of the classification model performance.

$$Precision = \frac{TP}{TP + FP} \tag{13.27}$$

$$Recall = \frac{TP}{TP + FN} \tag{13.28}$$

$$F1\ score = 2 \times \frac{Precision \times Recall}{Precision + Recall} \tag{13.29}$$

13.4 A comparative analysis

13.4.1 Methods selected for comparison

This section presents various texture feature analysis methods selected for the evaluation of their performance. The implementation details of all texture analysis methods are discussed in detail. This work intends to select only those methods that are vital and effective for extracting features from MRI images.

13.4.1.1 First-order histogram-based texture analysis

The first-order histogram describes the intensity distribution in the image without making any correlation between the intensity values. For each image, a 256-bin histogram is generated and the six features (Eqs. 13.3–13.8) [49] are extracted. Input images and their corresponding histograms are shown in Fig. 13.8. Each feature is normalized to range between 0 to 1 before passing to the classifiers. Each classifier receives the same set of features.

13.4.1.2 Gray-level cooccurrence matrix

GLCM is a second-order statistical texture analysis method. It examines the spatial relationship among pixels and defines how frequently a combination of pixels are present in an image in a given direction Θ and distance d. Each image is quantized into 16 gray levels (0–15) and 4 GLCMs (M) each for $\Theta = 0, 45, 90$, and 135 degrees with $d = 1$ are obtained. From each GLCM, five features (Eq. 13.30–13.34) are extracted. Thus, there are 20 features for each image. Each feature is normalized to range between 0 to 1 before passing to the classifiers, and each classifier receives the same set of features.

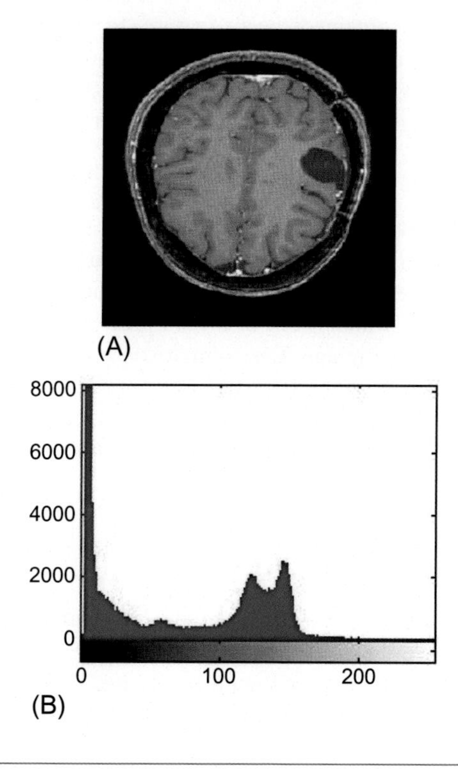

FIG. 13.8

(A) Input image. (B) Histogram of input image.

$$\text{Energy} = \sqrt{\Sigma_{i,j=0}^{G-1}(M(i,j))^2} \tag{13.30}$$

$$\text{Homogeneity} = \Sigma_{i,j=0}^{G-1}\frac{M(i,j)}{1+(i+j)^2} \tag{13.31}$$

$$\text{Correlation} = \Sigma_{i,j=0}^{G-1}M(i,j)\left[\frac{(i-\bar{i})(j-\bar{j})}{\sqrt{(\sigma_i^2)(\sigma_j^2)}}\right] \tag{13.32}$$

$$\text{Energy} = \Sigma_{i,j=0}^{G-1}M(i,j)\times(i,j)^2 \tag{13.33}$$

$$\text{Entropy} = -\Sigma_{b=0}^{G-1}P(b)\log_2\{P(b)\} \tag{13.34}$$

13.4.1.3 Local binary pattern

LBP is a local descriptor of the image based on the neighborhood for any given pixel. The neighborhood of a pixel is given in the form of P number of neighbors within a radius of R. It is a very powerful descriptor that detects all the possible edges in the image. The proposed work used $P = 8$ and $R = 1$ with uniform LBP (Eq. 13.12). Once an LBP descriptor is obtained, the six features from the histogram of the image are extracted using Eqs. (13.3)–(13.8). Each feature is normalized to range between 0 to 1 before passing to the classifiers, and each classifier receives the same set of features. Fig. 13.9 shows an input image and corresponding detected $LBP_{8,1}^{ri}$ uniform LBP.

13.4.1.4 Discrete wavelet transform

The advantage of wavelet transformation is that it captures both frequency and location information, which results in more accurate transformation. The Haar wavelet is used as a mother wavelet, which is one of the simplest wavelets and is very good for the analysis of edges in the image. For comparison purposes, the discrete wavelet transformation [16, 80] with three stages of transformation has been implemented. Then, the third-stage approximation coefficient is taken as a final feature vector, ignoring all other coefficients. As a result, the obtained feature vector contains 1024 features.

13.4.2 Comparative results

For the evaluation of the classification model, the hold-out policy is used. Out of 819 images, 535 images are used for the training of the model and the rest of the images are used for testing the efficiency of the model. Using 10-fold cross-validation, the model is fitted with the best parameters to test its effectiveness over the testing

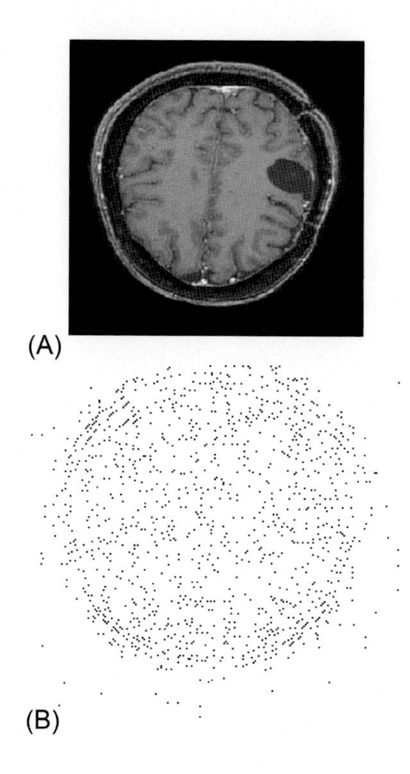

FIG. 13.9

(A) Input image. (B) Output of $LBP_{8,1}^{riu}$ method applied on input image.

dataset. The model is run 10 *times* over the testing dataset and the average of the 10 runs is exhibited as the overall performance of the model.

The precision, recall, and F1 score of the model are presented in Table 13.2. From the table, it is clear that the DWT and GLCM perform much better than the LBP. Also, the performance of the SVM classifier is far better than the performance of the *k*-NN and ANN classifiers.

The obtained ROC for each run is plotted, and the corresponding AUC is also calculated from the ROC. The ROC and corresponding AUC for 10 runs of first-order histogram texture analysis method for *k*-NN, ANN, and SVM classifier is shown in Fig. 13.10. Fig. 13.11–13.13 show the performance of LBP, GLCM, and DWT with *k*-NN, ANN, and SVM classifiers, respectively.

13.5 Conclusion

This chapter presents a study on MRI image texture analysis with the hope of illustrating the effectiveness of each method and its applicabilities. Statistical texture analysis methods are widely used for the texture analysis of MRI images. The

Table 13.2 Performance of texture analysis methods with *k*-NN, ANN, and SVM

Texture analysis method	Accuracy	*k*-NN	ANN	SVM
First-order histogram	Precision (%)	97.8	87.23	94.88
	Recall (%)	93	86.6	94.71
	F1 score (%)	93.87	86.85	94.76
LBP	Precision (%)	79.17	80.71	88.24
	Recall (%)	75.58	80.84	87.95
	F1 score (%)	77.23	80.55	88
GLCM	Precision (%)	97.7	94.35	94.44
	Recall (%)	96.53	94.83	97.37
	F1 score (%)	97.1	94.55	97.87
DWT	Precision (%)	99.11	99.58	99.47
	Recall (%)	98.39	99.1	99.59
	F1 score (%)	98.57	99.34	99.53

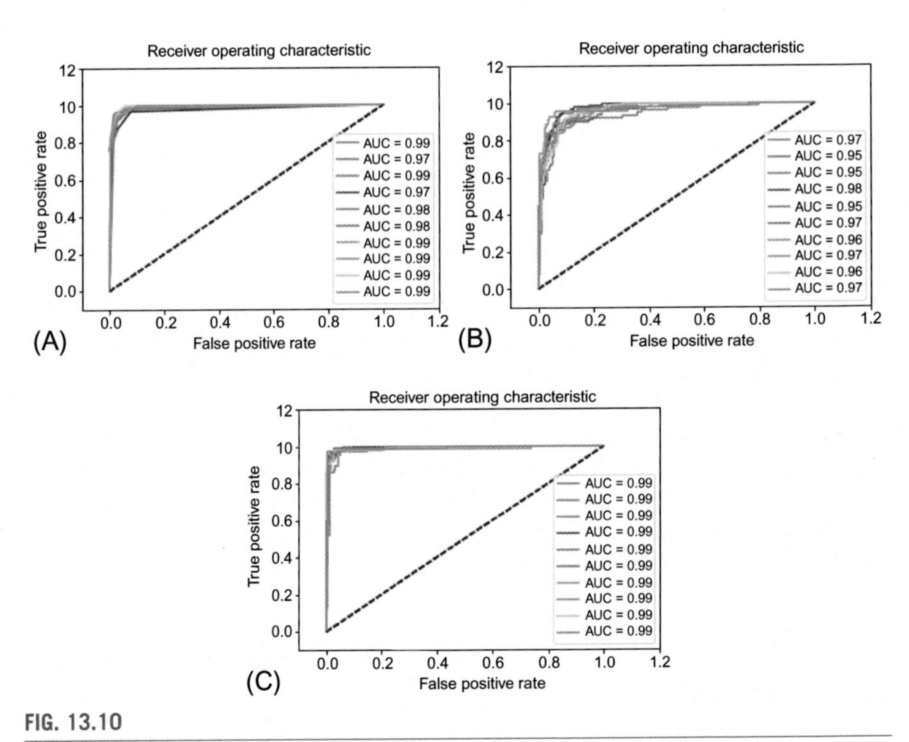

FIG. 13.10

ROC curve and corresponding AUC for the first-order histogram texture analysis method. (A) Performance of the *k*-NN classifier. (B) Performance of the ANN classifier. (C) Performance of the SVM classifier.

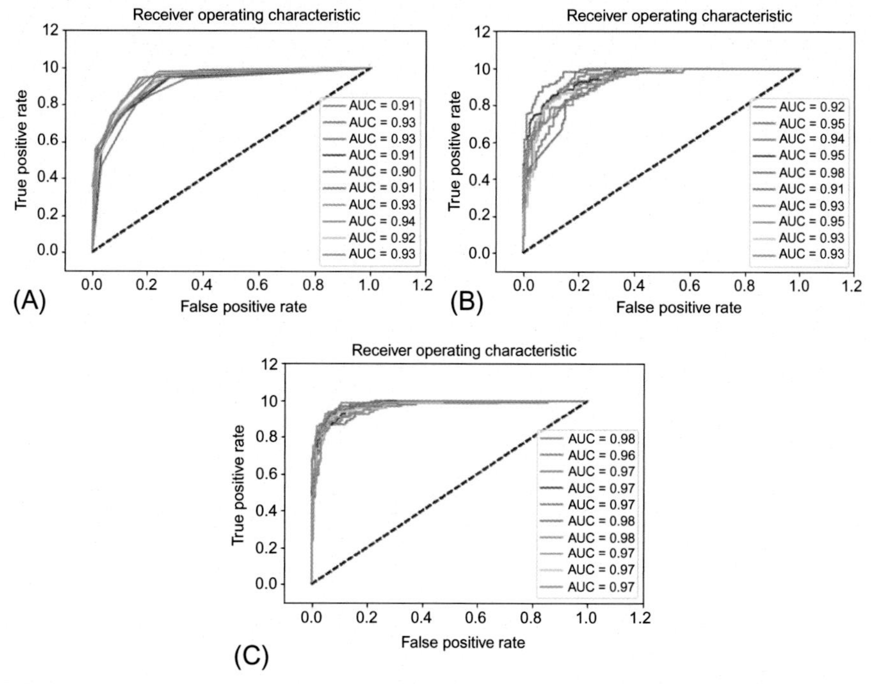

FIG. 13.11

ROC curve and corresponding AUC for the LBP texture analysis method. (A) Performance of the *k*-NN classifier. (B) Performance of the ANN classifier. (C) Performance of the SVM classifier.

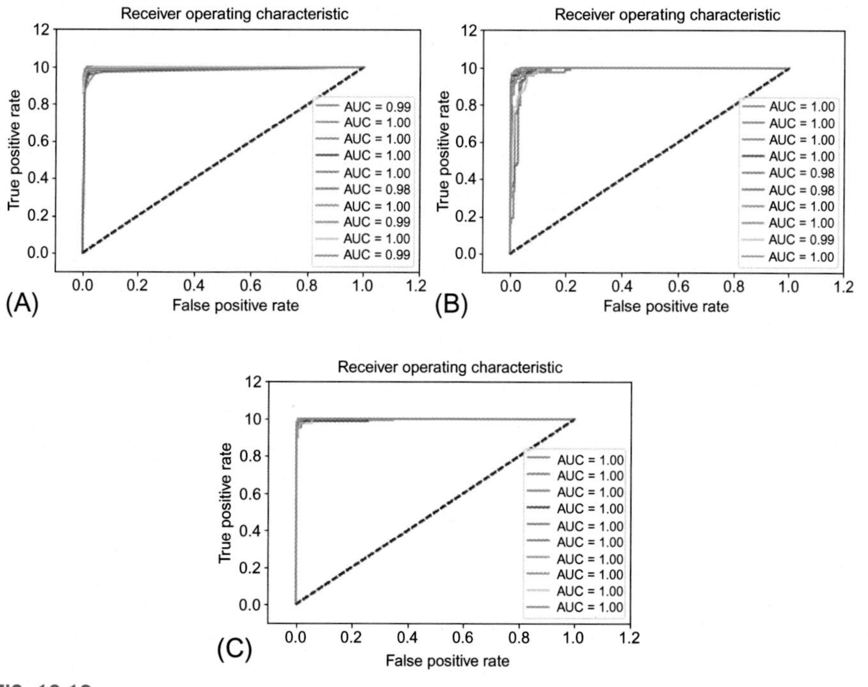

FIG. 13.12

ROC curve and corresponding AUC for the GLCM texture analysis method. (A) Performance of the *k*-NN classifier. (B) Performance of the ANN classifier. (C) Performance of the SVM classifier.

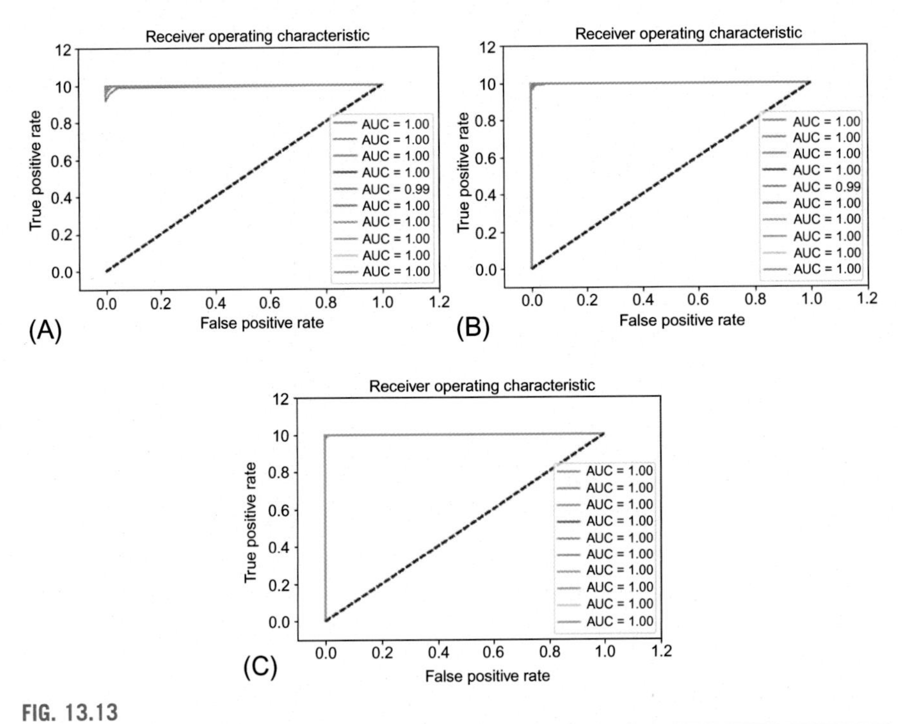

FIG. 13.13

ROC curve and corresponding AUC for the DWT texture analysis method. (A) Performance of the *k*-NN classifier. (B) Performance of the ANN classifier. (C) Performance of the SVM classifier.

recently proposed signal processing method DWT has shown very promising results for texture analysis of MRI images. However, model-based and geometrical texture analysis methods are not used widely for analysis of texture from medical images because of the absence of regular symmetric patterns in MRI images and the complexity of the model.

Comparing the four methods implemented under the common framework, DWT produced the best results for the detection of tumor images over our dataset. Among the three classifiers, the SVM classifier produced very impressive results. We believe that these texture analysis methods can be used for the classification of a tumor in various grades and that remains our plan.

References

[1] S. Chakraborty, S. Chatterjee, A.S. Ashour, K. Mali, N. Dey, Intelligent computing in medical imaging: a study, Advancements in Applied Metaheuristic Computing, IGI Global, Hershey, PA, 2018, pp. 143–163.

[2] J. Chaki, N. Dey, F. Shi, R.S. Sherratt, Pattern mining approaches used in sensor-based biometric recognition: a review, IEEE Sens. J. 19 (2019) 3569–3580.

[3] J. Chaki, N. Dey, Pattern analysis of genetics and genomics: a survey of the state-of-art, Multimed. Tools Appl. (2019) 1–32.
[4] P.K. Bhagat, P. Choudhary, Multiclass segmentation of brain tumor from MRI images, in: H. Malik, S. Srivastava, Y.R. Sood, A. Ahmad (Eds.), Applications of Artificial Intelligence Techniques in Engineering, Springer, Singapore, 2019, pp. 543–553.
[5] P.K. Bhagat, P. Choudhary, Image annotation: then and now, Image Vis. Comput. 80 (2018) 1–23.
[6] P. Brodatz, Textures: A Photographic Album for Artists and Designers, Dover Publications, New York, USA, 1999.
[7] A.W.M. Smeulders, M. Worring, S. Santini, A. Gupta, R. Jain, Content-based image retrieval at the end of the early years, IEEE Trans. Pattern Anal. Mach. Intell. 22 (12) (2000) 1349–1380.
[8] G. Leboucher, G.E. Lowitz, What a histogram can really tell the classifier, Pattern Recogn. 10 (5) (1978) 351–357.
[9] R.M. Haralick, K. Shanmugam, I. Dinstein, Textural features for image classification, IEEE Trans. Syst. Man Cybern. SMC-3 6 (1973) 610–621.
[10] V. Kovalev, M. Petrou, Multidimensional co-occurrence matrices for object recognition and matching, Graph. Models Image Process. 58 (3) (1996) 187–197.
[11] K. Valkealahti, E. Oja, Reduced multidimensional co-occurrence histograms in texture classification, IEEE Trans. Pattern Anal. Mach. Intell. 20 (1) (1998) 90–94.
[12] T. Ojala, M. Pietikainen, T. Maenpaa, Multiresolution gray-scale and rotation invariant texture classification with local binary patterns, IEEE Trans. Pattern Anal. Mach. Intell. 24 (7) (2002) 971–987.
[13] X. Tan, B. Triggs, Enhanced local texture feature sets for face recognition under difficult lighting conditions, IEEE Trans. Image Process. 19 (6) (2010) 1635–1650.
[14] T. Jabid, M.H. Kabir, O. Chae, Local directional pattern (LDP) for face recognition, in: 2010 Digest of Technical Papers International Conference on Consumer Electronics (ICCE), 2010, pp. 329–330.
[15] M.H. Kabir, T. Jabid, O. Chae, A local directional pattern variance (LDPv) based face descriptor for human facial expression recognition, in: 2010 7th IEEE International Conference on Advanced Video and Signal Based Surveillance, 2010, pp. 526–532.
[16] I. Daubechies, Ten Lectures of Wavelets, second ed., Society for Industrial and Applied Mathematics, Philadelphia, PA, USA, 1992.
[17] H. Bouyerbou, S. Oukid, N. Benblidia, K. Bechkoum, Hybrid image representation methods for automatic image annotation: a survey, in: 2012 International Conference on Signals and Electronic Systems (ICSES), 2012, pp. 1–6.
[18] M.S. Singh, P. Choudhary, Stroke prediction using artificial intelligence, in: 2017 8th Annual Industrial Automation and Electromechanical Engineering Conference (IEMECON), 2017, pp. 158–161.
[19] D. Nandi, A.S. Ashour, S. Samanta, S. Chakraborty, M.A.M. Salem, N. Dey, Principal component analysis in medical image processing: a study, Int. J. Image Mining 1 (1) (2015) 65–86.
[20] A.N. Karahaliou, I.S. Boniatis, S.G. Skiadopoulos, F.N. Sakellaropoulos, N.S. Arikidis, E.A. Likaki, G.S. Panayiotakis, L.I. Costaridou, Breast cancer diagnosis: analyzing texture of tissue surrounding microcalcifications, IEEE Trans. Inf. Technol. Biomed. 12 (6) (2008) 731–738.
[21] H. Yu, C. Caldwell, K. Mah, D. Mozeg, Coregistered FDG PET/CT-based textural characterization of head and neck cancer for radiation treatment planning, IEEE Trans. Med. Imaging 28 (3) (2009) 374–383.

[22] S.M.S. Reza, R. Mays, K.M. Iftekharuddin, Multi-fractal detrended texture feature for brain tumor classification, Proc. SPIE Int. Soc. Opt. Eng. 9414 (2015) 941410.
[23] Q.-U.-A. Qurat-Ul-Ain, G. Latif, S.B. Kazmi, M.A. Jaffar, A.M. Mirza, Classification and segmentation of brain tumor using texture analysis, in: Proceedings of the 9th WSEAS International Conference on Artificial Intelligence, Knowledge Engineering and Data Bases, World Scientific and Engineering Academy and Society (WSEAS), Stevens Point, WI, 2010, pp. 147–155.
[24] G.E. Lowitz, Can a local histogram really map texture information? Pattern Recogn. 16 (2) (1983) 141–147.
[25] B. Julesz, Experiments in the visual perception of texture, Sci. Am. 232 (1975) 34–43.
[26] S.R. Kodituwakku, S. Selvarajah, Comparison of color features for image retrieval, Indian J. Comput. Sci. Eng. 1 (3) (2010) 207–211.
[27] A.K. Jain, A. Vailaya, Image retrieval using color and shape, Pattern Recogn. 29 (8) (1996) 1233–1244.
[28] Z. Shi, Y. Yang, T.M. Hospedales, T. Xiang, Weakly-supervised image annotation and segmentation with objects and attributes, IEEE Trans. Pattern Anal. Mach. Intell. 39 (12) (2017) 2525–2538.
[29] S. Zheng, M.M. Cheng, J. Warrell, P. Sturgess, V. Vineet, C. Rother, P.H.S. Torr, Dense semantic image segmentation with objects and attributes, in: 2014 IEEE Conference on Computer Vision and Pattern Recognition, 2014, pp. 3214–3221.
[30] G. Singh, J. Kosecka, Nonparametric scene parsing with adaptive feature relevance and semantic context, in: 2013 IEEE Conference on Computer Vision and Pattern Recognition, 2013, pp. 3151–3157.
[31] A. Vezhnevets, V. Ferrari, J.M. Buhmann, Weakly supervised structured output learning for semantic segmentation, in: 2012 IEEE Conference on Computer Vision and Pattern Recognition, 2012, pp. 845–852.
[32] M. Rubinstein, C. Liu, W.T. Freeman, Annotation propagation in large image databases via dense image correspondence, in: A. Fitzgibbon, S. Lazebnik, P. Perona, Y. Sato, C. Schmid (Eds.), Computer Vision—ECCV 2012, Springer, Berlin, Heidelberg, 2012, pp. 85–99.
[33] H. Zhang, J.E. Fritts, S.A. Goldman, Image segmentation evaluation: a survey of unsupervised methods, Comput. Vis. Image Underst. 110 (2) (2008) 260–280.
[34] E. Shelhamer, J. Long, T. Darrell, Fully convolutional networks for semantic segmentation, IEEE Trans. Pattern Anal. Mach. Intell. 39 (4) (2017) 640–651.
[35] L.C. Chen, G. Papandreou, I. Kokkinos, K. Murphy, A.L. Yuille, DeepLab: semantic image segmentation with deep convolutional nets, atrous convolution, and fully connected CRFs, IEEE Trans. Pattern Anal. Mach. Intell. (99) (2017) 1.
[36] N. Sebe, Q. Tian, E. Loupias, M.S. Lew, T.S. Huang, Evaluation of salient point techniques, Image Vis. Comput. 21 (13) (2003) 1087–1095. British Machine Vision Computing 2001.
[37] K.S. Pedersen, M. Loog, P. Dorst, Salient point and scale detection by minimum likelihood, in: N.D. Lawrence, A. Schwaighofer, J.Q. Candela (Eds.), Proceedings of Machine Learning Research, Gaussian Processes in Practice, , vol. 1, PMLR, Bletchley Park, UK, 2007, pp. 59–72.
[38] D.G. Lowe, Distinctive image features from scale-invariant keypoints, Int. J. Comput. Vis. 60 (2) (2004) 91–110.
[39] N. Dalal, B. Triggs, Histograms of oriented gradients for human detection, in: 2005 IEEE Computer Society Conference on Computer Vision and Pattern Recognition (CVPR'05), vol. 1, 2005, pp. 886–893.

[40] H. Bay, T. Tuytelaars, L.V. Gool, SURF: speeded up robust features, in: A. Leonardis, H. Bischof, A. Pinz (Eds.), Computer Vision—ECCV 2006, Springer, Berlin, Heidelberg, 2006, pp. 404–417.
[41] E. Rosten, T. Drummond, Machine learning for high-speed corner detection, in: A. - Leonardis, H. Bischof, A. Pinz (Eds.), Computer Vision—ECCV 2006, Springer, Berlin, Heidelberg, 2006, pp. 430–443.
[42] M. Calonder, V. Lepetit, C. Strecha, P. Fua, BRIEF: binary robust independent elementary features, in: K. Daniilidis, P. Maragos, N. Paragios (Eds.), Computer Vision—ECCV 2010, Springer, Berlin, Heidelberg, 2010, pp. 778–792.
[43] E. Rublee, V. Rabaud, K. Konolige, G. Bradski, ORB: an efficient alternative to SIFT or SURF, in: 2011 International Conference on Computer Vision, 2011, pp. 2564–2571.
[44] N. Dey, P. Das, S. Biswas, A. Das, S.S. Chaudhuri, Feature analysis for the reversible watermarked electrooculography signal using low distortion prediction-error expansion, in: 2012 International Conference on Communications, Devices and Intelligent Systems (CODIS), 2012, pp. 624–627.
[45] L. Saba, N. Dey, A.S. Ashour, S. Samanta, S.S. Nath, S. Chakraborty, J. Sanches, D. Kumar, R. Marinho, J.S. Suri, Automated stratification of liver disease in ultrasound: an online accurate feature classification paradigm, Comput. Methods Prog. Biomed. 130 (2016) 118–134.
[46] S. Cheriguene, N. Azizi, N. Zemmal, N. Dey, H. Djellali, N. Farah, Optimized tumor breast cancer classification using combining random subspace and static classifiers selection paradigms, in: Applications of Intelligent Optimization in Biology and Medicine: Current Trends and Open Problems, Springer International Publishing, Cham, 2016, pp. 289–307.
[47] M. Tuceryan, A.K. Jain, Texture analysis, in: C.H. Chen, L.F. Pau, P.S.P. Wang (Eds.), Handbook of Pattern Recognition & Computer Vision, World Scientific Publishing Co., Inc., River Edge, NJ, USA, 1993, ISBN 981-02-113,6-8pp. 235–276.
[48] X. Xie, A galaxy of texture features, in: M. Mirmehdi, X. Xie, J. Suri (Eds.), Handbook of Texture Analysis, ICP, Berlin, 2008, pp. 375–406.
[49] W.K. Pratt, Digital Image Processing, fourth ed., John Wiley & Sons, Inc., USA, 2007.
[50] A. Papoulis, S.U. Pillai, Probability Random Variables, and Stochastic Processes, fourth ed., McGraw-Hill Inc., New York, 2001.
[51] W.-K. Lam, C.-K. Li, Rotated texture classification by improved iterative morphological decomposition, IEE Proc. Vis. Image Signal Process. 144 (3) (1997) 171–179.
[52] M.D. Levine, Vision in Man and Machine, McGraw-Hill Inc., New York, 1985.
[53] R.A. Lerski, K. Straughan, L.R. Schad, D. Boyce, S. Bluml, I. Zuna, VIII. MR image texture analysis—an approach to tissue characterization, Magn. Reson. Imaging 11 (6) (1993) 873–887.
[54] L. Alparone, F. Argenti, G. Benelli, Fast calculation of co-occurrence matrix parameters for image segmentation, Electron. Lett. 26 (1) (1990) 23–24.
[55] J. Sachdeva, V. Kumar, I. Gupta, N. Khandelwal, C.K. Ahuja, Segmentation, feature extraction, and multiclass brain tumor classification, J. Digit. Imaging 26 (2013) 1141–1150.
[56] D.M. Joshi, N.K. Rana, V.M. Misra, Classification of brain cancer using artificial neural network, in: 2010 2nd International Conference on Electronic Computer Technology, 2010, pp. 112–116.
[57] D.B. Kadam, S.S. Gade, M.D. Uplane, R.K. Prasad, An artificial neural network approach for brain tumor detection based on characteristics of GLCM texture features, Int. J. Innov. Eng. Technol. 2 (2013) 193–199.

[58] P. Mohanaiah, P. Sathyanarayana, Detection of tumour using grey level co-occurrence matrix and lifting based DWT with radial basis function, Int. J. Eng. Res. Technol. 2 (2013) 1677–1683.
[59] P. Mohanaiah, P. Sathyanarayana, Brain cancer classification using GLCM based feature extraction in artificial neural network, Int. J. Comput. Sci. Eng. Technol. 4 (2013) 966–970.
[60] N.V. Chavan, B.D. Jadhav, P.M. Patil, Detection and classification of brain tumors, Int. J. Comput. Appl. 112 (2015) (2015) 48–53.
[61] A. Fahrurozi, S. Madenda, Ernastuti, D. Kerami, Wood texture features extraction by using GLCM combined with various edge detection methods, J. Phys. Conf. Ser. 725 (1) (2016) 012005.
[62] A. Fahrurozi, S. Madenda, Ernastuti, D. Kerami, Texture fusion for batik motif retrieval system, Int. J. Electr. Comput. Eng. 6 (6) (2016) 3174–3187.
[63] K. Sharma, J. Virmani, A decision support system for classification of normal and medical renal disease using ultrasound images: a decision support system for medical renal diseases, Int. J. Ambient Comput. Intell. 8 (2) (2017) 52–69.
[64] T. Ojala, M. Pietikäinen, D. Harwood, A comparative study of texture measures with classification based on featured distributions, Pattern Recogn. 29 (1) (1996) 51–59.
[65] D.-C. He, L. Wang, Texture unit, texture spectrum, and texture analysis, IEEE Trans. Geosci. Remote Sens. 28 (4) (1990) 509–512.
[66] L. Wang, D.-C. He, Texture classification using texture spectrum, Pattern Recogn. 23 (8) (1990) 905–910.
[67] D. Unay, A. Ekin, M. Cetin, R. Jasinschi, A. Ercil, Robustness of local binary patterns in brain MR image analysis, in: 2007 29th Annual International Conference of the IEEE Engineering in Medicine and Biology Society, 2007, pp. 2098–2101.
[68] M. Pietikainen, G. Zhao, Two decades of local binary patterns: a survey, in: E. Bingham, S. Kaski, J. Laaksonen, J. Lampinen (Eds.), Advances in Independent Component Analysis and Learning Machines, Academic Press, Cambridge, MA, USA, 2015, pp. 175–210.
[69] K.H. Ghazali, M.F. Mansor, M.M. Mustafa, A. Hussain, Feature extraction technique using discrete wavelet transform for image classification, in: 2007 5th Student Conference on Research and Development, 2007, pp. 1–4.
[70] E.A. El-Dahshan, T. Hosny, A.M. Salem, Hybrid intelligent techniques for MRI brain images classification, Digit. Signal Process. 20 (2010) 433–441.
[71] L. Mercier, R.F. Del Maestro, K. Petrecca, D. Araujo, C. Haegelen, D.L. Collins, Online database of clinical MR and ultrasound images of brain tumors, Med. Phys. 39 (6 Part 1) (2012) 3253–3261.
[72] N. Dey, A.S. Ashour, S. Beagum, D.S. Pistola, M. Gospodinov, E.P. Gospodinova, J.M. R.S. Tavares, Parameter optimization for local polynomial approximation based intersection confidence interval filter using genetic algorithm: an application for brain MRI image de-noising, J. Imaging 1 (1) (2015) 60–84.
[73] A. Ashour, S. Samanta, N. Dey, N. Kausar, W. Abdessalemkaraa, A. Hassanien, Computed tomography image enhancement using cuckoo search: a log transform based approach, J. Signal Inf. Process. 1 (6) (2015) 244–257.
[74] D.D. Conge, M. Kumar, R.L. Miller, J. Luo, H. Radha, Improved seam carving for image resizing, in: 2010 IEEE Workshop on Signal Processing Systems, 2010, pp. 345–349.

[75] T. Cover, P. Hart, Nearest neighbor pattern classification, IEEE Trans. Inf. Theory 13 (1) (1967) 21–27.

[76] B.E. Boser, I.M. Guyon, V.N. Vapnik, A training algorithm for optimal margin classifiers, in: Proceedings of the Fifth Annual Workshop on Computational Learning Theory, ACM, New York, NY, 1992, pp. 144–152.

[77] S. Haykin, Neural Networks: A Comprehensive Foundation, second, Prentice Hall PTR, Upper Saddle River, NJ, 1998.

[78] A. Bhattacherjee, S. Roy, S. Paul, P. Roy, N. Kausar, N. Dey, Classification approach for breast cancer detection using back propagation neural network: a study, in: Biomedical Image Analysis and Mining Techniques for Improved Health Outcomes, IGI Global, Hershey, PA, 2016, pp. 210–221.

[79] D.M.W. Powers, Evaluation: from precision, recall and F-measure to ROC, informedness, markedness & correlation, J. Mach. Learn. Technol. 2 (1) (2011) 37–63.

[80] A.N. Sarlashkar, M. Bodruzzaman, M.J. Malkani, Feature extraction using wavelet transform for neural network based image classification, in: Proceedings of Thirtieth Southeastern Symposium on System Theory, 1998, pp. 412–416.

Index

Note: Page numbers followed by *f* indicate figures, *t* indicate tables, and *b* indicate boxes.

D

E

F

G

H

I

J

P

Q

R

S

Printed in the United States
By Bookmasters